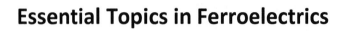

Essential Topics in Ferroelectrics

Essential Topics in Ferroelectrics

Edited by **Sharon Tatum**

New York

Published by NY Research Press,
23 West, 55th Street, Suite 816,
New York, NY 10019, USA
www.nyresearchpress.com

Essential Topics in Ferroelectrics
Edited by Sharon Tatum

© 2015 NY Research Press

International Standard Book Number: 978-1-63238-186-6 (Hardback)

Contents

Permissions

List of Contributors

Preface

Every book is initially just a concept; it takes months of research and hard work to give it the final shape in which the readers receive it. In its early stages, this book also went through rigorous reviewing. The notable contributions made by experts from across the globe were first molded into patterned chapters and then arranged in a sensibly sequential manner to bring out the best results.

Ferroelectricity is one of the most studied phenomena in the scientific community because of the vitality of ferroelectric materials in a broad spectrum of applications comprising of high dielectric constant capacitors, pyroelectric devices and transducers for medical diagnostic, piezoelectric sonars, electro-optic light valves, electromechanical transducers and ferroelectric random access memories. Ferroelectricity at nanoscale draws huge attention to the advancement of novel technologies. The need for ferroelectric systems with particular applications led to detailed research along with the betterment of processing and characterization techniques. This book provides an updated outlook of current research into ferroelectricity, covering several formulations, their forms (bulk, thin films, and ferroelectric liquid crystals), and ferroelectricity at nanoscale.

It has been my immense pleasure to be a part of this project and to contribute my years of learning in such a meaningful form. I would like to take this opportunity to thank all the people who have been associated with the completion of this book at any step.

Editor

Electronic Structures of Tetragonal ABX3: Role of the B-X Coulomb Repulsions for Ferroelectricity and Piezoelectricity

Kaoru Miura and Hiroshi Funakubo

Additional information is available at the end of the chapter

1. Introduction

Since Cohen proposed an origin for ferroelectricity in perovskites (ABX_3) [1], investigations of ferroelectric materials using first-principles calculations have been extensively studied [2-20]. Currently, using the pseudopotential (PP) methods, most of the crystal structures in ferroelectric ABX_3 can be precisely predicted. However, even in $BaTiO_3$, which is a well-known ferroelectric perovskite oxide with tetragonal structure at room temperature, the optimized structure by the PP methods is strongly dependent on the choice of the Ti PPs as illustrated in Fig. 1; preparation for Ti 3s and 3p semicore states in addition to Ti 3d, 4s, and 4p valence states is essential to the appearance of the tetragonal structure. This is an important problem for ferroelectricity, but it has been generally recognized for a long time that this problem is within an empirical framework of the calculational techniques [21].

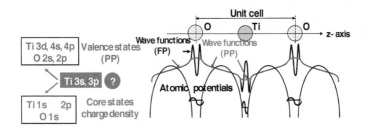

Figure 1. Illustration of the choice of Ti 3s and 3p states in pseudopotentials.

It is known that ferroelectric state appears when the long-range forces due to the dipole-dipole interaction overcome the short-range forces due to the Coulomb repulsions. Investigations about the relationship between the Ti-O Coulomb repulsions and the appearance of ferroelectricity in $ATiO_3$ (A = Ba, Pb) were reported both theoretically and experimentally. Theoretically, Cohen first proposed the hybridization between Ti 3d state and O 2p state (Ti 3d-O 2p) as an origin for ferroelectricity in $BaTiO_3$ and $PbTiO_3$ [1]. On the other hand, we investigated [20] the influence of the Ti-O_z Coulomb repulsions on Ti ion displacement in tetragonal $BaTiO_3$ and $PbTiO_3$, where O_z denotes the O atom to the z-axis (Ti is displaced to the z-axis). Whereas the hybridization between Ti 3d state and O_z $2p_z$ state stabilize Ti ion displacement, the strong Coulomb repulsions between Ti 3s and $3p_z$ states and O $2p_z$ states do not favourably cause Ti ion displacement. Experimentally, on the other hand, Kuroiwa *et al.* [22] showed that the appearance of ferroelectric state is closely related to the total charge density of Ti-O bonding in $BaTiO_3$. As discussed above, investigation about a role of Ti 3s and 3p states is important in the appearance of the ferroelectric state in tetragonal $BaTiO_3$, in addition to the Ti 3d-O 2p hybridization as an origin of ferroelectricity [1].

It seems that the strong B-X Coulomb repulsions affect the most stable structure of ABX_3. It has been well known that the most stable structure of ABX_3 is closely related to the tolerance factor t,

$$t = \frac{r_A + r_X}{\sqrt{2}\,(r_B + r_X)} \,, \tag{1}$$

where r_A, r_B, and r_X denote the ionic radii of A, B, and X ions, respectively [23]. In general ferroelectric ABX_3, the most stable structure is tetragonal for $t \gtrsim 1$, cubic for $t \approx 1$, and rhombohedral or orthorhombic for $t \lesssim 1$. In fact, $BaTiO_3$ with $t = 1.062$ shows tetragonal structure in room temperature. However, recently, $BiZn_{0.5}Ti_{0.5}O_3$ (BZT) with $t = 0.935$ was experimentally reported [24] to show a tetragonal $PbTiO_3$-type structure with high c/a ratio (1.211). This result is in contrast to that of $BiZn_{0.5}Mg_{0.5}O_3$ (BMT) with $t = 0.939$, *i.e.*, the most stable structure was reported to be the orthorhombic or rhombohedral structure [25, 26]. Several theoretical papers of BZT have been reported [4-6], but the role of the Zn-O Coulomb repulsions in the appearance of the tetragonal structure has not been discussed sufficiently.

Piezoelectric properties in ABX_3 are also closely related to the crystal structure. Investigations of the relationship between piezoelectric properties and the crystal structure of ABX_3 by first-principles calculations have been extensively studied [2-19]. Moreover, phenomenological investigations of the piezoelectric properties have been also performed [27, 28]. However, it seems that the piezoelectric properties in the atomic level have not been sufficiently investigated. Therefore, further theoretical investigation of the relationship between piezoelectric properties and the crystal structure of ABX_3, especially the B-X Coulomb repulsions, should be needed.

Recently, we investigated the roles of the Ti-O Coulomb repulsions in the appearance of a ferroelectric state in tetragonal $BaTiO_3$ by the analysis of a first-principles PP method [11-15]. We investigated the structural properties of tetragonal and rhombohedral $BaTiO_3$ with two kinds of Ti PPs, and propose the role of Ti 3s and 3p states for ferroelectricity. We also investigated the role of the Zn-O Coulomb repulsions in the appearance of a ferroelectric state in tetragonal BZT [10, 13]. Moreover, we also investigated the structural, ferroelectric, and piezoelectric properties of tetragonal ABX_3 and discussed the piezoelectric mechanisms based on the B-X Coulomb repulsions [12, 14, 15, 18, 19].

In this chapter, based on our recent papers and patents [10-19], we discuss a general role of B-X Coulomb repulsions for the appearance of the ferroelectric state in ABX_3. Then, we also discuss the relationship between the B-X Coulomb repulsions and the piezoelectric properties of tetragonal ABX_3.

2. Methodology

The calculations for ABX_3 were performed using the ABINIT code [29], which is one of the norm-conserving PP (NCPP) methods. Electron-electron interaction was treated in the local-density approximation (LDA) [30]. Pseudopotentials were generated using the OPIUM code [31]:

i. In BaTiO$_3$, 5s, 5p and 6s electrons for Ba PP, and 2s and 2p electrons for O PP were treated as semicore or valence electrons, respectively. Moreover, in order to investigate the role of Ti 3s and 3p states, two kinds of Ti PPs were prepared: the Ti PP with 3s, 3p, 3d and 4s electrons treated as semicore or valence electrons (Ti3spd4s PP), and that with only 3d and 4s electrons treated as valence electrons (Ti3d4s PP). In both PPs, the differences between the calculated result and experimental one are within 1.5 % of the lattice constant and within 10 % of the bulk modulus in the optimized calculation of bulk Ti. The cutoff energy for plane-wave basis functions was set to be 50 Hartree (Hr). The number of atoms in the unit cell was set to be five, and a 6×6×6 Monkhorst-Pack k-point mesh was set in the Brillouin zone of the unit cell. Positions of all the atoms were optimized within the framework of the tetragonal ($P4mm$) or rhombohedral ($R3m$) structure.

ii. In BZT and BMT, 5d, 6s, and 6p electrons for Bi PP, and 2s and 2p electrons for O PP were treated as semicore or valence electrons, respectively. Moreover, in order to investigate the roles of Zn and Ti 3s and 3p states, and Mg 2s and 2p states, two types of PPs were prepared: the PPs with only Zn and Ti 3d and 4s states, and Mg 3s states, considered as valence electrons (Case I), Zn and Ti 3s, 3p, 3d, and 4s states, and Mg 2s, 2p, and 3s states considered as semicore or valence electrons (Case II). The cutoff energy for plane-wave basis functions was set to be 70 Hr for Case I and 110 Hr for Case II. A 4×4×4 Monkhorst-Pack k-point mesh was set in the Brillouin zone of the unit cell. The calculated results can be discussed within 0.02 eV per formula unit (f.u.) using the above conditions. The present calculations were performed for the monoclinic, rhombohedral, and A-, C-, and G-type tetragonal structures. The number of atoms in the unit cell was set to be 10 for the rhombohedral and monoclinic structures, and 20 for the A-, C-, and G-type tetragonal structures. Positions of all the atoms were optimized within the framework of the rhombohedral ($R3$), monoclinic (Pm), and tetragonal ($P4mm$) structures.

iii.   Relationship between the B-X Coulomb repulsions and the piezoelectric properties
            in tetragonal ABX_3 is investigated. The pseudopotentials were generated using the
            opium code [31] with semicore and valence electrons (e.g., Ti3spd4s PP), and the
            virtual crystal approximation [32] were applied to several ABX_3.

Spontaneous polarizations and piezoelectric constants were also evaluated, due to the Born
effective charges [33]. The spontaneous polarization of tetragonal structures along the [001]
axis, P_3, is defined as

$$P_3 = \sum_k \frac{ec}{\Omega} Z_{33}^*(k) u_3(k) \; , \tag{2}$$

where e, c, and Ω denote the charge unit, lattice parameter of the unit cell along the [001]
axis, and the volume of the unit cell, respectively. $u_3(k)$ denotes the displacement along the
[001] axis of the kth atom, and $Z_{33}^*(k)$ denotes the Born effective charges [33] which contributes to the P_3 from the $u_3(k)$.

The piezoelectric e_{33} constant is defined as

$$e_{3j} = \left(\frac{\partial P_3}{\partial \eta_3} \right)_u + \sum_k \frac{ec}{\Omega} Z_{33}^*(k) \frac{\partial u_3(k)}{\partial \eta_j} \qquad (j = 3,1), \tag{3}$$

where e and Ω denote the charge unit and the volume of the unit cell. P_3 and c denote the
spontaneous polarization of tetragonal structures and the lattice parameter of the unit cell
along the [001] axis, respectively. $u_3(k)$ denotes the displacement along the [001] axis of the
kth atom, and $Z_{33}^*(k)$ denotes the Born effective charges which contributes to the P_3 from the
$u_3(k)$. η_3 denotes the strain of lattice along the [001] axis, which is defined as $\eta_3 \equiv (c - c_0)/c_0$; c_0
denotes the c lattice parameter with fully optimized structure. On the other hand, η_1 denotes
the strain of lattice along the [100] axis, which is defined as $\eta_1 \equiv (a - a_0)/a_0$; a_0 denotes the a
lattice parameter with fully optimized structure. The first term of the right hand in Eq. (3)
denotes the clamped term evaluated at vanishing internal strain, and the second term denotes the relaxed term that is due to the relative displacements.

The relationship between the piezoelectric d_{33} constant and the e_{33} one is

$$d_{33} \equiv \sum_{j=1}^{6} s_{3j}^E \times^T (e_{3j}), \tag{4}$$

where s_{3j^E} denotes the elastic compliance, and ``T'' denotes the transposition of matrix elements. The suffix j denotes the direction-indexes of the axis, i.e., 1 along the [100] axis, 2
along the [010] axis, 3 along the [001] axis, and 4 to 6 along the shear directions, respectively.

3. Results and discussion

3.1. Ferroelectricity

3.1.1. Role of Ti 3s and 3p states in ferroelectric BaTiO₃

Figures 2(a) and 2(b)show the optimized results for the ratio c/a of the lattice parameters and the value of the Ti ion displacement (δ_{Ti}) as a function of the a lattice parameter in tetragonal BaTiO₃, respectively. Results with arrows are the fully optimized results, and the others results are those with the c lattice parameters and all the inner coordination optimized for fixed a. Note that the fully optimized structure of BaTiO₃ is tetragonal with the Ti3spd4s PP, whereas it is cubic with the Ti3d4s PP. This result suggests that the explicit treatment of Ti 3s and 3p semicore states is essential to the appearance of ferroelectric states in BaTiO₃.

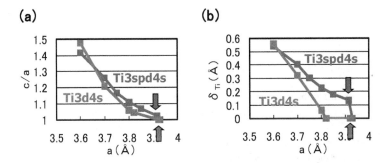

Figure 2. Optimized calculated results in tetragonal BaTiO₃. Results with arrows are the fully optimized results [11].

The calculated results shown in Fig. 2 suggest that the explicit treatment of Ti 3s and 3p semicore states is essential to the appearance of ferroelectric states in BaTiO₃. In the following, we investigate the role of Ti 3s and 3p states for ferroelectricity from two viewpoints.

One viewpoint concerns hybridizations between Ti 3s and 3p states and other states. Figure 3(a) and 3(b) shows the total density of states (DOS) of tetragonal BaTiO₃ with two Ti PPs. Both results are in good agreement with previous calculated results [7] by the full-potential linear augmented plane wave (FLAPW) method. In the DOS with the Ti3spd4s PP, the energy ``levels'', not bands, of Ti 3s and 3p states, are located at -2.0 Hr and -1.2 Hr, respectively. This result suggests that the Ti 3s and 3p orbitals do not make any hybridization but only give Coulomb repulsions with the O orbitals as well as the Ba orbitals. In the DOS with the Ti3d4s PP, on the other hand, the energy levels of Ti 3s and 3p states are not shown because Ti 3s and 3p states were treated as the core charges. This result means that the Ti 3s and 3p orbitals cannot even give Coulomb repulsions with the O orbitals as well as the Ba orbitals.

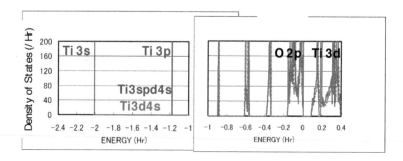

Figure 3. Total density of states (DOS) of fully optimized tetragonal BaTiO$_3$ with the Ti3spd4s PP (solid line) and cubic BaTiO$_3$ with the Ti3d4s PP (red dashed line) [11].

Another viewpoint is about the Coulomb repulsions between Ti 3s and 3p$_{x(y)}$ states and O$_{x(y)}$ 2s and 2p$_{x(y)}$ states in tetragonal BaTiO$_3$. Figure 4(a) and 4(b) show two-dimensional electron-density contour map on the xz-plane. These are the optimized calculated results with a fixed to be 3.8 Å, and the electron density in Fig. 4(a) is quantitatively in good agreement with the experimental result [22]. The electron density between Ti and O$_x$ ions in Fig. 3(a) is larger than that in Fig. 4(b), which suggests that Ti ion displacement is closely related to the Coulomb repulsions between Ti 3s and 3p$_{x(y)}$ states and O$_{x(y)}$ 2s and 2p$_{x(y)}$ states; the Ti-O Coulomb repulsion is an important role in the appearance of the ferroelectric state in BaTiO$_3$.

The present discussion of the Coulomb repulsions is consistent with the previous reports. A recent soft mode investigation [8] of BaTiO$_3$ shows that Ba ions contribute little to the appearance of Ti ion displacement along the [001] axis. This result suggests that Ti ion displacement is closely related to the structural distortion of TiO$_6$ octahedra. In the present calculations, on the other hand, the only difference between BaTiO$_3$ with the Ti3spd4s PP and with the Ti3d4s PP is the difference in the expression for the Ti 3s and 3p states, *i.e.*, the explicit treatment and including core charges. However, our previous calculation [20] shows that the strong Coulomb repulsions between Ti 3s and 3p$_z$ states and O$_z$ 2s and 2p$_z$ states do not favor Ti ion displacement along the [001] axis. This result suggests that the Coulomb repulsions between Ti 3s and 3p$_{x(y)}$ states and O$_{x(y)}$ 2s and 2p$_{x(y)}$ states would contribute to Ti ion displacement along the [001] axis, and the suggestion is consistent with a recent calculation [9] for PbTiO$_3$ indicating that the tetragonal and ferroelectric structure appears more favorable as the a lattice constant decreases.

(a)   **(b)**

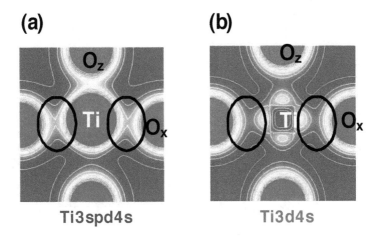

Ti3spd4s   **Ti3d4s**

Figure 4. Two-dimensional electron-density contour map on the xz-plane for tetragonal BaTiO$_3$: (a) with the Ti3spd4s PP, and (b) with the Ti3d4s PP. The optimized calculated results with a fixed to be 3.8 Å are shown in both figures. The electron density increases as color changes from blue to red via white. Contour curves are drawn from 0.4 to 2.0 $e/Å^3$ with increments of 0.2 $e/Å^3$ [11].

Considering the above investigations, we propose the mechanism of Ti ion displacement as follows: Ti ion displacement along the z-axis appears when the Coulomb repulsions between Ti 3s and 3 $p_{x\,(y)}$ states and O$_{x\,(y)}$ 2s and 2 $p_{x\,(y)}$ states, in addition to the dipole-dipole interaction, overcome the Coulomb repulsions between Ti 3s and 3p$_z$ states and O$_z$ 2s and 2p$_z$ states. An illustration of the Coulomb repulsions is shown in Fig. 5(a). In fully optimized BaTiO$_3$ with the Ti3spd4s PP, the Ti ion can be displaced due to the above mechanism. In fully optimized BaTiO$_3$ with the Ti3d4sPP, on the other hand, the Ti ion cannot be displaced due to the weaker Coulomb repulsions between Ti and O$_{x\,(y)}$ ions. However, since the Coulomb repulsion between Ti and O$_z$ ions in BaTiO$_3$ with the Ti3d4s PP is also weaker than that in BaTiO$_3$ with the Ti3spd4s PP, the Coulomb repulsions between Ti and O$_{x\,(y)}$ ions in addition to the log-range force become comparable to the Coulomb repulsions between Ti and O$_z$ ions both in Ti PPs, as the a lattice parameter becomes smaller. The above discussion suggests that the hybridization between Ti 3d and O$_z$ 2s and 2p$_z$ stabilizes Ti ion displacement, but contribute little to a driving force for the appearance of Ti ion displacement.

It seems that the above proposed mechanism for tetragonal BaTiO3 can be applied to the mechanism of Ti ion displacement in rhombohedral BaTiO3, as illustrated in Fig. 5(b). The strong isotropic Coulomb repulsions between Ti 3s and 3px (y, z) states and Ox (y, z) 2s and 2px (y, z) states yield Ti ion displacement along the [111] axis. On the other hand, when the isotropic Coulomb repulsions are weaker or stronger, the Ti ion cannot be displaced and therefore it is favoured for the crystal structure to be cubic.

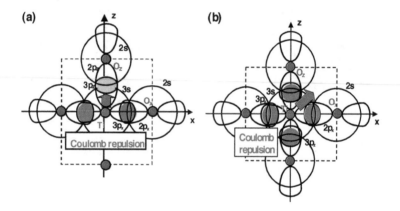

Figure 5. Illustrations of the proposed mechanisms for the Coulomb repulsions between Ti 3s and 3p states and O 2s and 2p states in BaTiO$_3$: (a) anisotropic Coulomb repulsions between Ti 3s and 3p$_{x (y)}$ states and O$_{x (y)}$ 2s and 2p$_{x (y)}$ states, and between Ti 3s and 3p$_z$ states and O$_z$ 2s and 2p$_z$ states, in the tetragonal structure. (b) isotropic Coulomb repulsions between Ti 3s and 3p$_{x (y, z)}$ states and O$_{x (y, z)}$ 2s and 2p$_{x (y, z)}$ states, in the rhombohedral structure [11].

Let us investigate the structural properties of rhombohedral BaTiO$_3$. Figures 6(a) and 6(b) show the optimized results of the 90-α degree and δ_{Ti} as a function of fixed volumes of the unit cells in rhombohedral BaTiO$_3$, respectively, where α denotes the angle between two lattice vectors. In these figures, α denotes the angle between two crystal axes of rhombohedral BaTiO$_3$, and δ_{Ti} denotes the value of the Ti ion displacement along the [111] axis. Results with arrows are the fully optimized results; V_{opt} denote the volume of the fully optimized unit cell with the Ti3spd4s PP. The other results are those with all the inner coordination optimized with fixed volumes of the unit cells. The proposal mechanisms about the Coulomb repulsions seem to be consistent with the calculated results shown in Fig.6: For $V/V_{opt} \lesssim 0.9$ or $\gtrsim 1.3$, the isotropic Coulomb repulsions are weaker or stronger, and the Ti ion cannot be displaced along the [111] axis and therefore the crystal structure is cubic for both Ti PPs. For $0.9 \lesssim V/V_{opt} \lesssim 1.3$, on the other hand, the isotropic Coulomb repulsions are strong enough to yield Ti ion displacement for both Ti PPs. However, since the magnitude of the isotropic Coulomb repulsion is different in the two Ti PPs, the properties of the 90-α degree and δ_{Ti} are different quantitatively.

(a)   **(b)**

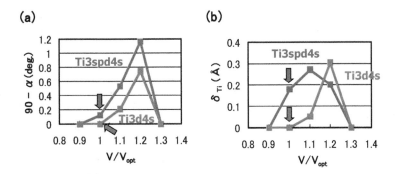

Figure 6. Optimized calculated results as a function of the fixed volumes of the unit cells in rhombohedral BaTiO₃: (a)
90-α degree and (b) δ$_{Ti}$ to the [111] axis. Blue lines correspond to the results with the Ti3spd4s PP, and red lines correspond to those with the Ti3d4s PP. V_{opt} denote the volume of the fully optimized unit cell with the Ti3spd4s PP. Results
with arrows are the fully optimized results, and the other results are those with all the inner coordination optimized
for fixed volumes of the unit cells [11].

3.1.2. Role of Zn 3s, 3p and 3d states in ferroelectric BiZn$_{0.5}$Ti$_{0.5}$O$_3$

As discussed in Sec. 3.1.1, the Coulomb repulsions between Ti 3s and 3p$_{x (y)}$ states and O$_{x (y)}$
2s and 2p$_{x (y)}$ states have an important role in the appearance of the ferroelectric state in tetragonal BaTiO₃. In this subsection, we discuss the role of Zn 3d (d^{10}) states in addition to 3s
and 3p states for ferroelectricity in tetragonal BZT.

Table 1 shows a summary of the optimized results of BZT in Cases I and II. ΔE_{total} denotes
the difference in total energy per f.u. between the rhombohedral and other structures. Although the lattice constant in each structure except the rhombohedral one seems to be quantitatively similar in both cases, properties of ΔE_{total} are different. In Case I, the rhombohedral
structure is the most stable, which is in disagreement with the experimental result [24]. In
Case II, on the other hand, the monoclinic structure, which is the ``pseudo-C-type-tetrago-
nal'' structure, is the most stable. Unfortunately, this result seems to be in disagreement with
the experimental result [24], but is in good agreement with the recent calculated result [6].
Note that the magnitude of ΔE_{total} in Case II is markedly smaller than that in Case I. In contrast to BZT, the rhombohedral structure is the most stable structure in both cases in BMT,
which is consistent with the experimental result [26].

Figures 7(a) and 7(b) show two-dimensional electron density contour maps of the C-type tetragonal BZT in Cases I and II, respectively. The Coulomb repulsion of Zn-O$_x$ in Case II is
larger than that in Case I, and the Coulomb repulsion favorably causes Zn ion displacement
to O$_z$ in Case II. This result is consistent with Sec. 3.1.1. In contrast to the properties of Zn-O
bonding, the inner coordination of the Ti ion is similar in both cases, although the electron
densities are markedly different. This result suggests that the Coulomb repulsion magnitude
of Ti-O$_z$ is the same as that of Ti-O$_x$ in small Ti-O bonding ($\approx$ 1.8 Å), in both Cases I and II.
Figures 7(c) and 7(d) show two-dimensional electron density contour maps of the C-type tet-

ragonal BMT in Cases I and II, respectively. Although the electron densities in both cases are markedly different, the inner coordination of the Mg ion are similar. This result suggests that the Coulomb repulsion between Mg and O is not strong sufficiently for inducing Mg ion displacement even in Case II.

Structure	a (Å)	c (Å)	c/a	α (deg.)	ΔE_{total} (eV/f.u.)
A-type Tetra.	3.748	4.579	1.222	90	0.316
C-type Tetra.	3.681	4.784	1.299	90	0.240
G-type Tetra.	3.725	4.574	1.228	90	0.158
Monoclinic	3.735	4.741	1.269	$\beta = 91.5$	0.193
Rhombohedral	5.560		1	59.93	0
Experiment [24]	3.822	4.628	1.211	90	---

(a)

Structure	a (Å)	c (Å)	c/a	α (deg.)	ΔE_{total} (eV/f.u.)
A-type Tetra.	3.711	4.662	1.256	90	0.135
C-type Tetra.	3.670	4.789	1.305	90	0.091
G-type Tetra.	3.684	4.698	1.275	90	0.047
Monoclinic	3.726	4.740	1.272	$\beta = 91.1$	-0.021
Rhombohedral	5.590		1	59.90	0
Experiment [24]	3.822	4.628	1.211	90	---

(b)

Table 1. Summary of the optimized results of BZT in (a) Case I and (b) Case II. a and c denote the lattice parameters, and α and β denote angles between two lattice axes. ΔE_{total} denotes the difference in total energy per f.u. between the rhombohedral and other structures [10].

Finally in this subsection, we discuss the difference in the electronic structures between the C-type tetragonal and the monoclinic BZT. Figures 8(a) and 8(b) show the electron density contour maps of the C-type tetragonal BZT and that of the monoclinic BZT in Case II, respectively. This result suggests that the strong Coulomb repulsion between Zn and O_z causes the small Zn ion displacement in the [110] direction in the monoclinic BZT, which makes the Coulomb repulsion of Zn-O_z weaker than that in the C-type tetragonal BZT. As a result, this small Zn ion displacement makes the monoclinic BZT more stable than the C-type tetragonal structure.

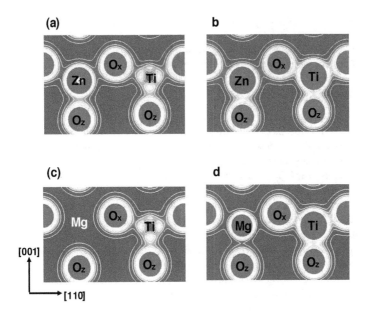

Figure 7. Two-dimensional electron density contour maps of monoclinic (a) BZT in Case I, (b) BZT in Case II, (c) BMT in Case I, and (d) BMT in Case II. The electron density increases as color changes from blue to red via white. Contour curves are drawn from 0.2 to 2.0 $e/Å^3$ with increments of 0.2 $e/Å^3$ [10].

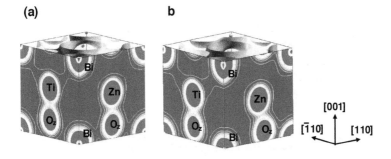

Figure 8. Two-dimensional electron density contour maps of BZT in Case II (a) C-type tetragonal and (b) monoclinic. The electron density increases as color changes from blue to red via white. Contour curves are drawn from 0.2 to 2.0 $e/Å^3$ with increments of 0.2 $e/Å^3$ [10].

3.2. Piezoelectricity

3.2.1. Role of the Ti-O Coulomb repulsions in tetragonal piezoelectric $SrTiO_3$ and $BaTiO_3$

As discussed in Sec. 3.1, the Coulomb repulsions between Ti 3s and $3p_{x\,(y)}$ states and $O_{x\,(y)}$ 2s and $2p_{x\,(y)}$ states have an important role in the appearance of the ferroelectric state in tetragonal $BaTiO_3$. In this subsection, we discuss the role of the Ti-O Coulomb repulsions for piezoelectric $SrTiO_3$ and $BaTiO_3$.

Figures 9(a) shows the optimized results for $c - c_{cub}$ as a function of the a lattice parameters in tetragonal $SrTiO_3$ and $BaTiO_3$, where c_{cub} denotes the c lattice parameter in cubic $SrTiO_3$ and $BaTiO_3$, respectively. These results are the fully optimized results and the results with the c lattice parameters and all the inner coordination optimized for fixed a. The fully optimized parameters of $SrTiO_3$ ($a = 3.84$ Å: cubic) and $BaTiO_3$ ($a = 3.91$ Å and $c = 4.00$ Å: tetragonal) are within 2.0 % in agreement with the experimental results in room temperature. Figure 9(b) shows the evaluated results for P_3 as a function of the a lattice parameters in tetragonal $SrTiO_3$ and $BaTiO_3$, where P_3, which is evaluated by Eq. (2), denotes the spontaneous polarization along the [001] axis. Note that the tetragonal and ferroelectric structures appear even in $SrTiO_3$ when the fixed a lattice parameter is compressed to be smaller than the fully-optimized a lattice parameter. As shown in Figs. 9(a) and 9(b), the tetragonal and ferroelectric structure appears more favorable as the fixed a lattice parameter decreases, which is consistent with previous calculated results [9, 11]. The results would be due to the suggestion discussed in the previous section that the large Coulomb repulsion of Ti-O bonding along the [100] axis (and the [010] axis) is a driving force of the displacement of Ti ions along the [001] axis, i.e., the large Coulomb repulsion along the [100] axis (and the [010] axis) is essential for the appearance of the tetragonal structure.

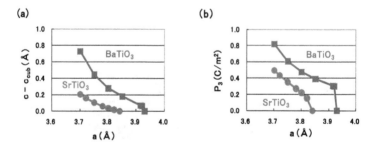

Figure 9. Optimized calculated results as a function of a lattice parameters in compressive tetragonal $SrTiO_3$ and Ba-TiO_3: (a) $c - c_{cub}$ and (b) P_3, i.e., spontaneous polarization along the [001] axis [12].

In the following, we use $c - c_{cub}$ as a functional parameter, because $c - c_{cub}$ is closely related to η_3. Figures 10(a) and 10(b) shows the piezoelectric properties of e_{33} and e_{31} as a function of $c - c_{cub}$ in tetragonal $SrTiO_3$ and $BaTiO_3$. The value $c - c_{cub}$ is optimized value as shown in Fig. 9(a) and e_{33} and e_{31} are evaluated values in their optimized structures. Note that e_{33} become larger at $c - c_{cub} \approx 0$, especially in SrTiO$_3$. These properties seem to be similar to the properties arond the Curie temperatures in piezoelectric ABO_3; Damjanovic emphasized the importance of the polarization extension as a mechanism of larger piezoelectric constants in a recent paper [28]. Contrary to e_{33}, on the other hand, the changes in e_{31} are much smaller than the changes in e_{33}, but note that e_{31} shows negative in SrTiO$_3$ while positive in BaTiO$_3$.

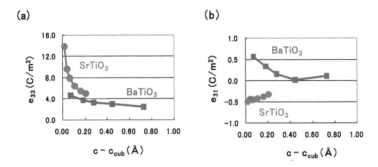

(a)   **(b)**

Figure 10. Evaluated piezoelectric constants as a function of $c - c_{cub}$ in optimized tetragonal SrTiO$_3$ and BaTiO$_3$: (a) e_{33} and (b) e_{31} [12].

As expressed in Eq. (3), e_{3j} is the sum of the contributions from the clamped term and the relaxed term. However, it has been generally known that the contribution to e_{3j} from the clamped term is much smaller than that from the relaxed term; in fact, the absolute values of the e_{33} clamped terms are less than 1 C/m^2 in both SrTiO$_3$ and BaTiO$_3$. We therefore investigate the contributions to the relaxed term of e_{33} and e_{31} in detail. As expressed in Eq. (3), the relaxed terms of e_{3j} are proportional to the sum of the products between the Z_{33}^* (k) and ∂u_3 $(k)/\partial \eta_j$ $(j = 3$ or $1)$ values. Let us show the evaluated results of Z_{33}^* (k), $\partial u_3(k)/\partial \eta_3$, and $\partial u_3(k)/\partial \eta_1$ in the following. Figures 11(a) and 11(b) show the Z_{33}^* (k) values in SrTiO$_3$ and BaTiO$_3$, respectively. Properties of the Z_{33}^* (k) values are quantitatively similar in both SrTiO$_3$ and BaTiO$_3$. Therefore, the difference in the properties of e_{33} and e_{31} between SrTiO$_3$ and BaTiO$_3$ must be due to the difference in the properties of $\partial u_3(k)/\partial \eta_j$. Figures 12(a) and 12(b) show the $\partial u_3(k)/\partial \eta_3$ values in SrTiO$_3$ and BaTiO$_3$, respectively. In these figures, O$_x$ and O$_z$ denote oxygen atoms along the [100] and [001] axes, respectively. Clearly, the absolute values of $\partial u_3(k)/\partial \eta_3$ are different in between SrTiO$_3$ and BaTiO$_3$. On the other hand, Figs. 13(a) and 13(b) show the $\partial u_3(k)/\partial \eta_1$ values in SrTiO$_3$ and BaTiO$_3$, respectively. The absolute values of $\partial u_3(k)/\partial \eta_1$, especially for Ti, O$_x$, and O$_z$ are different in between SrTiO$_3$ and BaTiO$_3$. As a result, the quantitative differences in e_{33} and e_{31} between SrTiO$_3$ and BaTiO$_3$ are due to the differences in the contribution of the $\partial u_3(k)/\partial \eta_3$ and $\partial u_3(k)/\partial \eta_1$ values, respectively.

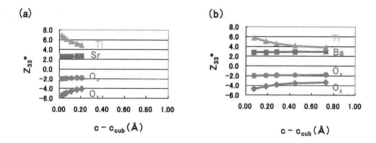

Figure 11. Evaluated Born effective charges $Z_{33}{}^*$ (k) as a function of $c - c_{cub}$: (a) SrTiO$_3$ and (b) BaTiO$_3$. O$_x$ and O$_z$ denote oxygen atoms along the [100] axis and the [001] axis, respectively [12].

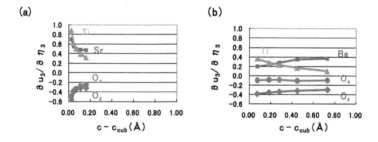

Figure 12. Evaluated values of $\partial u_3(k)/\partial \eta_3$ as a function of $c - c_{cub}$: (a) SrTiO$_3$ and (b) BaTiO$_3$ [12].

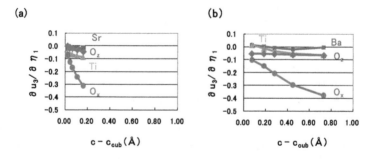

Figure 13. Evaluated values of $\partial u_3(k)/\partial \eta_1$ as a function of $c - c_{cub}$: (a) SrTiO$_3$ and (b) BaTiO$_3$ [12].

Let us discuss the reasons of the quantitative differences in e_{33} between SrTiO$_3$ and BaTiO$_3$. Figure 14(a) shows the difference between the Ti-O$_z$ distance ($R_{\text{Ti-Oz}}$) and the sum of the r_{Ti}

and r_{O_z} ($r_{Ti} + r_{O_z}$) along the [001] axis as a function of $c - c_{cub}$. Note that R_{Ti-O_z} is smaller than r_{Ti} $+ r_{O_z}$ in both SrTiO$_3$ and BaTiO$_3$. However, the difference in absolute value between R_{Ti-O_z} and $r_{Ti} + r_{O_z}$ in SrTiO$_3$ is smaller than the difference in BaTiO$_3$ for $0 \lesssim c - c_{cub} \lesssim 0.20$. This result suggests that the Ti-O$_z$ Coulomb repulsion along the [001] axis in SrTiO$_3$ is smaller than that in BaTiO$_3$ and that therefore the Ti ion of SrTiO$_3$ can be displaced more easily along the [001] axis than that of BaTiO$_3$. This would be a reason why the absolute values of $\partial u_3(k)/\partial \eta_3$ of Ti and O$_z$ ions in SrTiO$_3$ are larger than that in BaTiO$_3$. Figure 14(b) shows the difference between the A-O$_x$ distance (R_{A-O_x}) and the sum of r_A and r_{O_x} ($r_A + r_{O_x}$) on the (100) plane as a function of $c - c_{cub}$, where the values of the ionic radii are defined as Shannon's ones [23]. Note that R_{A-O_x} is smaller than $r_A + r_{O_x}$ in both SrTiO$_3$ and BaTiO$_3$. However, the difference in absolute value between R_{A-O_x} and $r_A + r_{O_x}$ in SrTiO$_3$ is much smaller than the difference in BaTiO$_3$ for $0 \lesssim c - c_{cub} \lesssim 0.20$. This result suggests that the Sr-O$_x$ Coulomb repulsion on the (100) plane in SrTiO$_3$ is much smaller than the Ba-O$_x$ Coulomb repulsion in BaTiO$_3$ and that therefore Sr and O$_x$ ions of SrTiO$_3$ can be displaced more easily along the [001] axis than Ba and O$_x$ ions of BaTiO$_3$. This would be a reason why the absolute values of $\partial u_3(k)/\partial \eta_3$ of Sr and O$_x$ ions in SrTiO$_3$ are larger than those of Ba and O$_x$ ions in BaTiO$_3$.

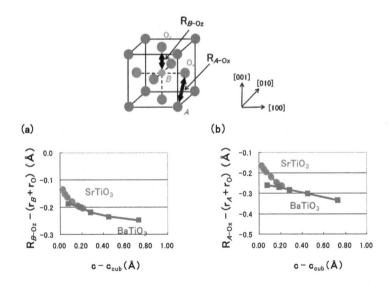

Figure 14. Evaluated values as a function of $c - c_{cub}$ in optimized tetragonal SrTiO$_3$ and BaTiO$_3$: (a) difference between the Ti-O$_z$ distance (R_{Ti-O_z}) and $r_{Ti} + r_{O_z}$. (b) difference between the A-O$_x$ distance (R_{A-O_x}) and $r_A + r_{O_x}$. R_{A-O_x} and R_{Ti-O_z} in ATiO$_3$ are also illustrated. Note that all the ionic radii are much larger and that A and Ti ions are displaced along the [001] axis in actual ATiO$_3$ [12].

Finally, in this subsection, we discuss the relationship between $\partial u_3(k)/\partial \eta_3$ and $c - c_{cub}$ in detail. Figure 15(a) shows the differences in the total energy (ΔE_{total}) as a function of u_3(Ti). In this figure, the properties of $SrTiO_3$ with $\eta_3 = 0.011$, $SrTiO_3$ with $\eta_3 = 0.053$ and fully opti-mized $BaTiO_3$ as a reference, are shown. Calculations of E_{total} were performed with the fixed crystal structures of previously optimized structures except Ti ions. Clearly, the magnitude of u_3(Ti) at the minimum points of the ΔE_{total} and the depth of the potential are closely relat-ed to the spontaneous polarization P_3 and the Curie temperature (T_C), respectively. On the other hand, e_{33} seems to be closely related to the deviation at the minimum points of the ΔE_{total}. Figure 15(b) shows illustrations of ΔE_{total} curves with deviations at the minimum points of the ΔE_{total} values, corresponding to the ΔE_{total} curves of $SrTiO_3$ in Fig. 15(a). Clearly, as η_3 becomes smaller, the deviated value at the minimum point of the ΔE_{total} values becomes smaller, i.e., the Ti ion can be displaced more favourably. On the other hand, as shown in Fig. 12(a), the absolute value of ∂u_3(Ti)$/\partial \eta_3$ becomes larger as η_3 becomes smaller.

Therefore, the Ti ion can be displaced more favourably as the deviated value at the mini-mum point of the ΔE_{total} values becomes smaller. The relationship between e_{33} and ∂u_3(Ti)$/\partial \eta_3$ is discussed in Sec. 3.2.3.

(a)   **(b)**

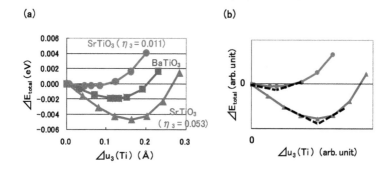

Figure 15. a) ΔE_{total} as a function of u_3(Ti) in tetragonal $SrTiO_3$ and $BaTiO_3$. (b) Illustration of the ΔE_{total} curves in tetrago-nal $SrTiO_3$ with $\eta_3 = 0.011$ and $SrTiO_3$ with $\eta_3 = 0.053$ with deviations at the minimum point of ΔE_{total} [14].

3.2.2. Proposal of new piezoelectric materials

The previous discussion in Sec. 3.2.1 suggests that the piezoelectric properties of e_{33} are closely related to the B-X Coulomb repulsions in tegtragonal ABX_3. In the viewpoint of the change of the B-X Coulomb repulsions, we recently proposed new piezoelectric materials [16, 17], i.e., $BaTi_{1-x}Ni_xO_3$ and $Ba(Ti_{1-3z}Nb_{3z})(O_{1-z}N_z)_3$.

It has been known that $BaNiO_3$ shows the 2H hexagonal structure as the most stable struc-ture in room temperature. Moreover, the ionic radius of Ni^{4+} (d^6) with the low-spin state in 2H $BaNiO_3$ is 0.48 Å, which is much smaller than that of Ti^{4+} (d^0), 0.605 Å, in $BaTiO_3$. There-fore, due to the drastic change in the $(Ti_{1-x}Ni_x)$-O Coulomb repulsions in tetragonal $BaTi_{1-x}Ni_xO_3$, the e_{33} piezoelectric values are expected to be larger than that in tetragonal $BaTiO_3$, especially around the morphotropic phase boundary (MPB). Figure 16(a) shows the total-energy difference ΔE_{total} between 2H and tetragonal structures of $BaTi_{1-x}Ni_xO_3$ as a function of x. The most stable structure changes at $x \approx 0.26$. Figure 16(b) shows $c - c_{cub}$ as a function of x. The $c - c_{cub}$ value shows 0 around $x = 0.26$, which suggests the appearance of the MPB, *i.e.*, the e_{33} piezoelectric value shows a maximum at $x \approx 0.26$.

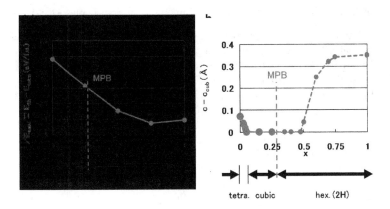

Figure 16. a) ΔE_{total} (total-energy difference between 2H and tetragonal structures), and (b) $c - c_{cub}$ of the tetragonal structure, as a function of x in $BaTi_{1-x}Ni_xO_3$ [16]. For or $0.26 \leq x \leq 1$, the tetragonal structure is not the most stable one.

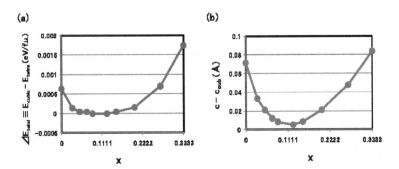

Figure 17. (a) ΔE_{total} (total-energy difference between cubic and tetragonal structures), and (b) $c - c_{cub}$, as a function of x in $Ba(Ti_{1-3z}Nb_{3z})(O_{1-z}N_z)_3$ [17].

Another proposal is tetragonal $Ba(Ti_{1-3z}Nb_{3z})(O_{1-z}N_z)_3$, which consists of $BaTiO_3$ and $BaNbO_2N$ [17]. Due to the change of $(Ti_{1-3z}Nb_{3z})$-$(O_{1-z}N_z)$ Coulomb repulsions, the e_{33} piezoelectric values are expected to be larger than that in tetragonal $BaTiO_3$. Recent experimental paper reported that the most stable structure of $BaNbO_2N$ is cubic in room temperature [34]. Contrary to the experimental result, however, our calculations suggest that the tetragonal structure will be more stable than the cubic one, as shown in Fig. 17(a). Figure 17(b) shows $c - c_{cub}$

as a function of x. The $c - c_{cub}$ value shows almost 0 at $x \approx 0.12$. Although the MPB does not appear in tetragonal $Ba(Ti_{1-3z}Nb_{3z})(O_{1-z}N_z)_3$, the e_{33} piezoelectric values are expected to show a maximum at $x \approx 0.12$.

3.2.3. Piezoelectric properties of in tetragonal ABX_3

In the following, we discuss the role of the B-X Coulomb repulsions in piezoelectric ABX_3.

Figures 18(a) and 18(b) show the piezoelectric properties of e_{33} as a function of the value $c - c_{cub}$ in tetragonal ABX_3, where c_{cub} denotes the c lattice parameter in cubic ABX_3; $c - c_{cub}$ is a closely related parameter to η_3. For ABX_3, $SrTiO_3$, $BaTiO_3$ and $PbTiO_3$ with the c lattice parameter and all the inner coordination optimized for fixed a, and $BaTi_{1-x}Ni_xO_3$ ($0 \leq x \leq 0.05$), $Ba(Ti_{1-3z}Nb_{3z})(O_{1-z}N_z)_3$ ($0 \leq z \leq 0.125$), $Ba_{1-y}Sr_yTiO_3$ ($0 \leq y \leq 0.5$), $BaTi_{1-x}Zr_xO_3$ ($0 \leq x \leq 0.06$), and $BiM'O_3$ (M' = Al, Sc) with fully optimized, were prepared [15]. Note that e_{33} becomes larger as $c - c_{cub}$ becomes smaller and that the trend of e_{33} is almost independent of the kind of A ions. Moreover, note also that e_{33} of $BaTi_{1-x}Ni_xO_3$ and that of $Ba(Ti_{1-3z}Nb_{3z})(O_{1-z}N_z)_3$ show much larger values than the other ABX_3.

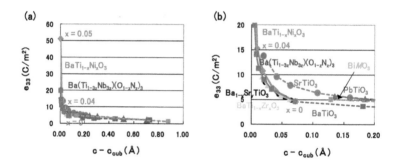

Figure 18. e_{33} as a function of $c - c_{cub}$ for different scales [15].

Let us discuss the relationship between $\partial u_3(k)/\partial \eta_3$ and $c - c_{cub}$ in $BaTi_{1-x}Ni_xO_3$ and $BaTiO_3$ in the following. Figures 19(a) and 19(b) show the $\partial u_3(k)/\partial \eta_3$ values. In these figures, O_x and O_z denote oxygen atoms along the [100] and [001] axes, respectively. Clearly, the absolute values of $\partial u_3(k)/\partial \eta_3$ in $BaTi_{1-x}Ni_xO_3$ are much larger than those in $BaTiO_3$. Moreover, in comparison with Fig.18, properties of e_{33} are closely related to those of $\partial u_3(k)/\partial \eta_3$. Figure 20(a) shows the difference between R_{B-Oz} and $r_B + r_{Oz}$ along the [001] axis, and Fig. 20(b) shows the difference between R_{A-Ox} and $r_A + r_{Ox}$ on the (100) plane for several ABO_3, as a function of $c - c_{cub}$. Clearly, the difference between R_{B-Oz} and $r_B + r_{Oz}$ is closely related to e_{33} shown in Fig. 18, rather than the difference between R_{A-Ox} and $r_A + r_{Ox}$. Moreover, note that the difference in absolute value between R_{B-Oz} and $r_B + r_{Oz}$ in $BaTi_{1-x}Ni_xO_3$ is much smaller than that in $BaTiO_3$. This result suggests that the $(Ti_{1-x}Ni_x)$-O_z Coulomb repulsion along the [001] axis in $BaTi_{1-x}Ni_xO_3$ is much smaller than the Ti-O_z Coulomb repulsion in $BaTiO_3$ and that therefore $Ti_{1-x}Ni_x$ ion of $BaTi_{1-x}Ni_xO_3$ can be displaced more easily along the [001] axis than Ti ion of

BaTiO$_3$. This must be a reason why the absolute value of $\partial u_3(k)/\partial \eta_3$ of Ti$_{1-x}$Ni$_x$ and O$_z$ ions in BaTi$_{1-x}$Ni$_x$O$_3$ is larger than those in BaTiO$_3$.

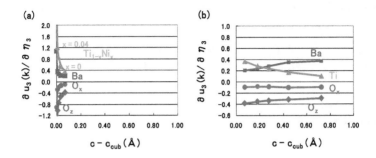

Figure 19. $\partial u_3(k)/\partial \eta_3$ as a function of $c - c_{cub}$: (a) BaTi$_{1-x}$Ni$_x$O$_3$ and (b) BaTiO$_3$ [15].

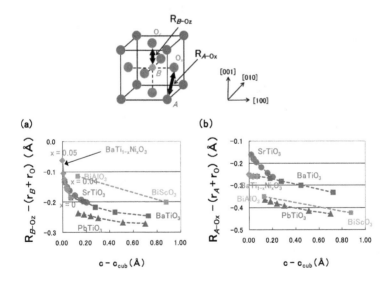

Figure 20. Evaluated values of in optimized tetragonal BaTi$_{1-x}$Ni$_x$O$_3$, BaTiO$_3$, and several ABO_3, as a function of $c - c_{cub}$: (a) $R_{B\text{-}Oz}$ - $(r_B + r_{Oz})$, and (b) $R_{A\text{-}Ox}$ - $(r_A + r_{Ox})$ [15].

Figure 21(a) shows ΔE_{total} as a function of the displacement of the Ti$_{1-x}$Ni$_x$ ions with fixed crystal structures of fully-optimized BaTi$_{1-x}$Ni$_x$O$_3$. Calculations of E_{total} were performed with the fixed crystal structures of previously optimized structures except Ti$_{1-x}$Ni$_x$ ions. The deviated value at the minimum point of ΔE_{total}, i.e., $\partial(\Delta E_{total})/\partial u_3(\text{Ti}_{1-x}\text{Ni}_x)$, becomes smaller as x becomes larger. Moreover, both e_{33} and $\partial u_3(Ti_{1-x}Ni_x)/\partial\eta_3$ become larger as x becomes larger, as shown in Figs. 18 and 19. This result is consistent with the result of $SrTiO_3$ shown in Fig. 15(a).

Let us discuss the above reasons in the following. $\partial(\Delta E_{total})/\partial u_3(Ti_{1-x}Ni_x)$ can be written as

$$\left(\frac{\partial\Delta E_{total}}{\partial u_3(Ti_{1-x}Ni_x)}\right) = \left(\frac{\partial\Delta E_{total}}{\partial\eta_3}\right) \times \left(\frac{\partial u_3(Ti_{1-x}Ni_x)}{\partial\eta_3}\right)^{-1}. \tag{5}$$

As shown in Fig. 21(b), $\partial(\Delta E_{total})/\partial\eta_3$ is almost constant, and therefore, $\partial(\Delta E_{total})/\partial u_3(Ti_{1-x}Ni_x)$ is almost proportional to $(\partial u_3(Ti_{1-x}Ni_x)/\partial\eta_3)^{-1}$, i.e.,

$$\left(\frac{\partial\Delta E_{total}}{\partial u_3(Ti_{1-x}Ni_x)}\right) \propto \left(\frac{\partial u_3(Ti_{1-x}Ni_x)}{\partial\eta_3}\right)^{-1}. \tag{6}$$

On the other hand, according to Eq. (3), e_{33} becomes larger as $\partial u_3(Ti_{1-x}Ni_x)/\partial\eta_3$ becomes larger. This is a reason why e_{33} becomes larger as $\partial(\Delta E_{total})/\partial u_3(Ti_{1-x}Ni_x)$ becomes smaller. This result is consistent with the result of $SrTiO_3$ discussed in Sec. 3.2.1.

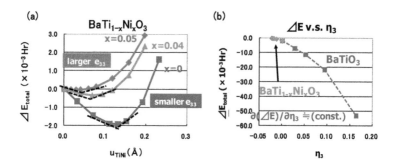

Figure 21. (a) ΔE_{total} as a function of $u_3(Ti_{1-x}Ni_x)$ in $BaTi_{1-x}Ni_xO_3$. Results with $x = 0.05, 0.04$, and 0 are shown. Dashed lines denote guidelines of $\partial(\Delta E_{total})/\partial u_3(Ti_{1-x}Ni_x)$ for each x. (b) ΔE_{total} as a function of η_3 for $BaTi_{1-x}Ni_xO_3$ and $BaTiO_3$ [15].

Finally, we comment on the difference in the properties between e_{33} and d_{33} in tetragonal ABX_3. Figures 22(a) and 22(b) show the piezoelectric properties of d_{33} as a function of $c - c_{cub}$. Note that the trend of d_{33} is closely dependent on the kind of A ions. This result is in contrast with the trend of e_{33} as shown in Fig. 18. As expressed in Eq. (4), d_{33} is closely related to the elastic compliance $s_{3j}{}^E$ as well as e_{3j}. In fact, the absolute value of $s_{3j}{}^E$ in $BiBX_3$ or $PbBX_3$ is generally larger than that in ABX_3 with alkaline-earth A ions. This result must be due to the larger Coulomb repulsion of Bi-X or Pb-X derived from 6s electrons in Bi (Pb) ion.

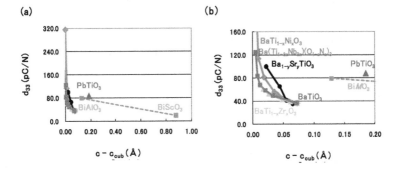

Figure 22. d_{33} as a function of $c - c_{cub}$ for different scales [15].

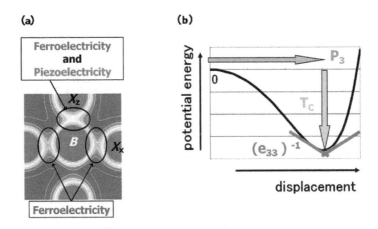

Figure 23. (a) Illustration of the relationship between the B-X Coulomb repulsions and the ferroelectric and piezoelectric states in tetragonal ABX_3. (b) Illustration of the relationship between e_{33} and the deviation. P_3 and T_C denote the spontaneous polarization and the Curie temperature, respectively.

4. Conclusion

We have discussed a general role of the B-X Coulomb repulsions for ferroelectric and piezoelectric properties of tetragonal ABX_y based on our recent papers and patents. We have found that both ferroelectric state and piezoelectric state are closely related to the B-X$_x$ Coulomb repulsions as well as the B-X$_z$ ones, as illustrated in Fig. 23(a). Moreover, as illustrated in Fig. 23(b), we have also found that e_{33} is closely related to the deviation at the minimum point of the ΔE_{total}.

Acknowledgements

We thank Professor M. Azuma in Tokyo Institute of Technology for useful discussion. We also thank T. Furuta, M. Kubota, H. Yabuta, and T. Watanabe for useful discussion and computational support. The present work was partly supported by the Elements Science and Technology Project from the Ministry of Education, Culture, Sports, Science and Technology.

Author details

Kaoru Miura[1] and Hiroshi Funakubo[2]

1 Canon Inc., Tokyo, Japan

2 Tokyo Institute of Technology, Yokohama, Japan

References

[1] Cohen R E. Origin of ferroelectricity in perovskite oxides. Nature 1992; 451 (6382) : 136-8.

[2] Wu Z, Cohen R E. Pressure-induced anomalous phase transitions and colossal enhancement of piezoelectricity in $PbTiO_3$. Phys. Rev. Lett. 2005; 95 (3): 037601, and related references therein.

[3] Deguez O, Rabe K M, Vanderbilt D. First-principles study of epitaxial strain in perovskites. Phys. Rev. B 2005; 72 (14): 144101, and related references therein.

[4] Qi T, Grinberg I, Rappe A M. First-principles investigation of the high tetragonal ferroelectric material $BiZn_{1/2}Ti_{1/2}O_3$. Phys. Rev. B 2010; 79 (13): 134113, and related references therein.

[5] Wang H, Huang H, Lu W, Chan H L W, Wang B, Woo C H. Theoretical prediction on the structural, electronic, and polarization properties of tetragonal Bi_2ZnTiO_6. J. Appl. Phys. 2009; 105 (5): 053713.

[6] Dai J Q, Fang Z. Structural, electronic, and polarization properties of Bi_2ZnTiO_6 supercell from first-principles. J. Appl. Phys. 2012; 111 (11): 114101.

[7] Khenata R, Sahnoun M, Baltache H, Rerat M, Rashek A H, Illes N, Bouhafs B. First-principle calculations of structural, electronic and optical properties of $BaTiO_3$ and $BaZrO_3$ under hydrostatic pressure. Solid State Commun. 2005; 136 (2): 120-125.

[8] Oguchi T, Ishii F, Uratani Y. New method for calculating physical properties from first principles-piezoelectric and multiferroics. Butsuri 2009; 64 (4): 270-6 [in Japanese].

[9] Uratani Y, Shishidou T, Oguchi T. First-principles calculations of colossal piezoelectric response in thin film PbTiO$_3$. Jpn. Soc. Appl. Phys. 2008: conference proceedings, 27-30 March 2008, Funabashi, Japan [in Japanese].

[10] Miura K, Kubota M, Azuma M, and Funakubo H. Electronic and structural properties of BiZn$_{0.5}$Ti$_{0.5}$O$_3$. Jpn. J. Appl. Phys. 2009; 48 (9): 09KF05.

[11] Miura K, Furuta T, Funakubo H. Electronic and structural properties of BaTiO$_3$: A proposal about the role of Ti 3s and 3p states for ferroelectricity. Solid State Commun. 2010; 150 (3-4): 205-8.

[12] Furuta T, Miura K. First-principles study of ferroelectric and piezoelectric properties of tetragonal SrTiO$_3$ and BaTiO$_3$ with in-plane compressive structures. Solid State Commun. 2010; 150 (47-48): 2350-3.

[13] Miura K, Azuma M, Funakubo H. Electronic and structural properties of ABO$_3$: Role of the B-O Coulomb repulsions for ferroelectricity. Materials 2011; 4 (1): 260-73. http://www.mdpi.com/1996-1944/4/1/260 (accessed 20 July 2012).

[14] Miura K. First-principles study of ABO$_3$: Role of the B-O Coulomb repulsions for ferroelectricity and piezoelectricity. Lallart M. (ed.) Ferroelectrics - Characterization and Modeling. Rijeka: InTech; 2012. p395-410. Available from http://www.intechopen.com/books/ferroelectrics-characterization-and-modeling/first-principles-study-of-abo3-role-of-the-b-o-coulomb-repulsions-for-ferroelectricity-and-piezoelec (accessed 20 July 2012).

[15] Miura K, Funakubo H. First-principles analysis of tetragonal ABO$_3$: Role of the B-O Coulomb repulsions for ferroelectricity and piezoelectricity. Proceedings of 15th US-Japan Seminar on Dielectric and Piezoelectric Ceramics, 6-9 November 2011, Kagoshima, Japan.

[16] Miura K, Ifuku T, Kubota M, Hayashi J. Piezoelectric Material. Japan Patent 2011-001257 [in Japanese].

[17] Miura K, Kubota M, Hayashi J, Watanabe T. Piezoelectric Material. submitted to Japan Patent [in Japanese].

[18] Miura K, Furuta T. First-principles study of structural trend of BiMO$_3$ and BaMO$_3$: Relationship between tetragonal and rhombohedral structure and the tolerance factors. Jpn. J. Appl. Phys. 2010; 49 (3): 031501 (2010).

[19] Miura K, Kubota M, Azuma M, and Funakubo H. Electronic, structural, and piezoelectric properties of BiFe$_{1-x}$Co$_x$O$_3$. Jpn. J. Appl. Phys. 2010; 49 (9): 09ME07.

[20] Miura K, Tanaka M. Electronic structures of PbTiO$_3$: I. Covalent interaction between Ti and O ions. Jpn. J. Appl. Phys. 1998; 37 (12A): 6451-9.

[21] LDA_TM_psp1_data: ABINIT.   http://www.abinit.org/downloads/psp-links/
      lda_tm_psp1_data/ (accessed 20 July 2012).

[22] Kuroiwa Y, Aoyagi S, Sawada A, Harada J, Nishibori E, Tanaka M, and Sakata M.
      Evidence for Pb-O covalency in tetragonal $PbTiO_3$. Phys. Rev. Lett. 2001; 87 (21):
      217601.

[23] Shannon R D. Revised effective ionic radii and systematic studies of interatomic dis-
      tances in halides and chalcogenides. Acta Crystallogr. Sect. A. 1976; 32 (5): 751-67.

[24] Suchomel M R, Fogg A M, Allix M, Niu H, Claridge J B, Rosseinsky M J. Bi_2ZnTiO_6:
      A lead-free closed-shell polar perovskite with a calculated ionic polarization of 150
      $\mu C\ cm^{-2}$. Chem. Mater. 2006; 18 (21), 4987-9.

[25] Khalyavin D D, Salak A N, Vyshatko N P, Lopes A B, Olekhnovich N M, Pushkarev
      A V, Maroz I I, Radyush T V. Crystal structure of metastable perovskite
      $Bi(Mg_{1/2}Ti_{1/2})O_3$: Bi-based structural analogue of antiferroelectric $PbZrO_3$. Chem. Ma-
      ter. 2006; 18 (21), 5104-10.

[26] Randall C A, Eitel R, Jones B, Shrout T R, Woodward D I, Reaney I M, Investigation
      of a high T_C piezoelectric system: $(1-x)Bi(Mg_{1/2}Ti_{1/2})O_3-(x)PbTiO_3$. J. Appl. Phys. 2004;
      95 (7): 3633-9.

[27] Rossetti Jr G A, Khachaturyan A G. Concepts of morphotropism in ferroelectric solid
      solutions. Proceedings of 13th US-Japan Seminar on Dielectric and Piezoelectric Ce-
      ramics, 4-7 November 2007, Awaji, Japan, and related references therein.

[28] Damjanovic D. A morphotropic phase boundary system based on polarization rota-
      tion and polarization extension. Appl. Phys. Lett. 2010; 97 (6): 062906, and related ref-
      erences therein.

[29] Gonze X, Beuken J-M, Caracas R, Detraux F, Fuchs M, Rignanese G-M, Sindic L, Ver-
      straete M, Zerah G, Jollet F, Torrent M, Roy A, Mikami M, Ghosez P, Raty J-Y, Allan
      D C. First-principles computation of material properties: the ABINIT software
      project. Comput. Mater. Sci. 2002; 25 (3), 478-92.

[30] Hohenberg P, Kohn W. Inhomogeneous electron gas. Phys. Rev. 1964; 136 (3B):
      B864-71.

[31] Opium - pseudopotential generation project. http://opium.sourceforge.net/
      index.html (accessed 20 July 2012).

[32] Ramer N J, Rappe A M. Application of a new virtual crystal approach for the study
      of disordered perovskites. J. Phys. Chem. Solids. 2000; 61 (2): 315-20.

[33] Resta R. Macroscopic polarization in crystalline dielectrics: the geometric phase ap-
      proach. Rev. Mod. Phys. 1994; 66 (3): 899-915.

[34] Kim Y-I, Lee E. Constant-wavelength neutron diffraction study of cubic perovskites
      $BaTaO_2N$ and $BaNbO_2N$. J. Ceram. Soc. Jpn. 2011; 119 (5): 371-4.

Chapter 2

Self Assembled Nanoscale Relaxor Ferroelectrics

Ashok Kumar, Margarita Correa, Nora Ortega,
Salini Kumari and R. S. Katiyar

Additional information is available at the end of the chapter

1. Introduction

Worldwide research on relaxor ferroelectric (RFE) has been carried out since 1950s. There are several schools in the world who have defined the evolution and origin of relaxor properties in the ferroelectric materials in their own way. One common consensus among the scientists is the presence of polar nano regions (PNRs) i.e. self assembled domains of short range ordering typically of less than 20-50 nm in the ferroelectric relaxor materials which causes the dielectric dispersion near the phase transition temperature. Another common approach has been also believed among the relaxor ferroelectric scientists i.e. the existence of random field at nanoscale. Random field model is considered on the experimental and theoretical facts driven from the different dielectric dispersion response under zero field cooled (ZFC) and field cooled (FC) among the relaxors.

Overall relaxor ferroelectrics have been divided into two main categories such as (i) classical relaxors (only short range ordering), the most common example is $PbMg_{1/3}Nb_{2/3}O_3$ (PMN), (ii) Semi-classical relaxor ferroelectrics (a combination of short and long range ordering). In the latter case, the relaxor properties can be arises due to compositional inhomogeneities, artificially induced strain, growth conditions (temperature, pressure, medium, etc.), and due to different ionic radii mismatched based chemical pressure in the matrix. The local domains (PNRs) reorientation induces polar-strain coupling which makes RFE the potentially high piezoelectric coefficient materials widely used in Micro/Nanoelectromechanical system (MEMS/NEMS). The basic features of the RFE over a wide range of temperatures and frequencies are as follows (i) dispersive and diffuse phase transition, (ii) partially disordered structure, (iii) existence of polar nano-ordered regions, etc. over a wide range of temperatures and frequencies. Physical and functional properties of relaxors are very different from normal ferroelectrics due to the presence of self assembled ordered regions in the configurational disorder matrix.

This article deals with the historical development of the relaxor ferroelectrics, microstructural origin of RFE, strain mediated conversion of normal ferroelectric to relaxor ferroelectrics, superlattice relaxors, lead free classical relaxor ferroelectrics, distinction of classical relaxors and semiclassical ferroelectric relaxors based on the polarization study, and their potential applications in microelectronic industry.

2. History

Relaxor ferroelectric materials were discovered in complex perovskites by Smolenskii [1] in early fifties of twentieth centuries. Since then a vast scientific communities have been working on relaxor materials and related phenomena but the plethora of mesoscopic and microscopic heterogeneities over a range of lengths and timescales made difficult to systematic observations [2]. The microscopic image and the dielectric response of relaxor are qualitatively different for that of normal ferroelectrics. The universal signature of RFE is as follows [3 - 18]: (i) the occurrence of broad frequency-dependent peak in the real and imaginary part of the temperature dependent dielectric susceptibility (χ') or permittivity (ε') which shifts to higher temperatures with increasing frequency, G. A. Smolenskii and A. I. Agranokskaya, Soviet Physics Solid State 1, 1429 (1959), (ii) Curie Weiss law is observed at temperatures far above dielectric maxima temperature (T_m), (iii) slim hysteresis loops because polar domains are nanosized and randomly oriented, (iv) existence of polar nano domains far above T_m, (v) no structural phase transition (overall structure remains pseudo-cubic on decreasing temperature but rhombohedral-type distortion occurs in crystal at local level) across T_m in relaxors in contrast with the normal ferroelectric in which phase transition implies a macroscopic symmetry change,(vii) history dependent functional properties and skin effects (surface behavior is quite different than the interior of the system) (vi) in most cases, relaxor crystal showed no optical birefringence (either far below the freezing temperature and/or under external electric field).

As we know ferroelectric and related phenomenon are synonyms of ferromagnetism, antiferromagnetism, spin glass, and random field model which have been originally studied for magnetic materials. Relaxors ferroelectrics possess a random field state, as initially proposed by Westphal, Kleemann and Glinchuk [16-18]. Near Burns temperature, small polar nano regions start originating in different directions inside the crystal at mesoscopic scale; however, these polar nano regions are not static in nature at Burns temperature. At low temperature these polar nanoregions (PNRs) become static and developed a more defined regions called random field, the nature of these fields are short range order. The evidence of the existence of nanosized polar domains at temperatures well above the dynamic transition temperature comes for experimental observations. The presence of these polar nanoregions (PNRs) was established by measuring properties that depend on square of the polarization (P^2) and by direct imaging with TEM studies. The first evidences came from a report by Burns and Dacol [5] who studied the electro-optical effect in a PMN crystal. In case of normal FE, the temperature dependence of the refraction index (n) show a linear decrease of n from the paraelectric phase down to phase transition temperature (T_c), below T_c n deviates from linearity. The deviation is proportional to the P^2 and it increases as the polarization changes with temperature. Burns and Dacol observed deviation from linear $n(T)$ in relaxor PMN crystal well above T_m, the onset of the deviation was observed at 620 K, almost 350 K above T_m. The temperature in which onset of deviation from linearity of $n(T)$ occurs (in any relaxor) had been identified in the literature as the Burns temperature $(T_B$ or $T_d)$.

Electrostriction is a property which depends on P^2. Thermal strain in a cubic perovskite has two components, one due to the linear coefficient of thermal expansion, α, and another due to electrostriction accounted by the electrostrictive coefficients Q_{ijkl}. Measures on the temperature dependence of thermal expansion in RFE crystals have shown that the contribution due to electrostriction vanishes only above T_B where P=0.

The existence of PNRs well above T_m and the growth of these domains with decreasing temperature have been demonstrated by TEM [8 - 12]. Transmission electron studies also account for the B-cation order in complex perovskites [11 - 12]. Ordering of the B-site cations occurs if there is sufficiently large interaction energy between neighboring cations.

3. Theory of relaxor ferroelectricity

The general formula of perovskite having A and B-site cations with different charge states can be written as $A'_x A''_{1-x} B'_y B''_{1-y} O_3$. The randomness occupancy of A', A'' and B', B'' in A, B site respectively, depends on the ionic sizes, distribution of cations ordering at sub lattices and charge of cations. If the charges of cations at B-sites are same it is unlikely to have the polar randomness at nanoscale. Long range order (LRO) is defined as a continuous and ordered distribution of the B cations on the nearest neighbor sites. The short coherence length of LRO occurs when the size of the ordered domains are in a range of 20 to 800 Å in diameter. The long coherence range of LRO occur when the size of the ordered domains are much greater than 1000 Å. Randall et al. made a classification of complex lead perovskite and their solid solutions based on B-cation order studied by TEM and respective dielectric, X-rays and optical properties [11]-[12]. This classification divides the complex lead perovskites into three subgroups; random occupation or disordered, nanoscale or short coherent long-range order and long coherent long-range order of B-site cations.

Diffuse phase transition behavior is characteristic of the disordered structures. In these structures, random lattice disorder introduces dipolar impurities and defects that influence the static and dynamic properties of these materials. The presence of the dipolar entities on a lattice site of the highly polarizable FE structure, induced dipoles in a region determined by the correlation length (r_c). The correlation length is a measure of the extent of dipoles that respond in a correlated manner. In normal FE, r_c is larger than the lattice parameter (a) and it is strongly temperature dependent. On decreasing the temperature, a faster increase of r_c promotes growing of polar domains yielding a static cooperative long-range ordered FE state at $T < T_c$. This is not the case for RFE where a small correlation length of dipoles leads to formation of polar nanodomains frustrating the establishment of long-range FE state. Therefore, the dipolar nanoregions form a dipolar-glass like or relaxor state at low temperature with some correlation among nanodomains.

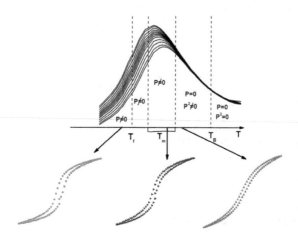

Figure 1. Temperature evolution of dielectric constant showing the characteristic temperatures in RFE. Representative hysteresis loops for each temperature interval are showed below.

These PNRs are dynamic and they experience a slowing down of their fluctuations at $T \leq T_m$. The dielectric relaxation does not fit the classical Debye relaxation model; instead there is a distribution of relaxation times related to the sizes of the nanodomains. The temperature dependence of dielectric constant as shown in Fig. 1 indentifies the main temperatures associated with relaxor ferroelectric behavior. These temperatures are T_B, T_m and T_f, we already defined T_B and T_m but the definition of T_f follow from the fit of the frequency dispersion of T_m with the Vogel-Fulcher law [15]. The dynamics of polar nanoregions does not follow Arrhenius type temperature dependence; instead nice fit of the frequency dispersion for each relaxor system is obtained with the Vogel-Fulcher law:

$$f = f_0 exp\left(\frac{-E_a}{k_B(T_m - T_f)}\right) \tag{1}$$

where f_0 is the attempt frequency which is related to the cut-off frequency of the distribution of relaxation times, E_a is the activation barrier to dipole reorientation, T_m is the dielectric maxima temperature, and T_f is the freezing temperature where polarization fluctuations "frozen-in".

The temperature dependence of the dielectric constant below T_B and in the vicinity of T_m is normally fitted by the empirical power law [19]:

$$\frac{1}{\varepsilon(\omega,\ T)} = \left(\frac{1}{\varepsilon_{max}(\omega,\ T)}\right)\left(1 + \frac{(T\ -\ T_{max})^{\gamma}}{2\delta^2}\right) \tag{2}$$

where γ and δ are parameters describing the degree of relaxation and diffuseness of the transition respectively. The parameter γ varies between 1 and 2, where values closer to 1 indicate normal ferroelectric behavior whereas values close to 2 indicate good relaxor behavior. Above T_B the inverse of dielectric permittivity is fitted by the Curie-Weiss law:

$$\chi = \frac{C}{T\ -\ \Theta} \tag{3}$$

where C and Θ are the Curie constant and Curie temperature respectively.

Last sixty years of extensive research in the field of relaxor ferroelectrics, several models have been proposed to explain the unusual dielectric behavior of these materials. Some of these models are: statistical composition fluctuations [20, 21], superparaelectric model [22], dipolar glass model [23], random field model [24], spherical random field model [25] etc. These models can explain much of the experimental observed facts but a clear understanding of the relaxor nature or even a comprehensive theory is not available yet. Despite the absence of a comprehensive theory of relaxor ferroelectricity, the literature agree to define relaxor materials in terms of the existence of polar nano regions as motioned above, these ordered regions exist in a disordered environment.

4. Strain induced relaxor phenomenon

The influence of epitaxial clamping on ferroelectric properties of thin films to rigid substrates has been largely studied and successfully explained in a Landau-Ginzburg-Devonshire (LGD) framework [26-30]. However, when the same treatment was applied to RFE, discrepancies among the predicted values and those observed in experiments were found [30]. For instance, a downward shift in T_m of PMN films in the presence of compressive in-plane strain was found, contrary to the expected upward shift. Catalan [30] et al. developed a model to analyze the influence of epitaxial strain on RFE. This model is based on LGD theory but assuming a quadratic dependence on $(T-T_m)$, rather than linear, for the first coefficient of the free energy. Catalan's model shows that the shift in T_m for relaxors thin films do not depend on the sign of the epitaxial mismatch strain, but rather on the thermal expansion mismatch between substrate and film. Reference [30] also provided a list of compounds which exhibited shift in T_m when they were grown in film forms. We have already observed in-plane strain in our $Pb(Sc_{0.5}Nb_{0.25}Ta_{0.25})O_3(PSNT)/La_{0.5}Sr_{0.5}CoO_3$ (LSCO)/MgO heterostructure, which may have caused the measured shift of T_m with respect to bulk values [31-33].

We have calculated shift in the peak position of the dielectric maxima of PSNT films compared to bulk values following the model developed by Catalan et.al. The model is as follow: In LGD formalism, the thermodynamical potential of a perovskite dielectric thin films is described as [27-28].

$$\Delta G = \left(\tfrac{1}{2}\alpha - Q_{i3}X_i\right)P_3^2 + \tfrac{1}{4}\beta P_3^2 - \tfrac{1}{2}s_{ij}(X_i X_j) \tag{4}$$

where α and β correspond to linear and nonlinear terms of the inverse permittivity, and s_{ij}, X_i, and Q_{i3} are the elastic compliances, the stress tensor, and the electrostriction tensor in Voigt notation; P_3 is out of the plane polarization. Strain gradients across the films are neglected. The inverse permittivity is the second derivative of the free energy with respect to the polarization:

$$\chi_3' = \frac{\partial^2 \Delta G}{\partial P^2} = \alpha - 4Q_{13}\frac{Y}{1-v}x_m + 3\beta P_3^2 \tag{5}$$

where x_m, Y, and v are the mismatch strain, Young's modulus, and Poisson's ratio of the film respectively. For conventional ferroelectrics, α is usually expanded in a power series of $(T-T_c)$:

$$\alpha = \frac{1}{C\varepsilon_0}(T - T_c) + f(T - T_c)^2 \tag{6}$$

which directly leads to Curie-Weiss behavior, where C is the Curie constant. Substituting α into the above equation showing that the strain shifts the critical temperature for a ferroelectric by:

$$\Delta T_c = 4C\varepsilon_0 Q_{13}\frac{Y}{1-v}x_m \tag{7}$$

Since Q_{13} is always negative, it has been observed that the shift in the critical temperature depends on the sign of in-plain strain: T_c decreases for the tensile strain and it increases for compressive strain. Equation 7 is not adequate to explain the shifts in the dielectric maximum temperature for the relaxor ferroelectric materials

In case of PSNT films, we observed a quadratic temperature dependence of the permittivity:

$$\frac{1}{\varepsilon(\omega, T)} - \frac{1}{\varepsilon_m(\omega, T)} = \frac{1}{2\varepsilon_m\varepsilon_0}(T - T_m)^2 \tag{8}$$

Since the inverse of the dielectric constant is the second derivative of free energy with respect to polarization (Equation 5), the measured dielectric constant can be related to the appropriate coefficients in the LGD thermodynamical potential. For P=0 and x_m=0 (in the absence of strain), χ correlates with the first coefficient of the Landau equation, i.e.

$$\chi(T, f) = \alpha(T, f) = \frac{1}{2\varepsilon_m \varepsilon_0}(T - T_m)^2 \tag{9}$$

We substituted the values of α in Equation 5 and calculated the dielectric maxima temperature under strain (T_m'). To calculate T_m', we differentiated the inverse of permittivity with respect to temperature at T_m':

$$\frac{\partial \chi'}{\partial T_m'} = 0 = \frac{1}{\varepsilon_0 \varepsilon_m \delta^2}(T_m' - T_m) - 4Q_{13}\frac{Y}{1-\nu}\frac{\partial x_m}{\partial T} \tag{10}$$

where it is assumed that Q_{13}, Y, and ν do not change substantially [34] with temperature in the dielectric maxima regions; only the mismatch strain changes due to differential thermal expansion:

$$\frac{\partial x_m}{\partial T} = \frac{\partial}{\partial T}\left[\frac{\left(a_{film}\left[1 - \lambda_{substrate}(T - T_0)\right] - a_{bulk}\right)}{a_{bulk}}\right] \tag{11}$$

where $\lambda_{substrate}$ is the coefficient of thermal expansion of the substrate, a_{film} and a_{bulk} are the lattice constant of the film and the bulk respectively. Using Equations, 9, 10, and 11 the expected shift in T_m is:

$$\Delta T_m = 4\delta^2 \varepsilon_m \varepsilon_0 Q_{13}\frac{Y}{1-\nu}\frac{a_{film}}{a_{bulk}}\lambda_{substrate} \tag{12}$$

We have used the above equation to calculate the shift in the position of dielectric maxima of PSNT films. Due to lack of experimental mechanical data in literature for PSNT films or ceramics, we have used the standard Q_{13} for PST thin films [34] and Y value for the PMN single crystal [35] other values we got from our experimental observation i.e. (δ, ε_m, a_{film}, a_{bulk}). For MgO substrate, $\lambda_{substrate} = 1.2\times10^{-5}$ K^{-1}. Putting all these data in Equation 12 we got a shift in the dielectric maximum temperature (T_m) of the same magnitude order than those of experimental observation.

5. Experimental observation of strain induced relaxor phenomenon

Fig. 2 shows the dielectric behavior and the microstructures of the PSNT thin films and their bulk matrix [31-33]. It has been shown that the two-dimensional (2D) clamping of films by the substrate may change profoundly the physical properties of ferroelectric heterostructure with respect to the bulk material. We observed a transfer and relaxing of epitaxial strain between the layers due to in-plane oriented heterostructure. However the mismatch strain due to different thermal expansion properties of each layer provokes a shift of 62 K in the temperature of the dielectric maxima.

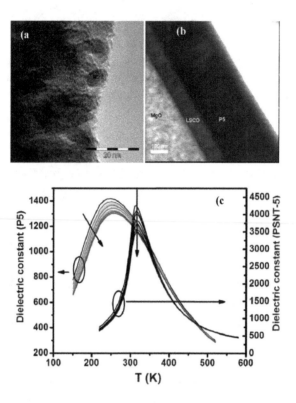

Figure 2. Microstructure of the PSNT bulk nanoceramics (a) and the thin films (b) grown on the LSCO coated MgO substrate, and a drastic shift in dielectric maxima temperature (c) in PSNT thin film with respect same compositions of their bulk counterpart. The microstructures of film and ceramic support the existence of strain induced relaxor phenomenon, as their microstructures- property correlation indicate the presence of some ordered polar nano-regions in the thin films whereas absence of such effect in bulk. Figure 2(a) was adapted from Ref. [31] and reproduced with permission (© 2008 John Wiley and Sons).

Transmission electron microscopy (TEM) images of PSNT in bulk (Fig. 2(a)) and thin film (Fig. 2(b)) form are shown in Fig. 2. The interlayer of about 20 nm between MgO and LSCO and the fibril structures along a single direction in the film confirm the in-plane strain state of the film. TEM image of bulk shows well ordered nanoregions in the grain matrix that meet a basic ingredient to produces relaxor behavior: the existence of nano-ordered regions surrounding by a disordered structure, however, it is unable to produce it. It has been observed from the neutron diffraction data that if the ordered nanoregions are in the range of ~ 5-10 nm, then it is capable of yielding frequency dispersion in the dielectric spectra [30]. But the relaxor state also depends on the coherency with what the dipoles respond to the probing field (and frequency). They can establish long range or short range coherent response. PSNT ceramic was incapable to produce the frequency dispersion behavior, but in thin films form the in plane strain and the breaking of long range ordering response, induced dielectric dispersion and shifting in dielectric maxima temperature towards lower temperature side. We have observed from the temperature evolution of the Raman spectra of bulk PSNT ceramic [31-33], the competitions between ordered domains of short and long-range order due to Nb- and Ta-rich regions respectively. The average disorder arrangement of $Sc^{3+}/Nb^{5+}/Ta^{5+}$ ions in the B-site octahedra leads to the observed diffuse phase transition. In film only short range ordering in B'/B" ions was developed within the disordered matrix. Although bulk matrix exhibited nano-ordered regions its average size must be higher than film. Stress effects change the ionic positions somehow favoring short-range ordering. These microstructural differences trigger the different observed dielectric responses. The microstructure-property relation of PSNT thin films and ceramics, one can build conclusions that perovskite with similar compositions have well defined relaxor behavior than their bulk matrix.

6. Birelaxors

We have discussed the origin and functional properties of relaxor, in nutshell the basic characteristic of relaxors are the existence of ordered PNRs in a disordered matrix [37-42]. Similarly, relaxor ferromagnets (synonyms: "Mictomagnets") are also known in the literature since the 1970s. A mictomagnet is described as having magnetic clusters (superparamagnetism) which form a spin glass and have a tendency to form short-range ordering [40, 41]. The materials hold both ferroelectric and magnetic orders at nanoscale are called "birelaxors". Birelaxors are suppose to possess only short range order over the entire temperature range with broken symmetry such as "global spatial inversion" and "time reversal symmetry" at nanoscale. Nanoscale broken symmetries and length scale coherency make birelaxors difficult to investigate microscopically; it is advisable to use optical tools for proper probing. Birelaxors do not usually show linear ME coupling ($a_{ij} P_i M_j$) but may have large nonlinear terms such as bP^2M^2. Since linear coupling is not allowed, local strain-mediated PNR-PNR, PNR-MNR (magnetic nanoregions), and MNR-MNR interactions may provide very strong ME effects [43-44].

A single-phase perovskite $PbZr_{0.53}Ti_{0.47}O_3$ (PZT) and $PbFe_{2/3}W_{1/3}O_3$ (PFW), (40% PZT-60 % PFW) solid solutions thin film grown by pulsed laser deposition system have shown interesting birelaxor properties. Raman spectroscopy, dielectric spectroscopy and temperature-dependent zero- field magnetic susceptibility indicate the presence of both ferroic orders at nanoscale [45].

Dipolar glass and spin glass properties are confirmed from the dielectric and magnetic response of the PZT-PFW system, the micro Raman spectra of this system also revealed the presence of polar nanoregions. Dielectric constant and tangent loss show the dielectric dispersion near the dielectric maxima temperature that confirms near-room-temperature relaxor behavior. Magnetic irreversibility is defined as ($M_{irr} = M_{FC}-M_{ZFC}$) and it represents the degree of spin glass behavior. PZT-PFW demonstrates special features in M_{irr} data at 100 Oe indicate that irreversibility persists above 220 K up to 4th or 5th order of the magnitude, T_{irr} (inset Fig. 3 (b)). The evidence of ZFC and FC splitting at low field ~100 Oe and even for high 1 kOe field suggests and supports glass-like behavior with competition be-tween long-range ordering and short-range order. The behavior of the ZFC cusp is the same for the entire range of field (>100 Oe), with a shift in blocking transition tempera-ture (T_B) (maximum value of magnetization in ZFC cusp) to lower temperature, i.e. from 48 K to 29.8 K with increasing magnetic field. The ZFC cusps disappeared in the FC proc-ess, suggesting that competing forces were stabilized by the field- cooled process. Magnet-ic irreversibilities were found over a wide range of temperature with a sharp cusp in the ZFC data, suggesting the presence of MNRs with spin-glass-like behavior. The detailed fabrication process, the complete characterization process and the functional properties are reported in reference [43]. This is not only the unique system to have birelaxor proper-ties, however, Levstik et al also observed the birelaxor properties in $0.8Pb(Fe_{1/2}Nb_{1/2})O_3$–$0.2Pb(Mg_{1/2}W_{1/2})O_3$ thin films [38].

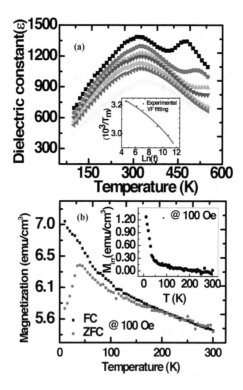

Figure 3. (a) Dielectric response and Vogel-Fulcher fitting (inset), (b) Zero field cooled and field cool magnetic re-sponse of PZT-PFW thin films indicate the presence of PNRs and MNRs at nanoscale. Adapted from Ref. [45] and repro-duced with permission (© 2011 AIP).

7. Superlattice relaxors and high energy density capacitors

Material scientists are looking for a system or some novel materials that possess high power density and high energy density or both. Relaxors demonstrate high dielectric constant, low loss, non linear polarization under external electric field, high bipolar density, nano dipoles (polar nano regions (PNRs) and moderate dielectric saturation, these properties support their potential candidature for the high power as well as high energy devices. At present high-k dielectric (dielectric constant less than 100, with linear dielectric) dominates in the high energy, high power density capacitor market. The high-k dielectrics show very high electric breakdown strength (> 3 MV/cm to 12 MV/cm) but their dielectric constant is relatively very low which in turns offer 1-2 J/cm^3 volume energy density. Recently, polymer ferroelectric, antiferroelectric, and relaxor ferroelectric have shown better potential and high energy density compare to the existing linear dielectrics [46-51].

BaTiO$_3$/Ba$_{0.30}$Sr$_{0.70}$TiO$_3$ (BT/BST) superlattices (SLs) with a constant modulation period of Λ= 80 Å were grown on (001) MgO substrate by pulsed laser deposition. The modulation periodicity $\Lambda/2$ was precisely maintained by controlling the number of laser shots; the total thickness of each SL film was ~6,000 Å = 0.6 μm. An excimer laser (KrF, 248 nm) with a laser energy density of 1.5 J/cm^2, pulse repetition rate of 10 Hz, substrate temperature 830 °C, and oxygen pressure at 200 mTorr was used for SL growth. The detailed growth and characterization techniques are presented elsewhere [52].

Dielectric responses of BT/BST SLs display the similar response as for normal relaxor ferroelectric which can be seen in Figure 4. It follows the non linear Vogel- Fulcher relationship (inset Figure. 4 (a)), frequency dispersion near and below the dielectric maxima temperature (T_m), merger of frequency far above the T_m, shift in T_m towards higher temperature side (about 50-60 K) with increase in frequencies, low dielectric loss, and about 60-70% dielectric saturation.

BT/BST SLs demonstrate a "in-built" field in as grown samples at low probe frequency (<1 kHz), whereas it becomes more symmetric and centered with increase in probe frequency system (>1 kHz) that ruled out the effect of any space charge and interfacial polarization. Energy density were calculated for the polarization-electric field (P-E) loops provide ~ 12.24 J/cm^3 energy density within the experimental limit, but extrapolation of this data in the energy density as function of applied field graph for different frequencies (see inset in Figure 4) suggests huge potential in the system such as it can hold and release more than 40 J/cm^3 energy density.

Experimental limitations restrict to proof the extrapolated data, however the current density versus applied electric field indicates exceptionally high breakdown field (5.8 to 6.0 MV/cm) and low current density (~ 10-25 mA/cm^2) near the breakdown voltage. Both direct and indirect measurements of the energy density indicate that it has ability to store very high energy density storage capacity (~ 46 J/cm^3).

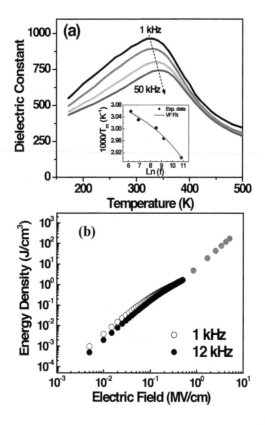

Figure 4. (a) Dielectric responses of BT/BST superlattice relaxors as function of temperature over wide range of frequencies, nonlinear Vogel-Fulcher relationship (inset), (b) Energy density capacity as function of applied electric filed, red dots show the extrapolated data. Adapted from Ref. [46] and reproduced with permission (© 2012 IOP).

The above experimental facts suggest the relaxor nature of BT/BST ferroelectric superlattices. It also indicates their potential to store and fast release of energy density which is comparable to that of high-k (<100) dielectrics, hopefully in the coming years it might be suitable for high power energy applications. These functional properties of relaxor superlattices make it plausible high energy density dielectrics capable of both high power and energy density applications.

8. Conclusions

Extensive studies on the relaxor ferroelectrics suggest the presence of localized nano size ordered regions in the disordered matrix or static random field are responsible for the dielectric dispersion near T_m. The nano regions and their coherence length are also critical for the relaxor behavior. A normal ferroelectric with diffuse phase transition system can be turn to relaxor ferroeletrics in their thin film forms under the suitable applications of strain, utilizing highly lattice mismatch substrate, growth conditions (thermal, oxygen partial pressure, atmosphere), etc.. It also indicates that the defects, oxygen vacancies, ordering of cations at A and B site of perovskite, tensile or comprehensive strain across the interface, etc. originate the ordered nano regions in films mainly responsible for dielectric dispersion. Materials with same compositions can have normal ferroelectric in bulk form whereas become relaxor in the thin film. Birelaxors hold both ferroics orders at nano scale with only short range ordering (SRO). These nanocale SRO make it potential candidates for non linear biquadratic magneto-electric coupling suitable for magnetic field sensors and non volatile memory applications. BT/BST superlattice has shown very high dielectric constant, high breakdown field, relaxor, and high energy density functional properties. A Relaxor superlattice with high functional properties is capable for both high power and energy density applications.

Acknowledgement

This work was partially supported by NSF-EFRI-RPI-1038272 and DOE-DE-FG02-08ER46526 grants.

Author details

Ashok Kumar[1,2], Margarita Correa[2], Nora Ortega[2], Salini Kumari[2] and R. S. Katiyar[2]

*Address all correspondence to: ashok553@gmail.com

1 National Physical Laboratory, Council of Scientific and Industrial Research (CSIR), New Delhi, India

2 Department of Physics and Institute for Functional Nanomaterials, University of Puerto Rico, San Juan, Puerto Rico

References

[1] G. A. Smolenskii and A. I. Agranokskaya, Soviet Physics Solid State 1, 1429 (1959)

[2] R. E. Cohen, Nature, 441, 941 (2006)

[3] Eric Cross, Ferroelectrics, 151, 305, (1994)

[4] G. A. Samara, J. Phys.: Condens. Matter 15, R367, (2003)

[5] G. Burns and F. H. Dacol, Solid State Communications, 48, 853, (1983)

[6] G. A. Samara, Solid State Physics, vol. 56, Edited by H. Ehrenreich, and R. Spaepen, New York: Academic, New York (2001)

[7] B. Krause, J. M. Cowley, and J. Wheatly, Acta Cryst. A35, 1015 (1979)

[8] C.A. Randall, D. J. Barber, and R. W. Whatmore, J. Micro. 145, 235 (1987)

[9] J. Chen, H. Chen, and M. A. Harmer, J. Am. Ceram. Soc. 72, 593 (1989)

[10] I. M. Brunskill, H. schmid, and P. Tissot, Ferroelectrics 37, 547 (1984)

[11] C. A. Randall, A. S. Bhalla, T. R. Shrout, and L. E. Cross, J. Mater. Res., 5, 829 (1990)

[12] C. A. Randall and A. S. Bhalla, Japanese Journal of Applied Physics, 29, 327 (1990)

[13] Ferroelectrics and Related Materials, G. A. Smolenskii (Ed.), V. A. Bokov, V. A. Isupov, N. N. Krainik, R. E. Pasynkov and A. I. Sokolov, Gordon & Breach Science Publisher, New York (1984)

[14] Z.-G Ye, Key Engineering Materials, 155-156, 81 (1998)

[15] H. Vogel. Z. Phys. 22, 645 (1921). G. Fulcher. J. Am. Ceram. Soc. 8, 339 (1925)

[16] V.Westphal,W. Kleeman, and M.D. Glinchuk, Phys. Rev. Lett. 68, 847 (1992).

[17] R. Pirc and R. Blinc, Phys. Rev. B. 60, 13470 (1999).

[18] R. Blinc, J. Dolinsek, A. Gregorovic, B. Zalar, C. Filipic, Z. Kutnjak, A. Levstik, and R. Pirc, Phys. Rev. Lett. 83, 424 (1999).

[19] H.T. Martirena and J.C. Burfoot. Ferroelectrics, 7, 151 (1974)

[20] G. A. Smolenskii, V. A. Isupov, A. I. Agranovskaya and S. N. Popov, Soviet Phys.-Solid State. 2, 2584 (1961).

[21] G. A. Smolenskii, J. Phys. Soc. Jpn. 28, Suppl. 26 (1970).

[22] L.E. Cross, Ferroelectrics 76, 241 (1987).

[23] D. Viehland, J. F. Li, S. J. Jang, and L. E. Cross and M. Wuttig. Phys. Rev. B. 43, 8316 (1991).

[24] R. Fisch. Phys. Rev. B. 67, 094110 (2003).

[25] R. Pirc and R. Blinc, Phys. Rev. B. 60, 13470 (1999)

[26] F. Devonshire, Philos. Mag. 40, 1040, (1949); 42, 1065, (1951)

[27] G.A. Rossetti Jr., L.E. Cross, and K. Kushida, Appl. Phys. Lett., 59, 2504 (1991)

[28] N.A. Perstev, A.G. Zembilgotov, and A. K. Tagantsev, Phys. Rev. Lett., 80, 1988 (1998)

[29] M. Dawber, K.M. Rabe, and J.F. Scott, Reviews of Modern Physics, 77, 1083, (2005)

[30] G. Catalan, M. H. Corbett, R. M. Bowman, and J. M. Gregg, J. App. Phys., 91, [4] 2295 (2002).

[31] Margarita Correa, A. Kumar and R.S. Katiyar, J. Am. Cer. Soc. 91 (6) 1788 (2008).

[32] Margarita Correa, Ashok Kumar, and R. S. Katiyar, Applied Physic Letters, 91, 082905 (2007).

[33] Margarita Correa, Ashok Kumar, and R. S. Katiyar, Integrated Ferroelectrics, 100, 297 (2008)

[34] K. Brinkman, A. Tagantsev, P. Muralt, and N. Setter, Jpn. J. App. Phys., 45, [9B] 7288 (2006)

[35] Q.M. Zhang and J. Zhao, Appl. Phys. Lett., 71, 1649 (1997)

[36] D. L. Orauttapong, J. Toulouse, J. L. Robertson, and Z. G. Ye, Physical Rev. B , 64, 212101 (2001).

[37] A. Kumar, N. M. Murari, and R. S. Katiyar, Appl. Phys. Lett. 90, 162903 (2007).

[38] A. Levstik, V. Bobnar, C. Filipič, J. Holc, M. Kosec, R. Blinc, Z. Trontelj, and Z. Jagličič, Appl. Phys. Lett. 91, 012905, (2007).

[39] V. V. Shvartsman, S. Bedanta, P. Borisov, W. Kleemann, A. Tkach, and P. M. Vilarinho, Phys. Rev. Lett. 101, 165704 (2008).

[40] G. A. Samara, in Solid State Physics, edited by H. Ehrenreich and R. Spaepen Academic, New York, 2001, Vol. 56, p. 239.

[41] J. Dho, W. S. Kim, and N. H. Hur, Phys. Rev. Lett. 89, 027202 (2002).

[42] R. Pirc and R. Blinc, Phys. Rev. B 76, 020101R (2007).

[43] A. Kumar, G. L. Sharma, R. S. Katiyar, R. Pirc, R. Blinc, and J. F. Scott, J. Phys.: Condens. Matter 21, 382204 (2009).

[44] R. Pirc, R. Blinc, and J. F. Scott, Phys. Rev. B 79, 214114 (2009).

[45] A. Kumar, G. L. Sharma, R. S. Katiyar, Appl. Phys. Lett. 99, 042907 (2011).

[46] J. H. Tortai, N. Bonifaci, A. Denat, J. Appl. Phys. 97, 053304 (2005).

[47] M. Rabuffi, G. Picci, IEEE Trans. Plasma Sci. 30, 1939 (2002).

[48] X. Hao, J. Zhai, and, X. Yao, J. Am. Ceram. Soc., 92, 1133 (2009).

[49] J. Parui, S. B. Krupannidhi, Appl. Phys. Lett., 92, 192901 (2008).

[50] R. M. Wallace, G. Wilk, MRS Bulletin. 27, 186 (2002).

[51] G. D. Wilk, R. M. Wallace, J. M. Anthony, J. Appl. Phys. 89, 5243 (2001).

[52] N. Ortega, A Kumar, J. F. Scott, Douglas B. Chrisey, M. Tomazawa, Shalini Kumari, D. G. B. Diestra and R. S. Katiyar. J. Phys.: Condens. Matter. 24 445901 (2012).

Chapter 3

Advances in Thermodynamics of Ferroelectric Phase Transitions

Shu-Tao Ai

Additional information is available at the end of the chapter

1. Introduction

The thermodynamics of ferroelectric phase transitions is an important constituent part of the phenomenological theories of them, as well as the interface dynamics of them. In particular, if we confine the thermodynamics to the equilibrium range, we can say that the Landau-Devonshire theory is a milestone in the process of the development of ferroelectric phase transition theories. This can be found in many classical books such as [1,2]. Many studies centering on it, especially the size-effects and surface-effects of ferroelectric phase transitions, have been carried out. For the reason of simplicity, we just cite a few [3,4]. But we think these are a kind of technical but not fundamental progress.

Why we think so is based on that the Landau-Devonshire theory is confined to the equilibrium range in essene so it can't deal with the outstanding irreversible phenomonon of first-order ferroelectric phase transitions strictly, which is the „thermal hysteresis". The Landau-Devonshire theory attributes the phenomenon to a series of metastable states existing around the Curie temperature T_C. In principle, the metastable states are not the equilibrium ones and can not be processed by using the equilibrium thermodynamics. Therefore, we believe the Landau-Devonshire theory is problematic though it is successful in mathematics. The real processes of phase transition were distorted. In this contribution, the Landau-Devonshire theory will be reviewed critically, then the latest phenomenological theory of ferroelectric phase transitions will be established on the basis of non-equilibrium or irreversible thermodynamics.

This contribution are organized as the follows. In Section 2, we will show the unpleasant consequence caused by the metastable states hypothesis, and the evidence for the non-existence of metastable states, i.e. the logical conflict. Then in Section 3 and 4, we will give the non-equilibrium (or irreversible) thermodynamic description of ferroelectric phase transitions, which eliminates the unpleasant consequence caused by the metastable states hypothesis. In Section 5, we will give the non-equilibrium thermodynamic explanation of the irreversibility of ferroelectric phase transitions, i.e. the thermal hysteresis and the domain occurrences in ferroelectrics. At last, in Section 6 we will make some concluding remarks and look forward to some possible developments.

2. Limitations of Landau-Devonshire theory and demonstration of new approach

The most outstanding merit of Landau-Devonshire theory is that the Curie temperature and the spontaneous polarization at Curie temperature can be determined simply. However, in the Landau-Devonshire theory, the path of a first-order ferroelectric phase transition is believed to consist of a series of metastable states existing around the Curie temperature. This is too difficult to believe because of the difficulties encounted (just see the follows) T_C.

2.1. Unpleasant consequence caused by metastable states hypothesis

Basing on the Landau-Devonshire theory, we make the following inference. Because of the thermal hysteresis, a first-order ferroelectric phase transition must occur at another temperature, which is different from the Curie temperature [5]. The state corresponding to the mentioned temperature (i.e. actural phase transition temperature) is a metastable one. Since the unified temperature and spontanous polarization can be said about the metastable state, we neglect the heterogeneity of system actually. In other words, every part of the system, i.e. either the surface or the inner part, is of equal value physically. When the phase transition occurs at the certain temperature, every part of the system absorbs or releases the latent heat simultaneously by a kind of action at a distance. (The concept arose in the electromagnetism first. Here it maybe a kind of heat transfer.) Otherwise, the heat transfer in system, with a finite rate, must destroy the homogeneity of system and lead to a non-equilibrium thermodynamic approach. The unpleasant consequence, i.e. the action at a distance should be eliminated and the lifeforce should be bestowed on the non-equilibrium thermodynamic approach.

In fact, a first-order phase transition process is always accompanied with the fundamental characteristics, called the co-existence of phases and the moving interface (i.e. phase boundary). The fact reveals that the phase transition at various sites can not occur at the same time. Yet, the phase transition is induced by the external actions (i.e. absorption or release of latent heat). It conflicts sharply with the action at a distance.

2.2. Evidence for non-existence of metastable states: logical conflict

In the Landau-Devonshire theory, if we neglect the influence of stress, the elastic Gibbs energy G_1 can be expressed with a binary function of variables, namely the temperature T and the electric displacement D (As G_1 is independent of the orientation of D, here we are interested in the magnitude of D only)

$$G_1 = g_1(T,D) \tag{1}$$

The long-standing, close correlation between analytical dynamics and thermodynamics implies that Equation (1) can be taken as a scleronomic constraint equation

$$f_1(G_1,T,D) = G_1 - g_1(T,D) = 0 \tag{2}$$

where G_1, T, D are the generalized displacements. The possible displacements dG_1, dT and dD satisfy the following equation

$$dG_1 - \frac{\partial g_1}{\partial T}dT - \frac{\partial g_1}{\partial D}dD = 0 \tag{3}$$

In the Landau-Devonshire theory, the scleronomic constraint equation, i.e. Equation (1) is expressed in the form of the power series of D (For simplicity, only the powers whose orders are not more than six are considered)

$$G_1 = G_{10} + \frac{1}{2}\alpha D^2 + \frac{1}{4}\beta D^4 + \frac{1}{6}\gamma D^6 \tag{4}$$

where α, β, γ are the functions of T, and G_{10} is the elastic Gibbs energy of paraelectric phase. The relation between G_1 and D at various temperature, which belongs to first-order phase transition ferroelectrics is represented graphically in Figure 1. The electric displacements which correspond to the bilateral minima of G_1 are identified as $\pm D^*$, and the electric displacement which corresponds to the middle minimum of G_1 equals zero. The possible electric displacements should be the above ones which correspond to the minima of G_1.

Equivalently, imposed on the generalized displacements G_1, T, D is a constraint, which is

$$\frac{\partial G_1}{\partial D} = 0 \tag{5}$$

So, the possible displacement dD should be the follows dD^*, $-dD^*$, $0-(\pm D^*) = \mp D^*$, $\pm D^*-0 = \pm D^*$, 0. After all, if our discussion are limited in the equilibrium thermodynamics strictly, there must be the third constraint, i.e. the equilibrium D and T should satisfy

$$h(D,T) = 0 \tag{6}$$

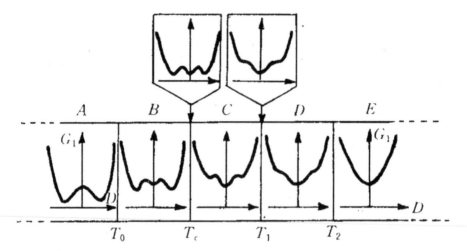

Figure 1. The relation between elastic Gibbs energy G_1 and zero field electric displacement D belongs to the ferroelectrics at various temperature, which undergoes a first-order phase transition [5]. T_0 :the lowest temperature at which the paraelectric phase may exist, i.e. the Curie-Weiss temperature; T_C :the phase transition temperature, i.e. the Curie temperature; T_1 : the highest temperature at which the ferroelectric phase may exist; T_2 : the highest temperature at which the ferroelectric phase may be induced by the external electric field.

where h is a binary function of the variables D and T. It can be determined by the principle of minimum energy

$$G_1 = \min \tag{7}$$

for certain T. Then, the metastable states are excluded. Thus, the thermal hysteresis does not come into being. The corollary conflicts with the fact sharply. This reveals that the first-order ferroelectric phase transition processes must not be reversible at all so as not to be dealt with by using the equilibrium thermodynamics.

How can this difficulty be overcome? An expedient measure adopted by Devonshire is that the metastable states are considered. However, do they really exist?

Because the metastable states are not the equilibrium ones, the relevant thermodynamic variables or functions should be dependent on the time t. In addition, the metastable states are close to equilibrium, so the heterogeneity of system can be neglected. Here, the elastic Gibbs energy G'_1 should be

$$G'_1 = g_2(T,D,t) \tag{8}$$

For the same reason as was mentioned above, Equation (8) can be regarded as a rheonomic constraint on the generalized displacements G'_1, T, D

$$f_2\left(G_1',T,D,t\right)=G_1'-g_2\left(T,D,t\right)=0 \tag{9}$$

In this case, the possible displacements dG'_1, dT and dD satisfy the following equation

$$dG_1'-\frac{\partial g_2}{\partial T}dT-\frac{\partial g_2}{\partial D}dD-\frac{\partial g_2}{\partial t}dt=0 \tag{10}$$

Comparing Equation (3) with Equation (10), we may find that the possible displacements here are not the same as those in the former case which characterize the metastable states for they satisfy the different constraint equations, respectively. (In the latter case, the possible displacements are time-dependent, whereas in the former case they are not.) Yet, the integral of possible displacement dD is the possible electric displacement in every case. The possible electric displacements which characterize one certain metastable state vary with the cases. A self-contradiction arises. So the metastable states can not come into being.

What are the real states among a phase transition process? In fact, both the evolution with time and the spatial heterogeneity need to be considered when the system is out of equilibrium [6-9]. Just as what will be shown in Section 2.3, the real states should be the stationary ones, which do not vary with the time but may be not metastable.

2.3. Real path: Existence of stationary states

The real path of a first-order ferroelectric phase transition is believed by us to consist of a series of stationary states. At first, this was conjectured according to the experimental results, then was demonstrated reliable with the aid of non-equilibrium variational principles.

Because in the experiments the ferroelectric phase transitions are often achieved by the quasi-static heating or cooling, we conjetured that they are stationary states processes [8]. The results on the motion of interface in ferroelectrics and antiferroelectrics support our opinion [10-12]. From Figure 2, we may find that the motion of interface is jerky especially when the average velocity v_a is small. A sequence of segments of time corresponding to the states of rest may be found. The experimental results about other materials such as $PbTiO_3$ are alike [11]. This reveals that in these segments of time (i.e. characteristic time of phase transition) the stationary distributions of temperature, heat flux, stress, etc. may be established. Otherwise, if the motion of interface is continuous and smooth, with the unceasing moving of interface (where the temperature is T_C) to the inner part, the local temperature of outer part must change to keep the temperature gradient ∇T of this region unchanged for it is determined by $\pm l\rho v'=J_q^{diff}=-\kappa\nabla T$, where l is the latent heat (per unit mass), ρ is the mass density, v' is the velocity of interface (where the phase transition is occurring), J_q^{diff} is the diffusion of heat, i.e. heat conduction, κ is the thermal conductivity (and maybe a tensor.) Then, the states are not stationary.

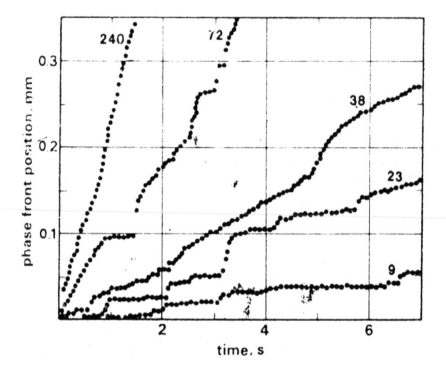

Figure 2. The position of phase boundary as a function of time for NaNbO$_3$ single crystal for the various values of average velocity v_a (the values in μm/s given against the curves) [12].

The non-equilibrium variational principles are just the analogue and generalization of the variational principles in analytical dynamics. The principle of least dissipation of energy, the Gauss's principle of least constraint and the Hamiltonian principle etc., in non-equilibrium thermodynamics play the fundamental roles as those in analytical dynamics. They describe the characteristics of stationary states or determine the real path of a non-equilibrium process.

For the basic characteristics of non-equilibrium processes is the dissipation of energy, the dissipation function φ is defined as

$$\varphi = \sigma_s - \pi \tag{11}$$

where σ_s is the rate of local entropy production and π is the external power supply (per unit volume and temperature). After the rather lengthy deducing and utilizing the thermodynamic Gauss's principle of least constraint which makes the system choose a real path [13], the evolution with time t of the deviation from a given non-equilibrium stationary state, $\xi(t)=\{\xi_i(t)\}$, was obtained in two cases. If no external power supply,

$$\xi_i(t) = \xi_i(0)e^{\frac{2S_i}{R_i}t} \tag{12}$$

where $\xi_i(0)$ is the initial value of $\xi_i(t)$, S_i is the coefficient between the linear variation in the thermodynamic force $\chi_{Ti}^{(1)}(\xi)$ and the deviation of the extensive pseudo-thermodynamic variable from a given non-equilibrium stationary state ξ_i, R_i is the coefficient between the linear variation in the dissipative force $\chi_{Di}^{(1)}(\xi)$ and the time-derivative of the deviation of the extensive pseudo-thermodynamic variable from a given non-equilibrium stationary state ξ_i. Equation (12) define the real path with the addition that R_i should be a suitable value R_i^*. If the external power supply exists, similarly the evolution of deviation ξ_i can be obtained

$$\xi_i(t) = \xi_i(0)e^{\frac{2(S_i - V_i)}{R_i}t} \tag{13}$$

where V_i is the coefficient between the linear variation in the force related to the external power $\chi_{Ei}^{(1)}$ and the deviation of the extensive pseudo-thermodynamic variable from a given non-equilibrium stationary state ξ_i. If the coefficients V_i, R_i assume the suitable values V_i^{**}, R_i^{**}, the system choose a real path.

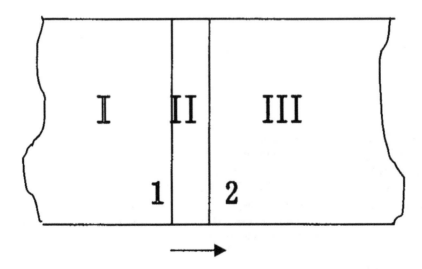

Figure 3. Three types of regions and their interfaces in the ferroelectric-paraelectric system in which a first-order phase transition is occurring [13].

Both the real paths in the two cases reveal that the deviations decrease exponentially when the system regresses to the stationary states. Stationary states are a king of attractors to non-equilibrium states. The decreases are steep. So the regressions are quick. It should be noted that we are interested in calculating the change in the generalized displacements during a macroscopically small time interval. In other words, we are concerned with the determination of the path of an irreversible process which is described in terms of a finite difference equation. In the limit as the time interval is allowed to approach zero, we obtain the variational equation of thermodynamic path.

So, if the irreversible process is not quick enough, it can be regarded as the one that consists of a series of stationary states. The ferroelectric phase transitions are usually achieved by the quasi-static heating or cooling in the experiments. So, the processes are not quick enough to make the states deviate from the corresponding stationary states in all the time. In Figure 3, three types of regions and their interfaces are marked with I, II, III, 1, 2 respectively. The region III where the phase transition will occur is in equilibrium and has no dissipation. In the region I where the phase transition has occurred, there is no external power supply, and in the region II (i.e. the paraelectric-ferroelectric interface as a region with finite thickness instead of a geometrical plane) where the phase transition is occurring, there exists the external power supply, i.e. the latent heat (per unit volume and temperature). According to the former analysis in two cases, we may conclude that they are in stationary states except for the very narrow intervals of time after the sudden lose of phase stability.

3. Thermo-electric coupling

In the paraelectric-ferroelectric interface dynamics induced by the latent heat transfer [6,7], the normal velocity of interface v_n was obtained

$$v_n = \frac{1}{l\rho}\left[k_{fer}\left(\nabla T\right)_{fer} - k_{par}\left(\nabla T\right)_{par}\right] \cdot \boldsymbol{n} \tag{14}$$

where l is the latent heat (per unit mass), ρ is the density of metastable phase (paraelectric phase), k_{fer} is the thermal conductivity coefficient of ferroelectric phase, k_{par} is the thermal conductivity coefficient of paraelectric phase, $(\nabla T)_{fer}$ is the temperature gradient in ferro-electric phase part, $(\nabla T)_{par}$ is the temperature gradient in paraelectric phase part, n is the unit vector in normal direction and directs from the ferroelectric phase part to the paraelectric phase part. The temperature gradients can be studied from the point of view that a ferroelectric phase transition is a stationary, thermo-electric coupled transport process [8].

3.1. Local entropy production

In the thermo-electric coupling case, the Gibbs equation was given as the following [8]

$$Tds = du - E \cdot dD - \sum_i \mu_i dn_i \tag{15}$$

where T, E, D is the temperature, the electric field intensity and the electric displacement within a random small volume, respectively; s, u, μ_i, n_i is the entropy density, the internal energy density, the chemical potential and the molar quantity density in the small volume, respectively. And there, it was assumed that the crystal system is mechanically-free (i.e. no force is exerted on it). Differentiating Equation (15) and using the following relations

$$\frac{\partial u}{\partial t} + \nabla \cdot J_u = 0 \tag{16}$$

$$\frac{\partial D}{\partial t} = \frac{\partial(\varepsilon_0 E + P)}{\partial t} = J_P \tag{17}$$

$$\frac{\partial n_i}{\partial t} + \nabla \cdot J_{ni} = 0 \tag{18}$$

we have

$$\frac{\partial s}{\partial t} = -\frac{1}{T}\nabla \cdot J_u - \frac{E}{T} \cdot J_P + \sum_i \frac{\mu_i}{T}\nabla \cdot J_{ni} \tag{19}$$

where J_u, J_P, J_{ni} is the energy flux, the polarization current and the matter flux; P is the polarization. J_u should consist of three parts: the energy flux caused by the heat conduction, the energy flux caused by the charge transport and the energy flux caused by the matter transport

$$J_u = J_q + \varphi J_P + \sum_i \mu_i J_{ni} \tag{20}$$

where J_q is the heat flux, φ is the electrical potential and satisfies

$$E = -\nabla \varphi \tag{21}$$

Then we deduce the following

$$\frac{\partial s}{\partial t} = -\nabla \cdot \left(\frac{J_q + \varphi J_P}{T} \right) + J_q \cdot \nabla \left(\frac{1}{T} \right) + J_P \cdot \nabla \left(\frac{\varphi}{T} \right) - \sum_i \frac{J_{ni}}{T} \cdot \nabla \mu_i \tag{22}$$

If we define a entropy flux J_s and a rate of local entropy production σ_s as

$$J_s = \frac{J_q + \varphi J_P}{T} \tag{23}$$

$$\sigma_s = J_q \cdot \nabla \left(\frac{1}{T} \right) + J_P \cdot \nabla \left(\frac{\varphi}{T} \right) - \sum_i \frac{J_{ni}}{T} \cdot \nabla \mu_i \tag{24}$$

Equation (22) can be written as

$$\frac{\partial s}{\partial t} + \nabla \cdot J_s = \sigma_s \tag{25}$$

This is the local entropy balance equation. We know, the system is in the crystalline states before and after a phase transition so that there is no diffusion of any kind of particles in the system. So, $J_{ni}=0$. The local entropy production can be reduced as

$$\sigma_s = J_q \cdot \nabla \left(\frac{1}{T} \right) + J_P \cdot \nabla \left(\frac{\varphi}{T} \right) \tag{26}$$

We know the existence of ferroics is due to the molecular field. It is an internal field. So we must take it into account. Here, the electric field should be the sum of the external electric field E_e and the internal electric field E_i

$$E = E_e + E_i \tag{27}$$

Correspondingly, there are the external electrical potential φ_e and the internal electrical potential φ_i and they satisfy

$$E_e = -\nabla \varphi_e \tag{28}$$

$$E_i = -\nabla \varphi_i \tag{29}$$

If the external electric field is not applied, φ_e can be a random constant. There is no harm in letting the constant equal zero. Then the entropy production equals

$$\sigma_s = J_q \cdot \nabla\left(\frac{1}{T}\right) + J_P \cdot \nabla\left(\frac{\varphi_i}{T}\right) \tag{30}$$

According to the crystal structures of ferroelectrics [2], we know the polarization current J_P originates from the displacement or ordering of ions in ferroelectrics. We may consider it as the transport of charges influenced by the internal electric field.

3.2. Description of phase transitions and verification of interface dynamics

Assume the external electric field is not applied. Here are the thermodynamic fluxex J_q, J_P and the corresponding thermodynamic forces X_q, X_P

$$X_q = \nabla\left(\frac{1}{T}\right) \tag{31}$$

$$X_P = \nabla\left(\frac{\varphi_i}{T}\right) \tag{32}$$

J_i can be expanded linearly with $X_j(i, \ j=q, \ P)$

$$J_q = L_{qq} \cdot X_q + L_{qP} \cdot X_P \tag{33}$$

$$J_P = L_{Pq} \cdot X_q + L_{PP} \cdot X_P \tag{34}$$

where L_{qq}, L_{qP}, L_{Pq} and L_{PP} are the transport coefficients, which are four second-order tensors. They should satisfy the generalized Onsager relations [14]
$L_{qq} = L_{qq}^T$, $L_{qP} = L_{Pq}^T$,

$$L_{PP} = L_{PP}^T \tag{35}$$

Because to a first-order ferroelectric phase transition the electric displacement changes suddenly and so does the internal electrical potential, the force X_P of the region where the phase transition is occurring can be regarded as a large constant roughly in the characteristic times of phase transition (i.e. the times in which the interface keeps rest). For the ferroelectric phase transition may be regarded as a stationary states process, the principle of minimum entropy production must be satisfied [15].

According to Equations (30)-(35), we have

$$\sigma_s = L_{qq} : X_q X_q + 2L_{qP} : X_q X_P + L_{PP} : X_P X_P \tag{36}$$

If there is no any restriction on X_q and X_P, according to the conditions on which the entropy production is a minimum

$$\left(\frac{\partial \sigma_s}{\partial X_q} \right)_{X_P} = 2L_{qq} \cdot X_q + 2L_{qP} \cdot X_P = 2J_q = 0 \tag{37}$$

$$\left(\frac{\partial \sigma_s}{\partial X_P} \right)_{X_q} = 2L_{Pq} \cdot X_q + 2L_{PP} \cdot X_P = 2J_P = 0 \tag{38}$$

We know the stationary states are equilibrium ones actually. If we let X_q (or X_P) be a constant, according to Equation (38) (or (37)) we know J_P (or J_q) which is corresponded to another force X_P (or X_q) should be zero.

Then, a first-order ferroelectric phase transition can be described by the second paradigm. Since the force X_P of the region where the phase transition is occurring is a large constant, the flux J_q of this region should be zero (but $J_P \neq 0$). This states clearly that the pure heat conduction and the heat conduction induced by the thermo-electric coupling cancel out each other so as to release or absorb the latent heat. It is certain that the latent heat passes through the region where the phase transition has occurred (at the outside of the region where the phase transition is occurring) and exchanges itself with the thermal bath. Accompanied with the change of the surface's temperature and the unceasing jerky moving of the region where the phase transition is occurring, a constant temperature gradient is kept in the region where the phase transition has occurred, i.e. the force X_q is a constant. So, the flux $J_P = 0$ (but $J_q \neq 0$). This states clearly that the electric displacement of the region where the phase transition has occurred will not change but keep the value at Curie temperature or zero until the phase transition finishes. Differently, the region where the phase transition will occur should be described by the first paradigm for there is no restriction on the two forces X_P, X_q. The states of this region are equilibrium ones. So the temperature gradient ∇T should be zero.

Considering that $(\nabla T)_{par} \approx 0$ for the region where the phase transition will occur (i.e. the paraelectric phase part) can be regarded as an equilibrium system, we modify Equation (14) as

$$v_n = \frac{k_{fer}(\nabla T)_{fer} \cdot \boldsymbol{n}}{l\rho} \tag{39}$$

In order to compare it with the experiments, we make use of the following values which are about PbTiO$_3$ crystal: ρ =7.1g/cm^3 [16], l =900cal/mol [17], k_{fer} =8.8 × 10^{5}erg/cmsK [18]. The value of the velocity of the interface's fast motion, which has been measured in the experiments, is 0.5mm/s [11]. According to Equation (39), we calculate the corresponding temperature gradient to be 57.35K/cm. However, in [19] it is reported that the experimental temperature gradient varies from 1.5 to 3.5K/mm while the experimental velocity of the interface's motion varies from 732 to 843μm/s. Considering the model is rather rough, we may conclude that the theory coincides with the experiments.

3.3. Relation between latent heat and spontaneous polarization

In the experiments, the latent heat and the spotaneous polarization are measured often for first-order ferroelectric phase transitions. So in the follows, we will establish the relation between latent heat and spontaneous polarization in the realm of non-equilibrium thermodynamics.

All the quantities of the region where the phase transition has occurred are marked with the superscript "I"; all the quantities of the region where the phase transition is occurring are marked with the superscript "II"; and all the quantities of the region where the phase transition will occur are marked with the superscript "III". Let's consider the heating processes of phase transition firstly. In the region where the phase transition has occurred,

$$J_q^{I-para} = L_{qq}^{I-para} \cdot X_q^{I-para} = l\rho v_a \tag{40}$$

where we have ignored the difference between the mass density of ferroelectric phase and that of paraelectric phase (almost the same) and denote them as ρ, v_a is the average velocity of interface. In the region where the phase transition is occurring,

$$J_q^{II} = L_{qq}^{II} \cdot X_q^{II} + L_{qP}^{II} \cdot X_P^{II} = 0 \tag{41}$$

$$J_P^{II} = L_{Pq}^{II} \cdot X_q^{II} + L_{PP}^{II} \cdot X_P^{II} \tag{42}$$

The heat which is transferred to the region where the phase transition is occurring is absorbed as the latent heat because the pure heat conduction and the heat conduction induced by the thermo-electric coupling cancel out each other. So,

$$l\rho v_a = \boldsymbol{L}_{qq}^{II} \cdot \boldsymbol{X}_q^{II} \qquad (43)$$

According to Eqations (41)-(43), we work out

$$\boldsymbol{J}_P^{II} = l\rho \left[\boldsymbol{L}_{Pq}^{II} \cdot \left(\boldsymbol{L}_{qq}^{II} \right)^{-1} - \boldsymbol{L}_{PP}^{II} \cdot \left(\boldsymbol{L}_{qP}^{II} \right)^{-1} \right] \cdot v_a \qquad (44)$$

where the superscript "-1" means reverse. While

$$\int \boldsymbol{J}_P^{II} \mathrm{d}t = -\boldsymbol{D}^{spon} = \varepsilon_0 \nabla \varphi_i^{III} - \boldsymbol{P}^{spon-III} = -\boldsymbol{P}^{spon-III} \qquad (45)$$

where we utilized the boundary condition of D and considered the region where the phase transition will occur is in equilibrium, the superscript "spon" means spontaneous. So,

$$\boldsymbol{P}^{spon} = \boldsymbol{P}^{spon-III} = -\int \boldsymbol{J}_P^{II} \mathrm{d}t \qquad (46)$$

The relation between latent heat and spontaneous polarization are obtained. In the cooling processes of phase transition,

$$-l\rho v_a = \boldsymbol{L}_{qq}^{I-ferr} \cdot \boldsymbol{X}_q^{I-ferr} + \boldsymbol{L}_{qP}^{I-ferr} \cdot \boldsymbol{X}_P^{I-ferr} = \boldsymbol{L}_{qq}^{II} \cdot \boldsymbol{X}_q^{II} \qquad (47)$$

$$\int \boldsymbol{J}_P^{II'} \mathrm{d}t = \boldsymbol{D}^{spon-I} = -\varepsilon_0 \nabla \varphi_i^I + \boldsymbol{P}^{spon-I} \qquad (48)$$

Repeating the above steps, we obtain

$$\boldsymbol{P}^{spon-I} = \int \boldsymbol{J}_P^{II'} \mathrm{d}t + \varepsilon_0 \nabla \varphi_i^I \qquad (49)$$

With

$$\boldsymbol{J}_P^{II'} = -l\rho \left[\boldsymbol{L}_{Pq}^{II} \cdot \left(\boldsymbol{L}_{qq}^{II} \right)^{-1} - \boldsymbol{L}_{PP}^{II} \cdot \left(\boldsymbol{L}_{qP}^{II} \right)^{-1} \right] \cdot v_a = -\boldsymbol{J}_P^{II} \qquad (50)$$

Then we find that $P^{spon-III}$ (serves as the equilibrium polarization) is not equal P^{spon-I} (serves as the non-equilibrium polarization). For the region where the phase transition will occur is stressed [9], there is some difference between $P^{spon}(=P^{spon-III})$ and the equilibrium spontaneous polarization without the affects of stress $P^{spon'}$ because of the piezoelectric effect.

4. Thermo-electro-mechanical coupling

The comprehensive thermo-electro-mechanical coupling may be found in ferroelectric phase transition processes. Because there exists not only the change of polarization but also the changes of the system's volume and shape when a ferroelectric phase transition occurs in it, the mechanics can not be ignored even if it is mechanically-free, i.e. no external force is exerted on it. To a first-order ferroelectric phase transition, it occurs at the surface layer of system firstly, then in the inner part. So, the stress may be found in the system.

Since one aspect of the nature of ferroelectric phase transitions is the thermo-electro-mechanical coupling, we take the mechanics into account on the basis of Section 3, where only the thermo-electric coupling has been considered. This may lead to a complete description in the sense of continuum physics.

4.1. Deformation mechanics

For a continuum, the momentum equation in differential form can be written as

$$\nabla \cdot \sigma + \rho f = \rho a \tag{51}$$

where σ, f, a, ρ is the stress, the volume force exerted on unit mass, the acceleration and the mass density, respectively. Let $k = \frac{1}{2}v^2$ be the local kinetic energy density (per unit mass), with v is the velocity. Then

$$\frac{dk}{dt} = v \cdot a \tag{52}$$

where t is the time. In terms of

$$\nabla \cdot (v \cdot \sigma) = v \cdot (\nabla \cdot \sigma) + (\nabla v) : \sigma \tag{53}$$

$$\sigma : (\nabla v) = \sigma : d \tag{54}$$

we can deduce the following balance equation of mechanical energy basing on Equations (51) (52)

$$\rho \frac{dk}{dt} - \nabla \cdot (v \cdot \sigma) = \rho v \cdot f - \sigma : d \tag{55}$$

which is in differential form and in Lagrangian form. Or in Eulerian form

$$\frac{\partial(\rho k)}{\partial t} + \nabla \cdot (\rho k v - v \cdot \sigma) = \rho v \cdot f - \sigma : d \tag{56}$$

where $d = \frac{1}{2}(\nabla v + v \nabla) = \frac{1}{2}[\nabla v + (\nabla v)^T]$ is the rate of deformation or strain rate (the super-script "T" means transposition).

To a ferroelectric phase transition, f and σ may be the nominal volume force and stress, which are the embodiments of the actions of thermo-electro-mechanical coupling and are two internal fields. Generally, they are the sums of real and nominal volume force or stress

$$f = f^{real} + f^{nom} \tag{57}$$

$$\sigma = \sigma^{real} + \sigma^{nom} \tag{58}$$

The nominal volume force and stress are not zero until the eigen (or free) deformation of system finishes in phase transitions. If they are zero, the eigen (or free) deformation finishes.

4.2. Local entropy production and description of phase transitions

The Gibbs equation was given as the following [9]

$$T ds + \sum_i \mu_i dn_i + \frac{1}{\rho} E \cdot dD + f \cdot dr + \frac{1}{\rho} \nabla \cdot (v \cdot \sigma) dt = du + dk \tag{59}$$

where T, E, D is the local temperature, the local electric field intensity and the local electric displacement, respectively; s, u, n_i, μ_i is the local entropy density (per unit mass), the local internal energy density (per unit mass), the local molar quantity density (per unit mass) and the chemical potential, respectively; r is the displacement vector. If the external electric field is not applied, the quantity E is the internal electric field E_i only [8].

Make the material derivative of Equation (59) with t, then obtain

$$\rho \frac{ds}{dt} + \frac{\rho}{T} \sum_i \mu_i \frac{dn_i}{dt} + \frac{1}{T} E_i \cdot \frac{dD}{dt} + \frac{\rho}{T} f \cdot \frac{dr}{dt} + \frac{1}{T} \nabla \cdot (v \cdot \sigma) = \frac{\rho}{T} \frac{de}{dt} \tag{60}$$

where e is the total energy with $e = u + k$. We know, the system is in the crystalline states before and after a phase transition so that there is no diffusion of any kind of particles in the system. So, n_i as the local molar quantity (per unit mass) does not change with t, i.e. $\dfrac{dn_i}{dt} = 0$. $\dfrac{dD}{dt}$ stands for the polarization current J_P, while $\dfrac{dr}{dt}$ stands for the velocity v, $E_i = -\nabla \varphi_i$ (φ_i stands for the internal electric potential).

After the lengthy and troublesome deduction [9], the local entropy balance equation in Lagrangian form can be obtained

$$\rho \frac{ds}{dt} = -\nabla \cdot J_s + \sigma_s \tag{61}$$

with the entropy flux J_s

$$J_s = \frac{J_q^{diff} + \varphi_i J_P - v \cdot \sigma}{T} \tag{62}$$

and the rate of local entropy production σ_s

$$\sigma_s = J_q^{diff} \cdot \nabla \left(\frac{1}{T}\right) + J_P \cdot \nabla \left(\frac{\varphi_i}{T}\right) - \sigma : \left[v \nabla \left(\frac{1}{T}\right)\right] - \rho f \cdot \frac{v}{T} \tag{63}$$

where J_q^{diff} is the diffusion of heat, i.e. heat conduction.

Here are the thermodynamic fluxes $J_q^{diff} (= J_q)$, J_P, $\sigma (= J_\sigma)$, $\rho f (= J_f)$ and the corresponding thermodynamic forces X_q, X_P, X_σ, X_f

$$X_q = \nabla \left(\frac{1}{T}\right) \tag{64}$$

$$X_P = \nabla \left(\frac{\varphi_i}{T}\right) \tag{65}$$

$$X_\sigma = -v \nabla \left(\frac{1}{T}\right) \tag{66}$$

$$X_f = -\frac{v}{T} \tag{67}$$

$J_i(i=q, P, \sigma, f)$ can be expanded linearly with $X_j(j=q, P, \sigma, f)$

$$J_q = L_{qq} \cdot X_q + L_{qP} \cdot X_P + L_{q\sigma} : X_\sigma + L_{qf} \cdot X_f \tag{68}$$

$$J_P = L_{Pq} \cdot X_q + L_{PP} \cdot X_P + L_{P\sigma} : X_\sigma + L_{Pf} \cdot X_f \tag{69}$$

$$J_\sigma = L_{\sigma q} \cdot X_q + L_{\sigma P} \cdot X_P + L_{\sigma\sigma} : X_\sigma + L_{\sigma f} \cdot X_f \tag{70}$$

$$J_f = L_{fq} \cdot X_q + L_{fP} \cdot X_P + L_{f\sigma} : X_\sigma + L_{ff} \cdot X_f \tag{71}$$

where L_{qq}, L_{PP}, L_{ff}, L_{qP}, L_{Pq}, L_{qf}, L_{fq}, L_{Pf}, L_{fP} are nine second-order tensors, $L_{q\sigma}$, $L_{\sigma q}$, $L_{P\sigma}$, $L_{\sigma P}$, $L_{\sigma f}$, $L_{f\sigma}$ are six third-order tensors, $L_{\sigma\sigma}$ is a fourth-order tensor. They should satisfy the generalized Onsager relations [14]

$$L_{\alpha\alpha} = L_{\alpha\alpha}^T$$

$$\left(\alpha = q, P, \sigma, f\right) \tag{72}$$

$$L_{\alpha\beta} = L_{\beta\alpha}^T$$

$$\left(\alpha, \beta = q, P, \sigma, f, \text{ and } \alpha \neq \beta\right) \tag{73}$$

So the rate of local entropy production can be written as

$$\begin{aligned}
\sigma_s = {}& L_{qq} : X_q X_q + L_{qP} \cdots X_P X_q + L_{q\sigma} \cdots X_\sigma^T X_q + L_{qf} \cdots X_f X_q \\
& + L_{Pq} \cdots X_q X_P + L_{PP} : X_P X_P + L_{P\sigma} \cdots X_\sigma^T X_P + L_{Pf} \cdots X_f X_P \\
& + L_{\sigma q} \cdots X_q X_\sigma^T + L_{\sigma P} \cdots X_P X_\sigma^T + L_{\sigma\sigma} \cdots X_\sigma^T X_\sigma^T + L_{\sigma f} \cdots X_f X_\sigma^T \\
& + L_{fq} \cdots X_q X_f + L_{fP} \cdots X_P X_f + L_{f\sigma} \cdots X_\sigma^T X_f + L_{ff} : X_f X_f
\end{aligned} \tag{74}$$

According to the condition on which the local entropy production is a minimum, from Equation (74) we can deduce the following

$$\left(\frac{\partial \sigma_s}{\partial X_q}\right)_{X_P, X_\sigma, X_f} = 2L_{qq} \cdot X_q + L_{qP} \cdot X_P + L_{q\sigma} : X_\sigma + L_{qf} \cdot X_f$$
$$+ L_{Pq}^T \cdot X_P + L_{\sigma q}^T : X_\sigma + L_{fq}^T \cdot X_f$$
$$= 2\left(L_{qq} \cdot X_q + L_{qP} \cdot X_P + L_{q\sigma} : X_\sigma + L_{qf} \cdot X_f\right)$$
$$= 2J_q$$
$$= 0$$

(75)

$$\left(\frac{\partial \sigma_s}{\partial X_P}\right)_{X_q, X_\sigma, X_f} = L_{qP}^T \cdot X_q + L_{Pq} \cdot X_q + 2L_{PP} \cdot X_P + L_{P\sigma} : X_\sigma$$
$$+ L_{Pf} \cdot X_f + L_{\sigma P}^T : X_\sigma + L_{fP}^T \cdot X_f$$
$$= 2\left(L_{Pq} \cdot X_q + L_{PP} \cdot X_P + L_{P\sigma} : X_\sigma + L_{Pf} \cdot X_f\right)$$
$$= 2J_P$$
$$= 0$$

(76)

$$\left(\frac{\partial \sigma_s}{\partial X_\sigma}\right)_{X_q, X_P, X_f} = L_{q\sigma}^T \cdot X_q + L_{P\sigma}^T \cdot X_P + L_{\sigma q} \cdot X_q + L_{\sigma P} \cdot X_P$$
$$+ 2L_{\sigma\sigma} : X_\sigma + L_{\sigma f} \cdot X_f + L_{f\sigma}^T \cdot X_f$$
$$= 2\left(L_{\sigma q} \cdot X_q + L_{\sigma P} \cdot X_P + L_{\sigma\sigma} : X_\sigma + L_{\sigma f} \cdot X_f\right)$$
$$= 2J_\sigma$$
$$= 0$$

(77)

$$\left(\frac{\partial \sigma_s}{\partial X_f}\right)_{X_q, X_P, X_\sigma} = L_{qf}^T \cdot X_q + L_{Pf}^T \cdot X_P + L_{\sigma f}^T : X_\sigma + L_{fq} \cdot X_q$$
$$+ L_{fP} \cdot X_P + L_{f\sigma} : X_\sigma + 2L_{ff} \cdot X_f$$
$$= 2\left(L_{fq} \cdot X_q + L_{fP} \cdot X_P + L_{f\sigma} : X_\sigma + L_{ff} \cdot X_f\right)$$
$$= 2J_f$$
$$= 0$$

(78)

This reveals that if the k forces among those are kept constant, i.e. $X_i = const(i = 1, 2, \cdots, k, k < 4)$, the fluxes corresponding to the left $4-k$ forces are zero. Of course, if there are no restrictions on $X_i(i = 1, 2, 3, 4)$, all the flues are zero. (For convenience, we have modified the superscripts q, P, σ, f to be 1,2,3,4).

We may describe a ferroelectric phase transition by using the two paradigms above similarly as we have done in Section 3. To a first-order ferroelectric phase transition, the forces X_p, X_σ, X_f of the region where the phase transition is occurring can be regarded as three large constants roughly in the characteristic times of phase transition (i.e. the times in which the interface keeps rest) because the electric displacement, the volume and the shape change suddenly. So, the flux J_q^{diff} of the region should be zero (but $J_p \neq 0$, $\sigma \neq 0$, $\rho f \neq 0$). This states clearly that the pure heat conduction and the heat conduction induced by the thermo-electric coupling and the thermo-mechanical coupling cancel out each other so as to release or absorb the latent heat. The phase transition occurs at the surface layer firstly, which is mechanically-free. So, when the phase transition occurs in this region, the flux σ maybe the nominal stress σ^{nom} only, which does work to realize the transformation from internal energy to kinetic energy. When the phase transition occurs in the inner part, the flux σ should be the sum of σ^{real} and σ^{nom} because the sudden changes of the inner part's volume and shape have to overcome the bound of outer part then σ^{real} arises. The region where the phase transition is occurring, i.e. the phase boundary is accompanied with the real stress σ^{real} usually, which does work to realize the transformation from kinetic energy to internal energy. This has been predicted and described with a propagating stress wave [20].

It is certain that the latent heat passes through the region where the phase transition has occurred (at the outside of the region where the phase transition is occurring) and exchange itself with the thermal bath. For $\pm l \rho v_a = J_q^{diff} = -\kappa \cdot \nabla T$, a constant temperature gradient ∇T is kept in the region where the phase transition has occurrd, i.e. the force X_q at every site is a constant (which does not change with the time but may vary with the position). So, the fluxes $J_p = \sigma = \rho f = 0$ (but the flux $J_q^{diff} \neq 0$). This states clearly that the electric displacement D will not change but keep the value at Curie temperature or zero until the phase transition finishes and $\sigma^{real} = -\sigma^{nom}$ in this region. Because the electric displacement D and the strain (or deformation) are all determined by the crystal structure of system, $J_p = 0$ reveals that D of this region does not change so does not the crystal structure then does not the strain (or deformation). According to [20], we know the region where the phase transition has occurred is unstressed, i.e. $\sigma^{real} = 0$, then $\sigma^{nom} = 0$. This reveals that the eigen (or free) strain (or deformation) of system induced by the thermo-electro-mechanical coupling of phase transition is complete and the change of it terminates before the phase transition finishes. The two deductions coincide with each other. σ^{real} may relaxes via the free surface.

The region where the phase transition will occur should be in equilibrium because there are no restrictions on the forces X_q, X_p, X_σ, X_f. Whereas, according to [20], the region is stressed, i.e. $\sigma^{real} \neq 0$. To the heating process of phase transition, this may lead to a change of the spontaneous polarization of this region because of the electro-mechanical coupling (piezo-electric effect).

An immediate result of the above irreversible thermodynamic description is that the action at a distance, which is the kind of heat transfer at phase transitions, is removed absolutely. The latent heat is transferred within a finite time so the occurrence of phase transition in the inner part is delayed. (Of course, another cause is the stress, just see Section 5) In other words, the various parts absorb or release the latent heat at the various times. The action at a distance does not affect the phase transition necessarily.

5. Irreversibility: Thermal hysteresis and occurrences of domain structure

5.1. Thermal hysteresis

The "thermal hysteresis" of first-order ferroelectric phase transitions is an irreversible phenomenon obviously. But it was treated by using the equilibrium thermodynamics for ferroelectric phase transitions, the well-known Landau-Devonshire theory [2]. So, there is an inherent contradiction in this case. The system in which a first-order ferroelectric phase transition occurs is heterogeneous. The occurrences of phase transition in different parts are not at the same time. The phase transition occurs at the surface layer then in the inner part of system. According to the description above, we know a constant temperature gradient is kept in the region where the phase transition has occurred. The temperature of surface layer, which is usually regarded as the temperature of the whole system in experiments, must be higher (or lower) than the Curie temperature. This may lead to the thermal hysteresis.

No doubt that the shape and the area of surface can greatly affect the above processes. We may conclude that the thermal hysteresis can be reduced if the system has a larger specific surface and, the thermal hysteresis can be neglected if a finite system has an extremely-large specific surface. So, the thermal hysteresis is not an intrinsic property of the system.

The region where the phase transition will occur can be regarded as an equilibrium system for there are no restrictions on the forces X_q, X_p, X_σ, X_f. In other words, the forces and the corresponding fluxes are zero in this region. To a system where a second-order ferroelectric phase transition occurs, the case is somewhat like that of the region where a first-order ferroelectric phase transition will occur. The spontaneous polarization, the volume and the shape of system are continuous at the Curie temperature and change with the infinitesimal magnitudes. This means X_q, X_p, X_σ, X_f and J_q^{diff}, J_p, σ, ρf can be arbitrary infinitesimal magnitudes. The second-order phase transition occurs in every part of the system simultaneously, i.e. there is no the co-existence of two phases (ferroelectric and paraelectric). So, there is no the latent heat and stress. The thermal hysteresis disappears.

The region where a first-order ferroelectric phase transition will occur is stressed. This reveals that the occurrences of phase transition in the inner part have to overcome the bound of outer part, where the phase transition occurs earlier. This may lead to the delay of phase transition in the inner part.

5.2. Occurrences of domain structure

Though the rationalization of the existence of domain structures can be explained by the equilibrium thermodynamics, the evolving characteristics of domain occurrences in ferroelectrics can not be explained by it, but can be explained by the non-equilibrium thermodynamics.

In the region where the phase transition is occurring, the thermodynamic forces X_P $\left(= \nabla \left(\dfrac{\varphi_i}{T}\right)\right)$, X_σ $\left(= -v\nabla\left(\dfrac{1}{T}\right)\right)$, X_f $\left(= -\dfrac{v}{T}\right)$ can be regarded as three large constants in the characteristic times of transition and the thermodynamic flux J_q^{diff} =0 (but $J_P \neq 0$, $\sigma \neq 0$, $\rho f \neq 0$). The local entropy production (cf. Equation (63)) reduces to

$$\sigma_s = J_P \cdot \nabla\left(\frac{\varphi_i}{T}\right) - \quad : \left[v\nabla\left(\frac{1}{T}\right)\right] - \rho f \cdot \frac{v}{T} \tag{79}$$

Now, we are facing a set of complicated fields of T, v and φ_i, respectively. Assume that the phase transition front is denoted by S. The points included in S stand for the locations where the transition is occurring. Because the transition occurs along all directions from the outer part to the inner part, we may infer that the orientations of $\nabla\left(\dfrac{\varphi_i}{T}\right)$, or of $-v\nabla\left(\dfrac{1}{T}\right)$ and or of $-\dfrac{v}{T}$ vary continuously such that they are differently oriented at different locations.

There are always several (at least two) symmetry equivalent orientations in the prototype phase (in most cases it is the high temperature phase), which are the possible orientations for spontaneous polarization (or spontaneous deformation or spontaneous displacement). Therefore, the spontaneous polarization, the spontaneous deformation and the spontaneous displacement must take an appropriate orientation respectively to ensure σ_s is a positive minimum when the system transforms from the prototype (paraelectric) phase to the ferroelectric (low temperature) phase. The underlying reasons are that

$$J_P = \frac{\mathrm{d}D}{\mathrm{d}t} = \frac{\mathrm{d}\left(\varepsilon_0 E_i + P\right)}{\mathrm{d}t} \tag{80}$$

$$= L \varepsilon \tag{81}$$

$$\rho f = \rho \frac{\mathrm{d}v}{\mathrm{d}t} - \nabla \cdot \tag{82}$$

where L, ε, and ε_0 are the modulus of rigidity, the strain and the permittivity of vacuum, respectively. Therefore, P, ε at different locations will be differently oriented. The domain structures in ferroelectrics thus occur.

It seems that the picture of domain occurrences for first-order ferroelectric phase transition systems should disappear when we face second-order ferroelectric phase transition systems. This is true if the transition processes proceed infinitely slowly as expounded by the equilibrium thermodynamics. But any actual process proceeds with finite rate, so it is irreversible. Then the above picture revives.

In [21], the domain occurrences in ferromagnetics can be described parallelly by analogy. And the case of ferroelastic domain occurrences is a reduced, simpler one compared with that of ferroelectrics or ferromagnetics.

It is well known that the Landau theory or the Curie principle tells us how to determine the symmetry change at a phase transition. A concise statement is as follows [22]: for a crystal undergoing a phase transition with a space-group symmetry reduction from G_0 to G, whereas G determines the symmetry of transition parameter (or vice versa), it is the symmetry operations lost in going from G_0 to G that determine the domain structure in the low-symmetry phase. The ferroic phase transitions are the ones accompanied by a change of point group symmetry [23]. Therefore, the substitution of "point group" for the "space group" in the above statement will be adequate for ferroic phase transitions. From the above statement, the domain structure is a manifestation of the symmetry operations lost at the phase transition. In our treatment of the domain occurrences in ferroics, we took into account the finiteness of system (i.e. existence of surface) and the irreversibility of process (asymmetry of time). The finiteness of system make the thermodynamic forces such as $\nabla\left(\dfrac{\varphi_i}{T}\right)$, $-v\nabla\left(\dfrac{1}{T}\right)$, $-\dfrac{v}{T}$ have infinite space symmetry. The infinite space symmetry, combined with the asymmetry of time, reproduces the symmetry operations lost at the phase transition in the ferroic phase. It can be viewed as an embodiment of time-space symmetry.

After all, for the domain structures can exist in equilibrium systems, they are the equilibrium structures but not the dissipative ones, for the latter can only exist in systems far from equilibrium [24].

6. Concluding remarks

In order to overcome the shortcoming of Landau-Devonshire theory, the non-equilibrium thermodynamics was applied to study the ferroelectric phase transitions. The essence of transitions is the thermo-electro-mechanical coupling. Moreover, the irreversibility, namely thermal hysteresis and domain occurrences can be explained well in the realm of non-equilibrium thermodynamics.

The non-equilibrium thermodynamic approach utilized here is the linear thermodynamic one actully. In order to get the more adequate approaches, we should pay attention to the new developments of non-equilibrium thermodynamics. The thermodynamics with internal variables [25] and the extended (irreversible) thermodynamics [26] are two current ones. They all expand the fundamental variables spaces to describe the irreversible processes more adequately. Whereas, the relevant theoretical processing must be more complicated undoubtly. This situation needs very much effort.

Acknowledgements

The work is supported by the Natural Science Foundation Program of Shandong Province, China (Grant No. ZR2011AM019).

Author details

Shu-Tao Ai*

School of Science, Linyi University, Linyi, People's Republic of China

References

[1] Grindlay, I. An Introduction to the Phenomenological Theory of Ferroelectricity. Oxford: Pergamon; (1970).

[2] Lines, M E, & Glass, A M. Principles and Applications of Ferroelectrics and Related Materials. Oxford: Clarendon; (1977).

[3] Tilley, D R, & Zeks, B. Landau Theory of Phase Transitions in Thick Films. Solid State Communications (1984). , 49(8), 823-828.

[4] Scott, J F, Duiker, H M, Beale, P D, Pouligny, B, Dimmler, K, Parris, M, Butler, D, & Eaton, S. Properties of Ceramic KNO_3 Thin-Film Memories. Physica B+C (1988).

[5] Zhong, W L. Physics of Ferroelectrics (in Chinese). Beijing: Science Press; (1996).

[6] Gordon, A. Finite-Size Effects in Dynamics of Paraelectric-Ferroelectric Interfaces Induced by Latent Heat Transfer. Physics Letters A (2001).

[7] Gordon, A, Dorfman, S, & Fuks, D. Temperature-Induced Motion of Interface Boundaries in Confined Ferroelectrics. Philosophical Magazine B (2002). , 82(1), 63-71.

[8] Ai, S T. Paraelectric-Ferroelectric Interface Dynamics Induced by Latent Heat Transfer and Irreversible Thermodynamics of Ferroelectric Phase Transitions. Ferroelectrics (2006). , 345(1), 59-66.

[9] Ai, S T. Mechanical-Thermal-Electric Coupling and Irreversibility of Ferroelectric Phase Transitions. Ferroelectrics (2007). , 350(1), 81-92.

[10] Yufatova, S M, Sindeyev, Y G, Garilyatchenko, V G, & Fesenko, E G. Different Kinetics Types of Phase Transformations in Lead Titanate. Ferroelectrics (1980). , 26(1), 809-812.

[11] Dec, J. Jerky Phase Front Motion in $PbTiO_3$ Crystal. Journal of Physics C (1988). , 21(7), 1257-1263.

[12] Dec, J, & Yurkevich, J. The Antiferroelectric Phase Boundary as a Kink. Ferroelectrics (1990). , 110(1), 77-83.

[13] Ai, S T, Xu, C T, Wang, Y L, Zhang, S Y, Ning, X F, & Noll, E. Comparison of and Comments on Two Thermodynamic Approaches (Reversible and Irreversible) to Ferroelectric Phase Transitions. Phase Transitions (2008). , 81(5), 479-490.

[14] Ai, S T, Zhang, S Y, Jiang, J S, & Wang, C C. On Transport Coefficients of Ferroelectrics and Onsager Reciprocal Relations. Ferroelectrics (2011). , 423(1), 54-62.

[15] Lavenda, B H. Thermodynamics of Irreversible Processes. London and Basingstoke: Macmillan; (1978).

[16] Chewasatn, S, & Milne, S T. Synthesis and Characterization of $PbTiO_3$ and Ca and Mn Modified $PbTiO_3$ Fibres Produced by Extrusion of Diol Based Gels. Journal of Material Science (1994). , 29(14), 3621-3629.

[17] Nomura, S, & Sawada, S. Dielectric Properties of Lead-Strontium Titanate. Journal of the Physical Society of Japan (1955). , 10(2), 108-111.

[18] Mante, A, & Volger, H. J. The Thermal Conductivity of $PbTiO_3$ in the Neighbourhood of Its Ferroelectric Transition Temperature. Physics Letters A (1967). , 24(3), 139-140.

[19] Dec, J. The Phase Boundary as a Kink. Ferroelectrics (1989). , 89(1), 193-200.

[20] Gordon, A. Propagation of Solitary Stress Wave at First-Order Ferroelectric Phase Transitions. Physics Letters A (1991).

[21] Ai, S T, Zhang, S Y, Ning, X F, Wang, Y L, & Xu, C T. Non-Equilibrium Thermodynamic Explanation of Domain Occurrences in Ferroics. Ferroelectrics Letters Section (2010). , 37(2), 30-34.

[22] Janovec, V. A Symmetry Approach to Domain Structures. Ferroelectrics (1976). , 12(1), 43-53.

[23] Wadhawan, V K. Ferroelasticity and Related Properties of Crystals. Phase Transitions (1982). , 3(1), 3-103.

[24] Glansdoff, P, & Prigogine, I. Thermodynamic Theory of Structure, Stability and Fluctuation. New York: Wiley-Interscience; (1971).

[25] Maugin, G A, & Muschik, W. Thermodynamics with Internal Variables (I) General Concepts, (II) Applications. Journal of Non-Equilibrium Thermodynamics (1994).

[26] Jou, D, Casas-vázquez, J, & Lebon, G. Extended Irreversible Thermodynamics (3rd edition). Berlin: Springer; (2001).

Chapter 4

Pb(Mg$_{1/3}$Nb$_{2/3}$)O$_3$ (PMN) Relaxor: Dipole Glass or Nano-Domain Ferroelectric?

Desheng Fu, Hiroki Taniguchi, Mitsuru Itoh and
Shigeo Mori

Additional information is available at the end of the chapter

1. Introduction

In sharp contrast to normal ferroelectric (for example BaTiO$_3$), relaxors show unusually large dielectric constant over a large temperature range (~100 K) (Fig. 1 (a)) [1,2]. Such large dielectric response is strongly dependent on the frequency. Its origin has been the focus of interest in the solid-state physics. Unlike the dielectric anomaly in BaTiO$_3$, which is associated with a ferroelectric phase transition, the maximum of dielectric response in relaxor does not indicate the occurrence of a ferroelectric phase transition. Such huge dielectric response suggests that local polarization might occur in the crystal. This was envisioned by Burns and Dacol from the deviation from linearity of refractive index $n(T)$ (Fig. 1(c)) around the so-called Burns temperature (T_{Burns}) [3] because the deviation Δn is proportional to polarization P_s. The local polarization is suggested to occur in a nano-region, and is generally called as polar nano-region (PNR). The existence of PNR in relaxor is well confirmed from the neutron scattering measurements [4] and transition electron microscopy (TEM) observation. However, it is still unknown how PNRs contribute to the large dielectric response. More recently, it is also suggested that a strong coupling between zone-center and zone-boundary soft-modes may play a key role in understanding the relaxor behaviors [5]. Clearly, the question "what is the origin of the giant dielectric constant over a broad temperature range?"is still unclear [6,7].

Another longstanding issue on relaxors is how PNRs interact at low temperature. There are two acceptable models: (1) dipole glass model [2,8-11], and (2) random-field model [12,13]. In spherical random-bond–random-field (SRBRF) model, Pirc and Blinc assumed that PNRs are spherical and interact randomly and proposed a frozen dipole glass state for relaxor (Fig. 2(a)) [10,11]. It predicts that the scaled third-order nonlinear susceptibility $a_3 = -\varepsilon_3/\varepsilon_1^4$ will shows a nearly divergent behavior at the freezing temperature T_f of the spherical glass phase. In sharp contrast, the random field model of Fisch [13], which assumes non-random two-spin exchange, predicts a ferroelectric or ferroelectric domain state in relaxor (Fig. 2(b)). This random Potts field model also predicts a broadening specific heat peak for the glass phase and shows that the latent heat at ferroelectric transition T_c is so small that it may be difficult to be detected, which reasonably explains the data reported by Moriya et al. [14].

Combining our recent results [15] from the electrical polarization, Raman scattering, TEM measurements and those reported in the literature, here, we propose a physical picture to understand the dielectric behaviors of Pb(Mg₁/₃Nb₂/₃)O₃ (PMN) relaxor.

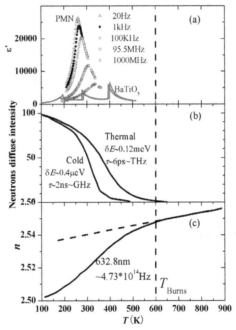

Figure 1. Temperature dependence of (a) linear dielectric constant [9], (b) cold [35] & thermal [4] neutrons diffuse intensities, and (c) refractive index [3] for PMN relaxor crystal.

2. Multiple inhomogeneities in relaxors

PMN is a prototypical relaxor with A(B′,B″)O₃ perovskite structure (Fig. 3), in which B-sites are occupied by two kinds of heterovalent cations. Such chemical inhomogeneity is a common feature of relaxor crystals. Although it remains an average centrosymmetric cubic structure down to 5 K [16], local structural inhomogeneity has been detected in PMN relaxor. In addition to PNR mentioned above, chemically ordering region (COR) [17-19] with size of several nm has been observed in PMN crystal by TEM. It should be noticed that PNR and COR belong to different symmetry groups and are considered to have non-centrosymmetry of$R3m$ and centrosymmetry of$Fm\bar{3}m$, respectively. Therefore, there is spontaneous polarization P_s along the $<111>_c$ direction of pseudocubic structure in PNRs [20,21], but none of P_s exists in CORs. In addition to chemical and structural inhomogeneities, we will show that PMN relaxor also has inhomogeneity of ferroelectric domain structure. Multiple inhomogeneities are thus considered to play a crucial role in inducing the intriguing behaviors in relaxors.

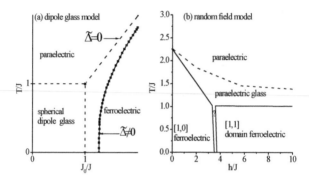

Figure 2. Two phase diagrams proposed for relaxor by (a) a dipole glass model [10], and (b) a random field model [13].

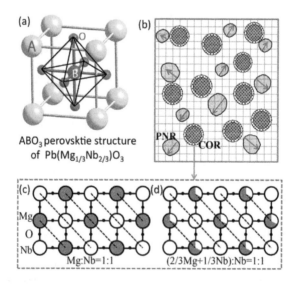

Figure 3. (a) ABO₃ perovskite structure. (b) Model for relaxor structure. PNR and COR represent the polar nano-region and chemically order region, respectively. (c) &(d) show two models of atom arrangement for COR. To maintain the electric neutrality, a Nb-rich layer is required for case (c).

3. Evolution of the electrical polarization and origin of the huge dielectric responses

In order to understand the nature of the huge dielectric response and the ground state of the electrical polarization in PMN, it is essential to know the polarization hysteresis of all states including virgin state in PMN crystal. Although there are many reports on the polarization hysteresis of PMN crystal, there is a lack of understanding of the polarization hysteresis of the virgin state. In our polarization measurements, in order to access the virgin state of the crystal at a temperature, it was firstly annealed at 360 K and then cooled to the desired temperature for the measurements.

Figure 4 shows the D-E hysteresis at three typical temperatures observed for (110)$_c$-cut PMN crystal At T=360 K that is greatly higher than the freezing temperature T_f=224 K assumed for PMN crystal [9],there is no remnant polarization within the experimental time scale of τ~10 ms (one cycle of the D–E loop) and the polarization is history-independent, indicating that the crystal is macroscopically paraelectric at this temperature. When temperature is lower than room temperature (for example, T=250 K), remnant polarization was observed but it generally disappears after removing the electric field.

Upon further cooling to temperatures lower than ~220 K,PMN shows polarization hysteresis similar to that of normal ferroelectric [22]. Fig. 4(a) shows an example of the characteristic hysteresis loop in this temperature range. In the virgin state as indicated by the thick red line, it appears that there is no remnant polarization in the crystal at zero electric field. However, as increasing the electric field, we can see the gradual growth of the polarization. This is a characteristic behavior of the polarization reversal (switching) in ferroelectric. When the applied field is larger than the coercive field E_c, ferroelectric domains are aligned along the direction of the electric field, leading to a stable macroscopic polarization in the crystal. This is evident from the fact that the remnant polarization is identical to the saturation polarization. These results clearly indicate that PMN is a ferroelectric rather than a dipole glass at temperature lower than 220 K.

There are many reports on the electric-field induced phase transition in PMN. On the basis of the change of dielectric constant under the application of a DC electric field, Colla et al. proposed an E-T phase diagram for PMN [8], which suggests a phase transition from a glass phase to a ferroelectric phase at a critical field of E_t=1. 5kV/cm in the temperature range of 160 K - 200 K. However, from our polarization results, we cannot find such a critical field except the coercive field E_c. If we assume that E_c is the critical field, then its value completely disagrees with the reported value. We observed that E_c increases rapidly with lowering the temperature, for example, E_c can reach a value of ~11 kV/cm at 180 K, which is about 7 times of E_t. Moreover, E_c is strongly dependent on frequency (Fig. 4b), and has an exponential form(1/f∝ exp (α/E_c) (Fig. 4c), which will result in an undefined critical field if we assume E_c=E_t because E_c is dependent on the measurement time-scale.

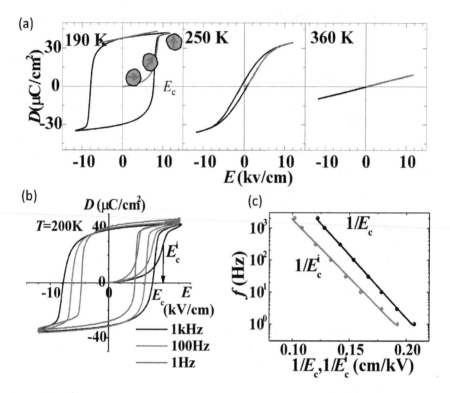

Figure 4. (a) Polarization hysteresis in PMN crystal at 190 K, 250 K, 360K. (b) Frequency dependence of polarization hysteresis. (c) Relationship between frequency and coercive field determined from the peak of switching currents. Superscript i denotes the value observed for the virgin state.

As discussed in following, ferroelectric micro-domain and soft-mode behaviors have also been observed at zero field in PMN in our measurements. Also, lowering of symmetry of local structure at zero field was also revealed around 210 K by a NMR study [21]. All these results direct to the fact of occurrence of a ferroelectric state at zero field in PMN crystal. We therefore consider that it is more rational to attribute E_c to the coercive field required for the domain switching rather than the critical field for a field-induced phase transition. In fact, the exponential relationship of coercive field with frequency is well-known as Merz's law ($f=1/\tau \propto \exp(-\alpha/E_c)$, τ=switching time, α=activation field) [23-28] in the normal ferroelectrics (for example, $BaTiO_3$, TGS) [23,24]. Using this relationship, the activation field is estimated to be 83. 5 kV/cm for PMN crystal at 200 K (Fig. 4c), which is one order of magnitude greater than that of the $BaTiO_3$ crystal. This indicates that the domain switching in PMN become more and more difficult as lowering temperatures. For example, when an electric field of 1 kV/cm is applied to the crystal at 200 K, an unpractical time of 2. 3×10^{29}s (~7. 3×10^{21} years) is required for the domain switching. Actually, it is impossible to observe the spontaneous polarization by this weak field within a limit time at this temperature.

The high-resolution data of the polarization obtained by a 14-bit oscilloscope allow us to cal-culate the linear and nonlinear dielectric susceptibilities (defined by the expansion $P=\varepsilon_0(\chi_1 E +\chi_3 E^3+$)) directly from the D-E hysteresis by differentiating the polarization with respective to the electric field. The calculated results are summarized in Fig. 5 for various electric fields in the virgin state and the zero electric field after the polarization reversal. These results al-low us to have a deep insight into the nature of abnormal dielectric behaviors and the phase transition in PMN relaxor. Fig. 5(a) shows the linear dielectric response for the virgin state in various electric fields. When comparing the response obtained at zero field with that ob-tained by LCR meter at the frequency corresponding to the sampling rate used in D-E hyste-resis measurements, one might find that they both behavior in the same way with the temperature. As mentioned above that the polarization response under the electric field in the virgin state is essentially due to the polarization reorientation, we therefore can reasona-bly attribute the dielectric anomaly usually observed in PMN relaxor to the polarization re-orientation. This indicates that the reorientation of the PNRs dominates the huge dielectric response in PMN relaxor. Fig. 5 (a) also shows that the peak of dielectric response shifts to lower temperature at a higher electric field. This means that the activation field required for domain switching increases with lowering the temperature.

The nonlinear dielectric susceptibility ε_{3v} and its scaled value a_{3v} for the virgin state are given in Fig. 5(c) and (d), respectively. ε_{3v}shows a broad peak around 255 K, which is in good agree-ment with those obtained by Levstik et al. using a lock-in to wave analyzer technique for vari-ous frequencies [9]. Levstik et al. attributed this behavior to the freezing of dipole glass at T_f=224 K in PMN. However, this picture is inconsistent with the results shown in the above polarization measurements, which indicates that PMN relaxor is ferroelectric but not glass at $T<T_f$. Such dipole glass picture is also excluded by the results of the scaled susceptibility a_3 shown in Fig. 5(d), and those reported in the previous studies [29, 30]. We can see that there is no divergent behavior of a_3 in PMN relaxor. This result again suggests that there is no freez-ing of dipole glass in PMN as predicted for the dipole glass model. The observed anomaly of nonlinear dielectric susceptibility around 255 K is more consistent with the phase diagram of the random field model proposed by Fisch [13], in which a glass phase occurs between the paraelectric phase and the ferroelectric phase in relaxors. Such anomaly of nonlinear dielec-tric susceptibility may be a manifestation of spherical dipole glass with random interaction used in SRBRF model, and its nature requires further theoretical investigations.

In the random field model, the ferroelectric phase transition is suggested to be smeared due to the quenched random fields, but it may be visible if the random fields are overcome by an external electric field [12]. This ferroelectric phase transition has been convincingly shown by the sharp peak of the linear dielectric susceptibility ε_{1p} (Fig. 5(b)) and the anomaly of its nonlinear components ε_{3p} and a_{3p} (solid circles in Fig. 5(c) and (d)) when ferroelectric do-mains are aligned by an external field. The sharp peak of linear susceptibility indicates that the ferroelectric phase transition occurs at $T=T_c$=225 K, around which Curie-Weise law was observed. The value of Curie constant is estimated to be $C = 2.05*10^5$ K, which is character-istic of that of the displacive-type phase transition, suggesting a soft mode-driven phase transition in this system. This conclusion is supported by the occurrence of soft–mode in the crystal observed by neutron [31] and Raman scattering measurements [32].

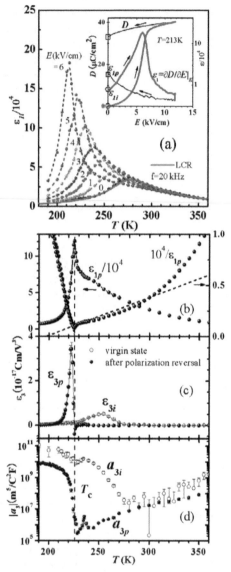

Figure 5. Dielectric responses in PMN crystal. (a) Chang of linear dielectric responses ($\varepsilon_1 = \partial D / \partial E|_E$) with the electric field for the virgin state. Inset shows an example of the polarization and dielectric responses, in which the thick lines indicate the virgin state. Dielectric constant obtained by LCR impedance measurements at an ac level of 1 V/cm (red solid line) is also shown for comparison. (b) Linear dielectric response ($\varepsilon_{1p} = \partial D / \partial E|_{E=0}$) at zero field after the polarization reversal (indicated by the square in inset of (a)), and its inverse. (c) Nonlinear dielectric constant($\varepsilon_3 = \partial^3 D / \partial^3 E|_{E=0}$) at zero field for the virgin state, and the state after the polarization reversal. (d) The corresponding $a_3 = \varepsilon_3 / \varepsilon_1^4|_{E=0}$ for these two states.

Here, we can see that there are two characteristic temperatures in relaxors: Burns temperature T_{Burns} and ferroelectric phase transition temperature T_c. At T_{Burns}, local polarizations (PNRs) begin to occur. Before the ferroelectric transition occurs, PNRs are dynamic, and more importantly, interactions among them are random. Consequently, the existence of PNRs between T_c and T_{Burns} can be considered as precursor phenomenon in a phase transition [33]. Actually, such precursor phenomenon has also been observed in the normal ferroelectric BaTiO$_3$ at temperatures far above T_c[34]. A difference between BaTiO$_3$ and PMN relaxor is that the temperate region of precursor existence in PMN is greatly lager than in BaTiO$_3$. To probe the precursor behavior, one has to consider the time scale used. For example, we cannot detect a spontaneous polarization at ms scale by D-E measurements for PMN at room temperature, but at the probe time scale of 2 ns, cold neutron high-flux backscattering spectrometer can detect PNRs up to ~400K [35] (Fig. 1b). In contrast, due to a shorter probe time scale of ~6 ps [4], thermal neutron scattering can probe PNRs up to 600 K, which is close to T_{Burns} determined by the optical measurements that have the shortest probe time scale.

4. Soft mode behaviors in PMN relaxor

In the displacive-type ferroelectrics, soft-modes should occur in the lattice dynamics of the crystal. Actually, in a study of neutron inelastic scattering, a FE soft mode was revealed to recovers, i. e., becomes underdamped, below 220 K, and from there its energy squared $(\hbar\omega_s)^2$ increases linearly with decreasing T as for normal FEs below T_c (see also Fig. 6(e)) [31]. This has long been a puzzle for PMN relaxor: how can this be, since it has been thought that PMN remains cubic to at least 5 K [16]? However, such soft-mode behavior is exactly consistent with our results from polarization measurements shown in previous section, which show a ferroelectric phase transition at T_c ~225 K. Our Raman scattering measurements also support the occurrence of FE softmode in PMN relaxor [32].

In Raman scattering studies for relaxors, the multiple inhomogeneities due to the coexistence of different symmetry regions such as the PNR and COR has been a tremendous barrier to clarify the dynamical aspect of relaxor behavior in PMN. In particular, the intense temperature-independent peak at 45 cm^{-1} (indicated by ↓ in Fig. 6(a)), which stems from the COR with $Fm\overline{3}m$ [32], always precludes a detailed investigation of low-wave number spectra of PMN crystal. Our angular dependence of the Raman spectra together with the results from the Raman tensor calculations clearly indicate that the strong F_{2g} mode located at 45 cm^{-1} can be eliminated by choosing a crossed Nicols configuration with the polarization direction of the incident laser along <110>$_c$ direction (see right panel in Fig. 6(c)).

Such special configuration allows us to observe the other low-wave number modes easily. Fig. 6 (d) shows the spectra obtained by this configuration. At the lowest temperature, a well-defined mode can be seen from the spectrum, which softens as increasing the temperature, indicating the occurrence of FE soft mode in PMN relaxor. Due to the multiple inhomogeneities of the system, the shape of soft-mode of PMN relaxor is not as sharp as that observed in normal displacive-type ferroelectrics. However, we still can estimate its frequency reliably from the careful spectrum analysis. Its temperature dependence is shown in Fig. 6(e) (indicated by ●) in comparison with the results obtained by neutron inelastic scattering (○) [31].

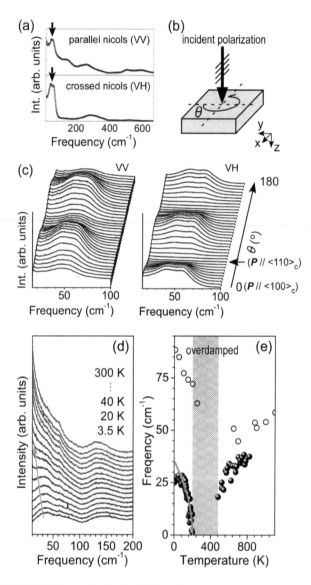

Figure 6. a) Room-temperature Raman spectra in PMN observed by the parallel (upper panel) and crossed (bottom panel) Nicols configurations, with the polarization direction of the incident laser parallel to <100>$_c$ (**P** // <100>$_c$). (b) Scattering configurations used in measurements on the angular dependence of the Raman spectra. (c) Angular dependence of the low-wave number Raman spectra obtained at room temperature. (d) Temperature dependence of Raman spectra observed by the crossed Nicols configuration with the polarization direction of the incident laser along <110>c direction (**P** // <110>$_c$). (e) Soft mode wave numbers obtained by Raman scattering (●) and by neutron inelastic scattering by Wakimoto et al. (O) [31].

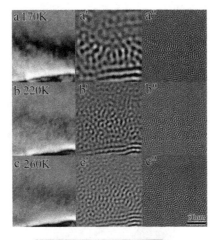

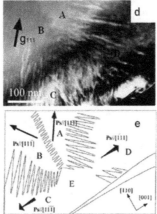

Figure 7. a)-(c) Temperature variation of TEM images observed for PMN relaxor. (a')-(c') & (a")-(c") show images of PNR and COR derived from (a)-(c). (d)Micrometric domain structure observed in the ferroelectric phase at 130K. (e) Schematic domain patterns shown in (d). Arrows and lines indicate the polarization directions and domain boundaries, respectively.

We find that the soft-mode exhibits softening towards T_c on heating and follows the conventional Curie–Weiss law (solid line) over a large temperature region. However, upon further heating, the soft mode becomes over damped in a temperature region extending over ~200 K, which does not allow us to estimate the frequency of the mode. At temperatures above 480 K, the soft mode recovers the under damped oscillation and hardens as the temperature increases. These phenomena are very similar to those revealed by neutron inelastic scattering [31]. A major difference is that the wave number of the soft mode in the present study is significantly lower than that observed by the neutron inelastic scattering. This can be reasonably understood by the splitting of the soft mode due to the lowering of symmetry as demonstrated in the NMR study [21]. According to previous results, the local symmetry in the PNR changes from cubic to rhombohedral. Therefore, the soft mode can be assumed to split from the F_{1u} mode to the A_1 and E modes. Generally, the A_1 mode is higher in wave number than the E mode due to the depolarization field effect.

In a short summary, we may say that the polarization in PMN is induced by the soft mode. This interpretation is essentially consistent with the results described in the polarization measurements, and the results obtained in previous neutron studies [36, 37], in which the crystallographic structure of PNR is attributed to the displacement pattern of the soft mode. The results of the Raman study also support that a ferroelectric state exists in PMN even at the zero-bias field.

5. Ferroelectric domain structures observed by TEM

In order to understand the microstructures of COR and PNR together with the domain structures and its evolution with temperature in the ferroelectric phase of PMN relaxor, we have carried out a detailed TEM observation. The typical results are summarized in Fig. 7. As shown in Fig. 7 (a")-(c"), COR was found to be spherical shape and has size less than 5 nm. It is very stable and remains unchanged within the temperature range of 130 K-675K. In the TEM observation, large amount of CORs were found to distribute in the PMN crystal. In a previous HRTEM study, its volume fraction has been estimated to be ~1/3 of the crystal [18]. CORs are thus considered to be the intense sources of the strong random fields.

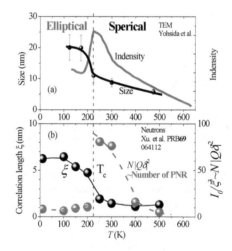

Figure 8. Temperature variation of PNR observed by (a) TEM [38]and (b) Neutron scattering [4]. In (b), ξ is the correlation length, and I_0 is the integrated diffuse scattering intensity and can be written to $N\xi^3|Q\delta|^2$, where N is the total number of PNR, δ is the average displacement of atoms within the PNR, and Q is the wave vector.

In contrast to CORs, PNRs exhibit remarkable change with temperature. As shown in Fig. 7(a')-(c') and Fig. 8, PNRs with size of several nm were found to occur in the crystal for $T<T_{Burns}$. These spherical PNRs show continuous growth as lowering the temperature. However, PNRs change from spherical shape to elliptical shape around T_c. Associating with the change in shape, its intensity was also found to drop rapidly. These results are consistent with those derived from neutron scattering (Fig. 8(b)) [4]. Neutron scattering study by Xu et al. [4] shows that the "correlation length" ξ, which is a direct measure of the length scale of the PNR, increases on cooling and changes remarkably around T_c. At the same time, the number of PNR increases on cooling from high temperatures and then drops dramatically at around T_c, remaining roughly constant below T_c.

Associating with the change in the number and the shape of PNR, micrometric ferroelectric domains were found to occur for $T<T_c$. It is because of growing into macroscopic domain that the number of PNR drop sharply in the ferroelectric phase. Figs. 7(d) and (e) show the structure of a FE micrometric domain in the FE phase of 130 Kand its schematic patterns, respectively. The micrometric domains are formed in the crystal with the spontaneous P_s along the <111>$_c$ direction. In comparison with the domain structure of normal ferroelectric such as BaTiO$_3$, domain size is relatively small and the domain boundaries blur in PMN relaxor.

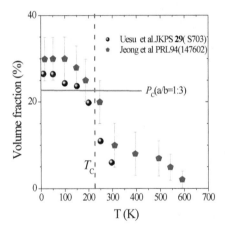

Figure 9. Volume fraction of PNR estimated from neutron scattering measurements [37,39]. Solid line denotes the threshold of percolation for elliptical shape [40].

The occurrence of micrometric domains, the soft-mode observed by Raman scattering, and the macroscopic polarization all direct to the same conclusion: PMN is essentially ferroelectric but not dipole-glass at $T<T_c$ although it exhibits some unique characteristic properties including broadening soft-mode, smearing domain wall, and very large activation field required for domain switching. Our TEM measurements clearly indicate that interactions among PNRs for $T<T_c$ are not random, but cooperative, which is different from the picture expected in the SRBRF model [10]. It is due to such non-random interaction, PNRs team up

together to form micrometric domain in the FE phase. Thus, our TEM observations support the random field model suggested by Fisch [13]. It should be emphasized that PNRs cannot merge together completely due to the blocking of the intense CORs in the crystal (Fig. 10).

Here, we made a discussion on the volume fraction of PNRs in PMN crystal. Neutron scattering technique has been used to estimate the volume fraction of PNRs. Fig. 9 replots two results reported by Jeong et al. [37] and Uesu et al. [39], respectively. Both studies indicate that PNRs occupy a volume fraction > 25% at the lowest measurement temperature. This volume fraction is larger than the threshold of 22% to form a percolated ferroelectric state with an ellipsoidal-shape [40], supporting again the picture of a ferroelectric state in PMN relaxor.

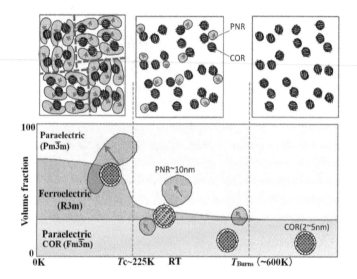

Figure 10. A physics picture of structure evolution in PMN relaxor.

6. A physics picture of relaxor

In summary, we propose a physics picture for relaxors. Figure 10 schematically shows a model of structure evolution in PMN relaxor. Since COR has been observed at $T>T_{Burns}$, it can be considered that there is a coexistence of paraelectric phase of COR with $Fm\bar{3}m$ symmetry and paraelectric $Pm\bar{3}m$ phasein this high temperature. Upon cooling, spherical PNRs occur from paraelectric $Pm\bar{3}m$ phase for $T<T_{Burns}$. Both number and size of PNR increase as lowering the temperature. Around room temperature, PNRs grow to a size of about 10 nm. For $T<T_c$ (~225K), neighboring PNRs merge together to form elliptical shape with anisotropy, associating with the reduction of its number. Due to the blocking by the intense CORs, individual PNR merely grows to a size of about 20 nm at $T<T_c$. However, PNRs with elliptical

shape tend to team up together to form a larger domain at $T<T_c$ at which a ferroelectric state occurs. The existence of a multi-scale inhomogeneity of ferroelectric domain structure provides a key point to understand the huge electromechanical coupling effects in relaxors and piezoelectrics with morphotropic phase boundary (MPB). This also gives idea to design new material with domain structure to have large elastic deformation with the application of an electric or magnetic field [41].

Acknowledgements

We thank Mr. M. Yoshida, Prof. N. Yamamoto and Prof. Shin-ya Koshihara of Tokyo Institute of Technology for their contributions in this work. We acknowledge the support of a Grant-in-Aid for Scientific Research, MEXT, Japan.

Author details

Desheng Fu[1*], Hiroki Taniguchi[2], Mitsuru Itoh[2] and Shigeo Mori[3]

*Address all correspondence to: ddsfu@ipc. shizuoka. ac. jp

1 Division of Global Research Leaders, Shizuoka University, Johoku, Naka-ku, Hamamatsu, Japan

2 Materials and Structures Laboratory, Tokyo Institute of Technology, Nagatsuta, Yokohama, Japan

3 Department of Materials Science, Osaka Prefecture University, Sakai, Osaka, Japan

References

[1] Smolenskii, G. A., Agranovskaia, A. I. *Sov. Phys. Tech. Phys.* (1958), 3, 1380.

[2] Cross, L. E. Relaxor ferroelectrics: An overview. *Ferroelectrics* (1994), 151, 305-320.

[3] Burns, G., Dacol, F. H. Glassy polarization behavior in ferroelectric compounds Pb(Mg$_{1/3}$Nb$_{2/3}$)O$_3$ and Pb(Zn$_{1/3}$Nb$_{2/3}$)O$_3$. *Solid State Commun.* (1983), 48, 853-856.

[4] Xu, G., Shirane, G., Copley, J. R. D. , Gehring, P. M. Neutron elastic diffuse scattering study of Pb(Mg$_{1/3}$Nb$_{2/3}$)O$_3$. *Phys. Rev. B* (2004), 69, 064112.

[5] Swainson, I. P., Stock, C., Gehring, P. M., Xu, G., Hirota, K., Qiu, Y., Luo, H., Zhao, X., Li, J.-F., Viehland, D., Soft phonon columns on the edge of the Brillouin zone in the relaxor PbMg$_{1/3}$Nb$_{2/3}$O$_3$. *Phys. Rev. B* (2009), 79, 224301.

[6] Vugmeister, B. E. Polarization dynamics and formation of polar nanoregions in relax-or ferroelectrics. *Phys. Rev. B* (2006), 73, 174117.

[7] Grinberg, I., Juhas, P., Davies, P. K., Rappe, A. M. Relationship between local structure and relaxor behavior in perovskite oxides. *Phys. Rev. Lett.* (2007), 99, 267603.

[8] Colla, E.V., Koroleva, E. Yu., Okuneva, N. M., Vakhrushev, S. B. Long-time relaxation of the dielectric response in lead magnoniobate. *Phys. Rev. Lett.* (1995), 74, 1681-1684.

[9] Levstik, A., Kutnjak, Z., Filipič, C., Pirc, R. Glassy freezing in relaxor ferroelectric lead magnesium niobate. *Phys. Rev. B* (1998), 57, 11204-11211.

[10] Pirc, R., Blinc, R. Spherical random-bond–random-field model of relaxor ferroelectrics. *Phys. Rev. B* (1999), 60, 13470-13478.

[11] Bobnar, V., Kutnjak, Z., Pirc, R., Blinc, R., Levstik, A. Crossover from glassy to inhomogeneous-ferroelectric nonlinear dielectric response in relaxor ferroelectrics. *Phys. Rev. Lett.* (2000), 84, 5892-5895.

[12] Westphal, V., Kleemann, W., Glinchuk, M. D. Diffuse phase transitions and random-field-induced domain states of the "relaxor" ferroelectric $PbMg_{1/3}Nb_{2/3}O_3$. *Phys. Rev. Lett.* (1992), 68, 847.

[13] Fisch, R. Random-field models for relaxor ferroelectric behavior. *Phys. Rev. B* 2003; 67, 094110.

[14] Moriya, Y., Kawaji, H., Tojo, T., Atake, T. Specific-heat anomaly caused by ferroelectric nanoregions in $Pb(Mg_{1/3}Nb_{2/3})O_3$ and $Pb(Mg_{1/3}Ta_{2/3})O_3$ relaxors. *Phys. Rev. Lett.* (2003), 90, 205901.

[15] Fu, D., Taniguchi, H., Itoh, M., Koshihara, S., Yamamoto, N., and Mori S. Relaxor $Pb(Mg_{1/3}Nb_{2/3})$ O_3: A ferroelectric with multiple inhomogeneities. *Phys. Rev. Lett.* (2009), 103, 207601.

[16] Mathan, N. de, Husson, E., Calvarn, G., Gavarri, J. R., Hewat, A. W., Morell, A. A structural model for the relaxor $PbMg_{1/3}Nb_{2/3}O_3$ at 5 K. *J. Phys. Condens. Matter* (1991), 3, 8159.

[17] Hilton, A. D., Barber, D. J., Randall, C.A., Shrout, T. R. On short range ordering in the perovskite lead magnesium niobate. *J. Mater. Sci.* (1990), 25, 3461-3466.

[18] Boulesteix, C., Varnier, F., Llebaria, A., Husson, E. Numerical determination of the local ordering of $PbMg_{1/3}Nb_{2/3}O_3$ (PMN) from high resolution electron microscopy images. *J. Solid State Chem.* (1994), 108, 141-147.

[19] Chen, J., Chan, H. M., Harmer, M. P. Ordering structure and dielectric properties of undoped and La/Na-doped $Pb(Mg_{1/3}Nb_{2/3})O_3$. *J. Am. Ceram. Soc.* (1989), 72, 593-598.

[20] Jeong, I. K., Darling, T. W., Lee, J. K., Proffen, Th., Heffner, R. H., Park, J. S., Hong, K. S., Dmowski, W., Egami, T. Direct observation of the formation of polar nanoregions

in Pb(Mg$_{1/3}$Nb$_{2/3}$)O$_3$ using neutron pair distribution function analysis. *Phys. Rev. Lett.* (2005), 94, 147602.

[21] Blinc, R., Laguta, V. V., Zalar, B. Field cooled and zero field cooled [207]Pb NMR and the local structure of relaxor PbMg$_{1/3}$Nb$_{2/3}$O$_3$. *Phys. Rev. Lett.* (2003), 91, 247601.

[22] Wieder, H. H. Electrical behavior of barium titanatge single crystals at low temperatures. *Phys. Rev.* (1955), 99, 1161-1165.

[23] Wieder, H. H. Activation field and coercivity of ferroelectric barium titanate. *J. Appl. Phys.* (1957), 28, 367-369.

[24] Wieder, H. H. Ferroelectric polarization reversal in rochelle salt. *Phys. Rev.* (1958), 110, 29-36.

[25] Merz, W. J. Domain formation and domain wall motions in ferroelectric BaTiO$_3$ single crystals. *Phys. Rev.* (1954), 95, 690-698.

[26] Fatuzzo, E., Merz, W. J. Switching mechanism in triglycine sulfate and other ferroelectrics. *Phys. Rev.* 1959; 116, 61-68.

[27] Fatuzzo, E. Theoretical Considerations on the Switching transient in ferroelectrics. *Phys. Rev.* (1962), 127, 1999-2005.

[28] Shin, Y.-H., Grinberg, I., Chen, I-W., Rappe, A. M. Nucleation and growth mechanism of ferroelectric domain-wall motion. *Nature* (2007), 449, 881-884.

[29] Glazounov, A. E., Tagantsev, A. K. Phenomenological model of dynamic nonlinear response of relaxor ferroelectrics. *Phys. Rev. Lett.* (2000), 85, 2192-2195.

[30] Dec, J., Miga, S., Kleemann, W., Dkhil, B. Nonlinear dielectric properties of PMN relaxor crystals within Landau-Ginzburg-Devonshire approximation. *Ferroelectrics* (2008), 363, 141-149.

[31] Wakimoto, S., Stock, C., Birgeneau, R. J., Ye, Z.-G., Chen, W., Buyers, W. J. L., Gehring, P. M., Shirane, G. Ferroelectric ordering in the relaxor Pb(Mg$_{1/3}$Nb$_{2/3}$)O$_3$ as evidenced by low-temperature phonon anomalies. *Phys. Rev. B* (2002), 65, 172105.

[32] Taniguchi, H., Itoh, M., Fu, D. Raman scattering study of the soft mode in Pb(Mg$_{1/3}$Nb$_{2/3}$)O$_3$. *J. Raman Spectrosc.* (2010), 42, 706-714.

[33] Bussmann-Holder, A., Beige, H., Völkel, G. Precursor effects, broken local symmetry, and coexistence of order-disorder and displacive dynamics in perovskite ferroelectrics. *Phys. Rev. B* (2009), 79, 184111.

[34] Tai, Z. R., Namikawa, K., Sawada, A., Kishimoto, M., Tanaka, M., Lu, P., Nagashima, K., Maruyama, H., Ando, M. Picosecond view of microscopic-scale polarization clusters in paraelectric BaTiO$_3$. *Phys. Rev. Lett.* (2004), 93, 087601.

[35] Gehring, P. M., Hiraka, H., Stock, C., Lee, S.-H., Chen, W., Ye, Z.-G., Vakhrushev, S. B., Chowdhuri, Z. Reassessment of the Burns temperature and its relationship to the

diffuse scattering, lattice dynamics, and thermal expansion in relaxor $Pb(Mg_{1/3}Nb_{2/3})O_3$. *Phys. Rev. B* (2009), 79, 224109.

[36] Hirota, K., Ye, Z. G., Wakimoto, S., Gehring, P. M., Shirane, G. Neutron diffuse scattering from polar nanoregions in the relaxor $Pb(Mg_{1/3}Nb_{2/3})O_3$. *Phys. Rev. B* (2002), 65, 104105.

[37] Jeong, I. K., Darling, T. W., Lee, J. K., Proffen, T., Heffner, R. H., Park, J. S., Hong, K. S., Dmowski, W., Egami, T. Direct observation of the formation of polar nanoregions in $Pb(Mg_{1/3}Nb_{2/3})O_3$ using neutron pair distribution function analysis. *Phys. Rev. Lett.* (2005), 94, 147602.

[38] Yoshida, M. Study of $Pb(Mg_{1/3}Nb_{2/3})O_3$ ferroelectric relaxor by Transmission Electron Microscopy, *Undergraduate thesis*, Tokyo Institute of Technology, Tokyo, (1996).

[39] Uesu, Y., Tazawa, H., Fujishiro, K., Yamada, Y. Neutron scattering and nonlinear-optical studies on the phase transition of ferroelectric relaxor $Pb(Mg_{1/3}Nb_{2/3})O_3$. *J. Korean Phys. Soc.* (1996), 29, S703-S705.

[40] Garboczi, E. I., Snyder, K. A., Douglas, J. F. Geometrical percolation threshold of overlapping ellipsoids. *Phys. Rev. E* (1995), 52, 819-828.

[41] "Ferroelectrics: Nanoregions team together", *NPG Asia Materials*, Feb. 1, 2010, doi: 10.1038/asiamat.2010.17.

Chapter 5

Electronic Band Structures and Phase Transitions of Ferroelectric and Multiferroic Oxides

Zhigao Hu, Junhao Chu, Yawei Li, Kai Jiang and Ziqiang Zhu

Additional information is available at the end of the chapter

1. Introduction

Perovskite ferroelectric (FE) materials have attracted considerable attention for a wide range of applications, such as dynamic random access memories (DRAM), microwave tunable phase shifters and second harmonic generators (SHGs). [1–3] Moreover, materials that have coupled electric, magnetic, and structural order parameters that result in simultaneous ferroelectricity, ferromagnetism, and ferroelasticity are known as multiferroics. [4–6] These multiferroics materials have attracted a lot of attention in recent years because they can potentially offer a whole range of new applications, including nonvolatile ferroelectric memories, novel multiple state memories, and devices based on magnetoeletric effects. Although there are some reports on the electrical and magnetic properties of perovskite-type ferroelectric and multiferroics materials, optical properties and electronic transitions have not been well investigated up to now. On the other hand, phase transition is one of the important characteristics for the ferroelectric/multiferroics system. As we know, the phase transition is strongly related to the structural variation, which certainly can result in the electronic band modifications. Therefore, one can study the phase transition of the above material systems by the corresponding spectral response behavior at different temperatures.

Among these materials, barium strontium titanate (BST) has been considered to be one of the most promising candidates for devices due to its excellent dielectric properties of high dielectric functions, low leakage current and an adjustable Curie temperature T_c through variation of the composition between barium titanate (BT) and strontium titanate (ST). However, the limited figure of merit at high frequency microwave region restricts the BST practical applications. In order to improve the physical properties of the BST materials, introducing small compositions of dopants has been used for several decades. Many experimental and theoretical studies have been performed on the dielectric properties of BST in the ferroelectric state by adding dopants such as Magnesium [7, 8], Aluminum [9], Manganese (Mn) [10–12], Samarium [13], and different rare earth [14], In particular, the Mn doping BST shows some advantages in reducing the dielectric loss, enhancing the resistivity, and increasing dielectric tunability. It can significantly improve the dielectric properties, which makes it a potential candidate for microwave elements. For example, the Mn doping can cause the variation of the oxygen vacancy, which is the crucial role in modifying the dielectric loss mechanism. On the other hand, the doping of Mn can also reduce the dielectric constant peak and broaden the dielectric phase transition temperature range, which results in a smaller temperature coefficient of capacitance in BSMT materials. [10] Therefore, it is important to further investigate the physical properties of BSMT materials in order to develop the potential applications.

Meanwhile, many research groups have focused on the doping effect on the fabrications and dielectric properties for strontium barium niobate $SrBi_2Nb_2O_9$ (SBN) materials. [15–20] The substitution of Ca ions in the Sr site for the SBN ceramic induced the T_C increasing, which is useful for the application in high-temperature resonators. [15, 18] However, the substitution of some rare earth ions such as La^{3+} or Pr^{3+} for Bi^{3+} in the Bi_2O_2 layers can result in a shift for the T_C to lower temperature. [16, 20, 21] It is found that the behavior of the Nd-doped SBN ceramic tends to change from a normal ferroelectrics to a relaxor type ferroelectrics owing to the introduction of Nd ions in the Bi_2O_2 layers. [16, 17, 20, 22] Up to now, a detailed understanding of the lattice dynamic properties and the phase transition behavior of Nd-doped SBN ceramics are still lacking. Raman spectroscopy is a sensitive technique for investigating the structure modifications and lattice vibration modes, which can give the information on the changes of lattice vibrations and the occupying positions of doping ions. Thus, it is a powerful tool for the detection of phase transition in the doping-related ferroelectric materials. [22, 23]

The optical properties such as the dielectric functions provide an important insight on dielectric and ferroelectric behaviors of the material and play an important role in design, optimization, and evaluation of optoelectronic devices. [24–27] In addition, the doping of Mn or Nd can induce more defects in the lattice structure, which can affect its electronic band structures and optoelectronic properties. Hence, the doping composition dependence of optical properties for BST and SBN ceramics is technically important for practical optoelectronic device development. Compared to film structure, the optical properties of bulk material (single crystals and dense ceramics) are not affected by interface layer, stress from clamping by the substrate, non-stoichiometry and lattice mismatch between film and substrate. Hence, it is desirable to carry out a delicate investigation regarding the optical properties of the BSMT and SBNN ceramics. Note that spectroscopic ellipsometry (SE), Raman scattering, and transmittance spectra are potentially valuable techniques for the studies of ferroelectric materials due to their high sensitivity of local structure and symmetry. Compared with the other techniques, they can provide dielectric functions of the materials. SE and transmittance spectra can provide optical band gap and optical conductivity, whereas Raman spectra can provide Raman-active modes of the materials. [28–31]

On the other hand, bismuth ferrite ($BiFeO_3$, BFO) is known to be the only perovskite material that exhibits multiferroic at room temperature (RT). At RT, it is a rhombohedrally distorted ferroelectric perovskite with the space group $R3c$ and a Curie temperature (T_C) of about 1100 K. [32–36] Since the physical properties of BFO films are related to their domain structure and phase states, which is sensitive to the applied stress, composition, and fabrication condition for BFO materials. PLD technique has the ability to exceed the solubility of magnetic impurity and to permit high quality film grown at low substrate temperature. Recent studies of photoconductivity, [35] photovoltaic effect, [37] and low open circuit voltage in a working solar device, [38] illustrate the potential of polar oxides as the active photovoltaic material.

In spite of the promising properties, there are no systematical reports focused on the optical properties of BFO films. In order to make BFO useful in actual electrical and optoelectronic devices, the physical properties, especially for electronic band structure and optical response behavior, need to be further clarified.

The objectives of the chapter will tentatively answer the interesting questions: (1) Is there an effective method to directly analyze electronic structure of FE materials by optical spectroscopy? (2) What kind of temperature dependence have FE oxides from band-to-band transitions? (3) Can spectral response at high-temperature be used to judge phase transition? Correspondingly, this chapter is arranged in the following way. In Sec. 2, detailed growths of $Ba_{0.4}Sr_{0.6-x}Mn_xTiO_3$ (BSMT), $SrBi_{2-x}Nd_xNb_2O_9$ (SBNN) ceramics and BFO films are described; In Sec. 3, solid state spectroscopic techniques are introduced; In Sec. 4, electronic band structures of BSMT ceramics are presented; In Sec. 5, phase transitions of SBNN ceramics are derived; In Sec. 6, temperature effects on electronic transitions of BFO films have been discussed; In Sec. 7, the main results and remarks are summarized.

2. Experimental

2.1. Fabrications of FE ceramics

The ceramics based on $Ba_{0.4}Sr_{0.6-x}Mn_xTiO_3$ (with x = 1, 2, 5 and 10%) specimens were prepared by the conventional solid-state reaction sintering. High purity $BaCO_3$ (99.8%), $SrCO_3$ (99.0%), TiO_2 (99.9%), and $MnCO_3$ were used as the starting materials. Weighted powers were mixed by ball milling with zirconia media in the ethanol as a solvent for 24 h and then dried at 110 °C for 12 h. After drying, the powders were calcined at 1200 °C for 4 h, and then remilled for 24 h to reduce the particle size for sintering. The calcined powders were added with 8 wt.% polyvinyl alcohol (PVA) as a binder. The granulated powders were pressed into discs in diameter of 10 mm and thickness of 1.0 mm. The green pellets were kept at 550 °C for 6 h to remove the solvent and binder, followed by sintering at 1400 °C for 4 h. More details of the preparation process can be found in Ref. [39]. On the other hand, the SBNN (x=0, 0.05, 0.1, and 0.2) ceramics were prepared by a similar method, and $SrCO_3$, Bi_2O_3, Nb_2O_5, and Nd_2O_3 were used as the starting materials. Details of the fabrication process for the ceramics can be found elsewhere. [17, 40]

2.2. Depositions of BFO films

The $BiFeO_3$ films were deposited on c-sapphire substrates by the PLD technique. The $BiFeO_3$ targets with a diameter of 3 cm were prepared through a conventional solid state reaction method using reagent-grade Bi_2O_3 (99.9%) and Fe_2O_3 (99.9%) powders. Weighed powders were mixed for 24 h by ball milling with zirconia media in ethanol and then dried at 100 °C for 12 h. The dried powders were calcined at about 680 °C in air for 6 h to form the desired phase, and followed by sintering at about 830 °C for 2 h. Before the deposition of the BFO films, c-sapphire substrates need to be cleaned in pure ethanol with an ultrasonic bath to remove physisorbed organic molecules from the surfaces, followed by rinsing several times with de-ionized water. Then the substrates were dried in a pure nitrogen stream before the film deposition. A pulsed Nd:YAG (yttrium aluminum garnet) laser (532 nm wavelength, 5 ns duration) operated with an energy of 60 mJ/pulse and repetition rate of 10 Hz was used as the ablation source. The films were deposited immediately after the target was

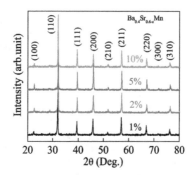

Figure 1. The XRD patterns of the BSMT ceramics with the Mn composition of 1, 2, 5 and 10%, respectively.

preablated in order to remove any surface contaminants. The distance between the target and the substrate was kept at 3 cm. The deposition time was set to about 30 min. Finally, the films were annealed at 600 °C in air atmosphere by a rapid thermal annealing process. A detailed preparation of the films can be found in Ref. [6].

3. Optical spectroscopy

The ellipsometric measurements were carried out in the photon energy range of 0.7-4.2 eV (300-1700 nm) with a spectral resolution of 5 nm by near-infrared-ultraviolet (NIR-UV) SE (SC630UVN by Shanghai Sanco Instrument, Co., Ltd.). The measurements were performed under the incident angle of 67° for all the ceramics corresponding to the experimental optimization near the Brewster angle of the BSMT. Raman scattering experiments were carried out using a Jobin-Yvon LabRAM HR 800 UV micro-Raman spectrometer, excited by a 632.8 nm He-Ne laser with a spectral resolution of 0.5 cm^{-1}. Temperature dependent measurements from 80 to 873 K were performed using the Linkam THMSE 600 heating stage, and the set-point stability is of better than 0.5 K. The normal-incident transmittance spectra were recorded using a double beam ultraviolet-infrared spectrophotometer (PerkinElmer Lambda 950) at the photon energy from 0.5 to 6.5 eV (190-2650 nm) with a spectral resolution of 2 nm. The samples at 5.3-300 K were mounted into an optical cryostat (Janis SHI-4-1) for variable temperature experiments. [41]

4. Electronic band structures of BSMT ceramics

The XRD patterns of the BSMT ceramics with different Mn composition are shown in Fig. 1 and no secondary phase appears within the detection limit of the XRD. Besides the strongest (110) peak, some weaker peaks (100), (111), (200), (210), (211), (220) can be also observed, which indicate that the ceramics are polycrystalline with single perovskite phase. The diffraction patterns are fitted by the Gaussian lineshape analysis to extract the peak positions and full width at half maximum (FWHM). The lattice constant a of the BSMT ceramics, which can be estimated from the (110) diffraction peak, is calculated to be about 3.954 Å. [39] The ionic radius of Mn^{2+} (1.27 Å) is smaller than that of Sr^{2+} (1.44 Å) and Ba^{2+} (1.61 Å), and is larger than that of Ti^{4+} (0.61 Å), which can be attributed to the change of the lattice constant. When the Mn composition is below 5%, the (110) diffraction peak positions shift from smaller angles to larger angles and the lattice constant slightly decreases, which can be

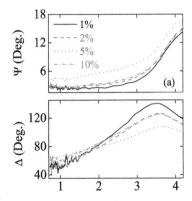

Figure 2. Experimental ellipsometric data (a) Ψ and (b) Δ for the BSMT ceramics from near-infrared to ultraviolet photon energy region at the incident angle of 67°. (Figure reproduced with permission from [39]. Copyright 2012, Springer.)

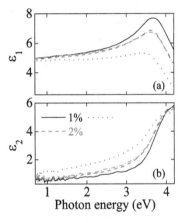

Figure 3. The (a) real and (b) imaginary parts of the NIR-UV dielectric functions for the BSMT ceramics with different Mn composition. (Figure reproduced with permission from [39]. Copyright 2012, Springer.)

ascribed to the fact that Mn occupies the A site together with Ba^{2+} and Sr^{2+}. In order to reduce distortion of the crystal lattice, the Mn mainly substitutes Sr^{2+} because the differences in ionic radius between Mn ion and Ba ion is larger than that between Mn and Sr. On the other hand, when the Mn composition is larger than 5%, the excess Mn can also substitute the Ti site, which results in the increase of the lattice constant and the smaller shift of the peak position. It can be concluded that the Mn ions substitute Sr sites of the BST lattice at first, then occupy Ti sites when the Mn composition is beyond 5%.

The experimental ellipsometric spectra of Ψ and Δ recorded at an incident angle of 67° for the BSMT ceramics are depicted in Figs. 2(a) and (b), respectively. The observed changes in the Ψ and Δ data for different Mn composition may be attributed to the lattice distortion and variation in atomic coordinate. Because the sample is bulk material with a thickness of several millimeters, the dielectric functions of the BSMT ceramics can be directly calculated

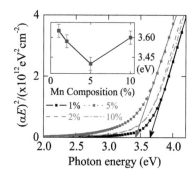

Figure 4. Absorption coefficient $vs.$ incident photon energy near the optical band gap of the BSMT ceramics. The insert is the optical band gap E_g with the different Mn composition. (Figure reproduced with permission from [39]. Copyright 2012, Springer.)

according to the ellipsometric spectra Figs. 3(a) and (b) show the real (ε_1) and imaginary (ε_2) parts of the dielectric functions in the photon energy range of 0.7-4.2 eV, respectively. The evolution of ε with the photon energy is a typical optical response behavior of ferroelectric and/or semiconductors. The optical band gap (E_g) of the BSMT ceramics is one of the important optical behaviors, which is calculated by considering a direct transition from the VB to the CB when the photon energy falls on the materials. In the BSMT system, the VB is mainly composed of the O $2p$ orbital and the CB is mainly composed of the Ti $3d$ orbital. It should be noted that because of the splitting of the Ti $3d$ conduction bands into t_{2g} and e_g subbands, the lowest CB arises from the threefold degenerate Ti $3d$ t_{2g} orbital, which has lower energy than the twofold degenerate Ti $3d$ e_g orbital. [42] The absorption coefficient related E_g of the BSMT ceramics can be determined according to the Tauc's law: $(\alpha E)^2 = A(E - E_g)$, where A is a constant, α and E are the absorption coefficient and incident photon energy, respectively. For the allowed direct transition, the straight line between $(\alpha E)^2$ and E will provide the value of the band gap, which is extrapolated by the linear portion of the plot to $(\alpha E)^2 = 0$, as seen in Fig. 4. The E_g is estimated to 3.65, 3.57, 3.40 and 3.60 eV corresponding to $x=1, 2, 5$ and 10% for the BSMT ceramics, respectively, as shown in the inset of Fig. 4. The results suggest that the band gap of the BSMT ceramics decreases and then increases with increasing Mn composition.

As we know, the optical band gap can be affected by some factors such as grain size, oxygen vacancy, stress and amorphous nature of the materials. [2] The decreasing trend of the band gap with the Mn composition below 5% can be attributed to the increase of the grain size and the smaller lattice constant, which are caused by the Mn introduction in the A sites. When the Mn composition is 10%, there is a sharp increase in the optical band gap because the excess Mn will substitute the Ti site at the Mn composition of 10%, causing the increase of the oxygen vacancies. In addition, the creation of an oxygen vacancy which is associated with the generation of free charge carriers can be described as the following: $MnO(-TiO_2) \rightarrow Mn_{Ti}'' + V_O^{\cdot\cdot} + O_O; O_O \rightarrow V_O^{\cdot\cdot} + 2e^- + 1/2O_2$, where $V_O^{\cdot\cdot}$ represents the doubly charged oxygen vacancy, O_O is an oxygen ion at its normal site, e^- is the free electronic charge generated through the vacancy formation. [9, 43] The heavy doping blocks the lowest states in the CB and the effective band-gap increases, which is known as the Burstein-Moss (BM) effect. [44] When a large number of vacancy-related charge carriers are

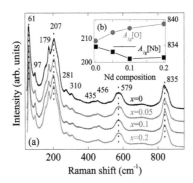

Figure 5. Raman scattering spectra of the SBNN ceramics with different Nd composition (x) recorded at RT. The dashed lines clearly indicate some Raman-active phonon modes. The inset shows the peak frequency variation of the A_{1g}[Nb] phonon mode at about 207 cm^{-1} and the A_{1g}[O] phonon mode at about 836 cm^{-1} as a function of Nd composition. (Figure reproduced with permission from [46]. Copyright 2012, Wiley.)

generated in the system, the lowest lying states in the CB are filled and the vertical distance needed for the optical transition increases. [44] Hence, it can be concluded that, when the Mn composition is 10%, the sharply increase in the band gap may be attributed to the BM shift caused by electrons generated by oxygen vacancies. Besides, the grain size for the ceramic decreases induced by the heavy doping with the Mn composition of 10%, which may also result in a sharply increase of the optical band gap. It can be concluded that the difference of the optical band gap could be due to the dopant composition, the oxygen vacancies, and the crystallinity of the BSMT ceramics.

5. Phase transitions of SBNN ceramics

The general formula of bismuth layer structure ferroelectrics (BLSFs) is given as $(Bi_2O_2)^{2+}$ $(A_{m-1}B_mO_{3m+1})^{2-}$, where A and B are the two types of cations that enter the perovskite unit, and m is the the number of perovskite unit cell between bismuth oxide layers. [20] SBN, which is known to be $m=2$ member of BLSFs family, has been regarded as a promising ferroelectric material due to low dielectric constants and excellent fatigue resistance. [40] Fig. 5(a) shows the Raman spectra of the SBNN ceramics with different Nd compositions at RT in the spectral range of $50 - 950$ cm^{-1}. The Raman selection rules allow 18 phonon modes $(4A_{1g} + 2B_{1g} + 6B_{2g} + 6B_{3g})$ for SBN ceramics at RT. [45] However, less than 10 phonon modes are observed because of the possible overlap of the same symmetry vibration or the weak feature of some vibration bands. [46, 47] According to the assignment of SrBi$_2$Nb$_2$O$_9$ single crystal, [45] the Raman phonon modes at about 61, 207 and 835 cm^{-1} can be assigned to the A_{1g} phonon mode, the vibrations at about 179 and 579 cm^{-1} can be assigned to the E_g phonon mode. However, the assignment of other phonon modes are still not clear now. The internal vibrations of NbO$_6$ octahedra occur in the high-frequency mode region above 200 cm^{-1} because the intragroup binding energy within the NbO$_6$ octahedra is much larger than the intergroup or crystal binding energy. [15] The composition dependence of the frequencies for two typical phonon modes is illustrated in Fig. 5(b). Note that the A_{1g}[Nb] phonon mode at 207 cm^{-1}, which arises from the distortion of NbO$_6$ octahedra, generally decreases with the Nd composition, whereas the A_{1g}[O] phonon mode at 835 cm^{-1} mode corresponding to the symmetric Nb$-$O stretching vibration, increases with the introduction of Nd ions.

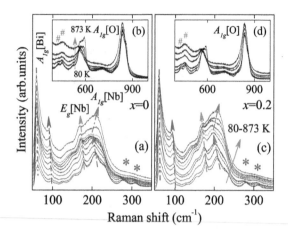

Figure 6. Temperature dependence of the SBNN ceramics with the composition of (a)(b) $x=0$ and (c)(d) $x=0.2$. The solid arrows indicate the temperature increasing from 80 K to 873 K, and the dash arrows show the shift of the frequency for the phonon modes with the temperature. The symbol asterisk ($*$) and pound sign ($\#$) indicate the two weak E_g phonon modes in the range of 281-310 cm^{-1} and 435-456 cm^{-1}, respectively. (Figure reproduced with permission from [46]. Copyright 2012, Wiley.)

In order to further understand the effect of Nd^{3+} ion substitution on the phonon modes, Fig. 6 presents the temperature dependence of the Raman spectra for the SBNN ceramics with two Nd compositions of $x=0$ and 0.2 in the temperature range from 80 to 873 K. It suggests that the intensities for all the phonon modes increase with the temperature except for the phonon mode at about 579 cm^{-1}, whose peak has been overwhelmed at high temperature. The broadening band can be assigned to a rigid sublattice mode, in which all the positive and negative ion displacements are equal and opposite. [48] It is found that a strong broadening peak can be observed at 80 K due to the combined effects of two modes splitting from the E_g character mode. However, with increasing the temperature, the frequency and intensity of the mode present a decreasing trend. Because the mode is assigned to the asymmetric Nb$-$O vibration, it can be concluded that the NbO$_6$ octahedra is sensitive to the temperature. On the other band, the phonon modes at about 281 and 310 cm^{-1} (labeled by $*$), which are associated with the O$-$Nb$-$O bending, become more difficult to be distinguished as the temperature increases and disappear at high temperature. Similar phenomena can be observed for the phonon modes in the range of 435-456 cm^{-1} (labeled by $\#$). The band at about 456 cm^{-1}, which is described to a Ti$-$O torsional mode, has been assigned as the E_g character and splits into two phonon modes centered at 435 and 456 cm^{-1} at lower temperature. As pointed out by Graves *et al.*, [45] it can be ascribed to the fact that the several E_g phonon modes split into the B_{2g} and B_{3g} phonon modes during the tetragonal to orthorhombic transition. Moreover, the splitting of the phonon modes reveals the structural changes in the SBNN ceramics with the temperature.

Considering that the phase transition temperature is related to the distortion extent of the NbO$_6$ octahedra for the SBNN ceramics, the temperature dependence of the Raman shift for the A_{1g}[Nb] phonon mode is plotted in Fig. 7(a)-(d). For all the SBNN ceramics, the decrease of the A_{1g}[Nb] phonon mode can be observed as the temperature is increased. Note that an obviously anomalous vibration occurs around the phase transition temperatures: the Raman shift sharply increases with increasing the temperature. In addition, the temperature

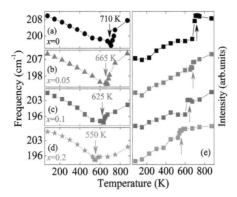

Figure 7. (a)-(d) Raman shift and (e) intensities of the A_{1g}[Nb] phonon mode as a function of the temperature for the SBNN ceramics. The arrows indicate that the anomalous vibration occurs around the ferroelectric to paraelectric phase transition temperatures. (Figure reproduced with permission from [46]. Copyright 2012, Wiley.)

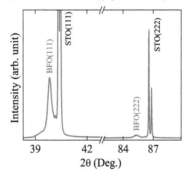

Figure 8. The XRD patterns of the BFO films deposited on STO (111) substrate. (Figure reproduced with permission from [5]. Copyright 2010, American Institute of Physics.)

of the anomalous points is different for the four ceramics: 710 K for $x=0$, 665 K for $x=0.05$, 625 K for $x=0.1$ and 550 K for $x=0.2$. The different anomalous points can be ascribed to the Nd^{3+} incorporation in the Bi_2O_2 layers. In addition, the phase transition from the orthorhombic to the tetragonal phase may occur at the temperature when the frequency of the A_{1g}[Nb] phonon mode shows the anomalous behavior. Therefore, the sharp change in the temperature dependence of both wavenumber and intensity of the A_{1g}[Nb] phonon mode was successfully applied to probe the phase transition of the SBNN ceramics.

6. Temperature effects on electronic transitions of BFO films

BFO film with the nominal thickness of about 330 nm was prepared on STO (111) substrate by pulsed laser deposition. [5] Fig. 8 shows the XRD pattern of the BFO film and there is no impurity phase. As can be seen, the film is well crystallized with the rhombohedral phase and presents a (111) single crystalline orientation. According to the known Scherrer's equation, the grain size from the (111) diffraction peak was evaluated to about 32 nm. A three-phase layered structure (air/film/substrate) was constructed to

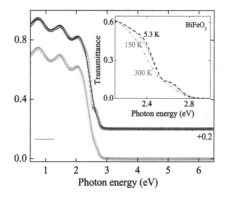

Figure 9. Experimental (dotted curves) and fitting (solid curves) transmittance spectra at temperatures of 300 and 5.3 K, respectively. The inset shows the enlarged band gap region of the BFO film at temperatures of 300, 200, and 5.3 K, respectively. (Figure reproduced with permission from [5]. Copyright 2010, American Institute of Physics.)

simulate the transmittance spectra of the BFO film. It should be emphasized that the normal-incident transmittance spectra cannot be sensitive to the thinner surface rough layer, which could be several nanometers and much less than the film thickness. Therefore, the surface rough layer can be reasonably neglected owing to a slight contribution in the evaluation of the optical properties. The optical constants of the BFO film can be expressed using four Tauc-Lorentz (TL) oscillators. [50] As an example, the experimental and fitting transmittance spectra of the BFO film at 300 and 5.3 K are shown in Fig. 9 with the dotted and solid curves, respectively. Note that the symmetrical interference period indicates that the film is of good uniformity and crystallization. From Fig. 9, it can be observed that the absorption edge remarkably shift toward the lower energy with increasing the temperature, suggesting that the OBG of the film has a negative temperature coefficient. Note that the shift at high temperature region (100-300 K) is larger than that at low temperature region (5.3-100 K). This is because the quantities of the conduction band downward and the valence band upward are different under the distinct temperature regions. Especially, two broadening shoulder structures appear and the intensities become much stronger with decreasing the temperature. The similar phenomena have been observed at 2.5 eV when the temperature decreases to about 4 K, which represents the onset of the optical absorption. [51] Furthermore, the shoulders are simply low-lying features of the electronic structure or evidence for excitonic character.

Based on the theoretical calculations and experimental observations, the four energy bands can be uniquely assigned to the following electronic transitions: (1) on-site Fe^{3+} d to d crystal field transition; (2) majority channel Fe $3d$ to O $2p$ charge transfer excitation; (3) minority channel dipole-allowed O p to Fe d charge transfer excitation; and (4) strong hybridized majority channel O p and Fe d to Bi p state excitation, respectively. [51–55] Within the experimental error bars, the energy positions shift toward the higher energy at the temperature of 5.3 K except for the second excitation, which can be attributed to the energy band variations. Nevertheless, the origin of the abnormal shift for the second excitation is unclear in the present work. Under the influence of the tetrahedral crystal field, the Fe $3d$ orbital states split into t_{2g} and e_g state and the t_{2g} state strongly hybrided with the O p orbital. [52] With decreasing the temperature, the t_{2g} and e_g states can be located at different level in

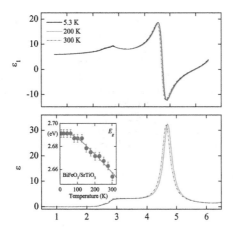

Figure 10. The dielectric functions of the BFO films in the photon energy range of 0.5-6.5 eV at 300, 200, and 5.3 K, respectively. The inset shows the temperature dependence of the E_g (dotted curve) and Bose-Einstein model fitting result (solid curve). (Figure reproduced with permission from [5]. Copyright 2010, American Institute of Physics.)

the energy space, which can affect the electronic excited ability of the Bi, Fe, and O states. On the other hand, the evaluated optical constants of the BFO film is presented in Fig. 10. The real and imaginary parts of dielectric functions increase with the temperature at the lower photon energy while decrease with further increasing the photon energy. With decreasing the temperature, both the ε_1 and ε_2 shift toward the higher energy. The phenomena are related to the modification of the electronic structure such as the fundamental band gap absorption under the lower temperature. From the inset of Fig. 10, it can be found that the E_g value increases from 2.65 ± 0.01 to 2.69 ± 0.01 eV, corresponding to decreasing the temperature from 300 to 5.3 K, which indicates that the total shift value of the E_g is about 40 meV. The observed decrease in the E_g with the temperature can be described using the Bose-Einstein model. It is widely recognized that the electron-phonon interaction and the lattice thermal expansion are responsible for the shrinkage in the optical band gap with the temperature.

7. Conclusion and remarks

In summary, electronic band structures and phase transitions of perovskite (ABO_3)-type Manganese (Mn) doped $Ba_{0.4}Sr_{0.6}TiO_3$ ceramics, $SrBi_{2-x}Nd_xNb_2O_9$, and $BiFeO_3$ materials have been investigated from infrared to ultraviolet transmittance, SE and temperature-dependent Raman scattering techniques. The interband electronic transitions and dielectric functions of these materials could be readily obtained in the wider photon energy region. Meanwhile, the phase transition temperature can be easily determined by phonon scattering measurements, indicating that the present solid state spectroscopy is useful to further clarify the physical phenomena for perovskite-type ferroelectric/multiferroics oxides.

It should be emphasized that optical properties and their related phenomena, such as phase transition and domain status have not been well investigated. The following factors must be addressed: (1) Growth of high-quality perovskite FE materials (crystal, ceramic and film); (2) Theoretical model and explanations; (3) Improved experimental methods. As for solid state spectroscopic technique, however, it can uniquely discover the electronic band structure of FE system. One can think that the electronic transition will be changed during the phase transition process owing to the crystal structure variation. Thus, we can check the status by recording the spectral response. Evidently, some optical setup at elevated temperatures are necessary because most of FE materials have the high T_C. In our research group, Transmittane/Reflectance, SE, and Raman systems from LHe temperature to about 800 K have been developed. Our next goal is to characterize the physical information of some typical FE materials in a wider spectral and temperature ranges, which play an important role in clarifying the structure transitions and the intrinsic origin.

Acknowledgments

The authors would like to thank Dr. Wenwu Li for valuable discussions. This work was financially supported by the Major State Basic Research Development Program of China (Grant Nos. 2011CB922200 and 2013CB922300), the Natural Science Foundation of China (Grant Nos. 11074076, 60906046 and 61106122), the Program of New Century Excellent Talents, MOE (Grant No. NCET-08-0192) and PCSIRT, the Projects of Science and Technology Commission of Shanghai Municipality (Grant Nos. 11520701300, 10DJ1400201 and 10SG28), and the Program for Professor of Special Appointment (Eastern Scholar) at Shanghai Institutions of Higher Learning.

Author details

Zhigao Hu[1,*], Junhao Chu[1,2,*],
Yawei Li[1], Kai Jiang[1] and Ziqiang Zhu[1]

1 Key Laboratory of Polar Materials and Devices, Ministry of Education, Department of Electronic Engineering, East China Normal University, Shanghai, People's Republic of China
2 National Laboratory for Infrared Physics, Shanghai Institute of Technical Physics, Chinese Academy of Sciences, Shanghai, People's Republic of China

References

[1] Nagaraj, B.; Sawhney, T.; Perusse, S.; Aggarwal, S.; Ramesh, R.; Kaushik, V. S.; Zafar, S.; Jones, R. E.; Lee, J. -H.; Balu, V. & Lee, J. (1999). (Ba,Sr)TiO$_3$ thin films with conducting perovskite electrodes for dynamic random access memory applications. *Appl. Phys. Lett.*, 74, 21, 3194-3196

[2] Roy, S. C.; Sharma, G. L. & Bhatnagar, M. C. (2007). Large blue shift in the optical band-gap of sol-gel derived Ba$_{0.5}$Sr$_{0.5}$TiO$_3$ thin films. *Solid State Commun.*, 141, 5, 243-247

[3] Kozyrev, A. B.; Kanareykin, A. D.; Nenasheva, E. A.; Osadchy, V. N. & Kosmin, D. M. (2009). Observation of an anomalous correlation between permittivity and tunability of a doped (Ba,Sr)TiO$_3$ ferroelectric ceramic developed for microwave applications. *Appl. Phys. Lett.*, 95, 01, 012908(1-3)

[4] Catalan, G. & Scott, J. F. (2009). Physics and applications of bismuth ferrite. *Adv. Mater.*, 21, 24, 2463-2485

[5] Li, W. W.; Zhu, J. J.; Wu, J. D.; Gan, J.; Hu, Z. G.; Zhu, M. & Chu, J. H. (2010). Temperature dependence of electronic transitions and optical properties in multiferroic $BiFeO_3$ nanocrystalline film determined from transmittance spectra. *Appl. Phys. Lett.*, 97, 12, 121102(1-3)

[6] Jiang, K.; Zhu, J. J.; Wu, J. D.; Sun, J.; Hu, Z. G. & Chu, J. H. (2011). Influences of oxygen pressure on optical properties and interband electronic transitions in multiferroic bismuth ferrite nanocrystalline films grown by pulsed laser deposition. *ACS Appl. Mater. Interfaces*, 03, 12, 4844-4852

[7] Su, B. & Button, T. W. (2004). Microstructure and dielectric properties of Mg-doped barium strontium titanate ceramics. *J. Appl. Phys.*, 95, 03, 1382-1385

[8] Xu, S.; Qu, Y. & Zhang, C. (2009). Effect of Mg^{2+} content on the dielectric properties of $Ba_{0.65-x}Sr_{0.35}Mg_xTiO_3$ ceramics. *J. Appl. Phys.*, 106, 01, 014107(1-5)

[9] Kim, S. S. & Park, C. (1999). Leakage current behaviors of acceptor- and donor-doped $(Ba_{0.5}Sr_{0.5})TiO_3$ thin films. *Appl. Phys. Lett.*, 75, 17, 2554-2556

[10] Yuan, Z.; Lin, Y.; Weaver, J.; Chen, X.; Chen, C. L.; Subramanyam, G.; Jiang, J. C. & Meletis, E. I. (2005). Large dielectric tunability and microwave properties of Mn-doped $(Ba,Sr)TiO_3$ thin films. *Appl. Phys. Lett.*, 87, 15, 152901(1-3)

[11] Zhang, J. J.; Zhai, J. W. & Yao, X. (2009). Dielectric tunable properties of low-loss $Ba_{0.4}Sr_{0.6}Ti_{1-y}Mn_yO_3$ceramics. *Scripta Mater.*, 61, 07, 764-767

[12] Liu, M.; Ma, C.; Collins, G.; Liu, J.; Chen, C.; Shui, L.; Wang, H.; Dai, C.; Lin, Y.; He, J.; Jiang, J.; Meletis, E. I. & Zhang, Q. (2010). Microwave dielectric properties with optimized Mn-doped $Ba_{0.6}Sr_{0.4}TiO_3$ highly epitaxial thin films. *Cryst. Growth Des.*, 10, 10, 4221-4223

[13] Li, Y. L. & Qu, Y. F. (2009). Dielectric properties and substitution mechanism of samarium-doped $Ba_{0.68}Sr_{0.32}TiO_3$ ceramics. *Mater. Res. Bull.*, 44, 01, 82-85

[14] Zhang, J. J.; Zhai, J. W.; Chou, X. J. & Yao, X. (2008). Influence of rare-earth addition on microstructure and dielectric behavior of $Ba_{0.6}Sr_{0.4}TiO_3$ ceramics. *Mater. Chem. Phys.*, 111, 02, 409-413

[15] Huang, S. M.; Li, Y. C.; Feng, C. D.; Gu, M. & Liu, X. L. (2008). Dielectric and structural properties of layer-structured $Sr_{1-x}Ca_xBi_2Nb_2O_9$. *J. Am. Ceram. Soc.*, 91, 09, 2933-2937

[16] Huang, S. M.; Feng, C. D.; Chen, L. D. & Wen, X. W. (2005). Dielectric properties of $SrBi_{2-x}Pr_xNb_2O_9$ ceramics (x=0, 0.04 and 0.2). *Solid State Commun.*, 133, 6, 375-379

[17] Sun, L.; Feng, C. D.; Chen, L. D. & Huang, S. M. (2007). Effect of substitution of Nd^3 for Bi^3 on the dielectric properties and structures of $SrBi_{2-x}Nd_xNb_2O_9$ bismuth layer-structured ceramics. *J. Appl. Phys.*, 101, 08, 084102(1-5)

[18] Wu, Y.; Forbess, M. J.; Seraji, S.; Limmer, S. J.; Chou, T. P.; Nguyen, C. & Cao, G. (2001). Doping effect in layer structured $SrBi_2Nb_2O_9$ ferroelectrics. *J. Appl. Phys.*, 90, 10, 5296-5302

[19] Liu, G. Z.; Wang, C.; Gu, H. S. & Lu, H. B. (2007). Raman scattering study of La-doped $SrBi_2Nb_2O_9$ ceramics. *J. Phys. D: Appl. Phys.*, 40, 24, 7817-7820

[20] Fang, P.; Fan, H.; Li, J. & Liang, F. (2010). Lanthanum induced larger polarization and dielectric relaxation in Aurivillius phase $SrBi_{2-x}La_xNb_2O_9$ ferroelectric ceramics. *J. Appl. Phys.*, 107, 06, 064104(1-4)

[21] Verma, M.; Sreenivas, K. & Gupta, V. (2009). Influence of La doping on structural and dielectric properties of $SrBi_2Nb_2O_9$ ceramics. *J. Appl. Phys.*, 105, 02, 024511(1-6)

[22] Speghini, A.; Bettinelli, M.; Caldiño, U.; Ramírez, M. O.; Jaque, D.; Bausá, L. E. & Solé, J. G. (2006). Phase transition in $Sr_xBa_{1-x}Nb_2O_6$ ferroelectric crystals probed by Raman spectroscopy. *J. Phys. D: Appl. Phys.*, 39, 23, 4930-4934

[23] Tenne, D. A.; Bruchhausen, A.; Kimura, N. D. L.; Fainstein, A.; Katiyar, R. S.; A. Cantarero,; Soukiassian, A.; Vaithyanathan, V.; Haeni, J. H.; Tian, W.; Schlom, D. G.; Choi, K. J.; Kim, D. M.; Eom, C. B.; Sun, H. P.; Pan, X. Q.; Li, Y. L.; Chen, L. Q.; Jia, Q. X.; Nakhmanson, S. M.; Rabe, K. M. & Xi, X. X. (2006). Probing nanoscale ferroelectricity by ultraviolet Raman spectroscopy. *Science*, 313, 5793, 1614-1616

[24] Zhu, J. J.; Li, W. W.; Xu, G. S.; Jiang, K.; Hu, Z. G. & Chu, J. H. (2011). A phenomenalogical model of electronic band structure in ferroelectric $Pb(In_{1/2}Nb_{1/2})O_3$-$Pb(Mg_{1/3}Nb_{2/3})O_3$-$PbTiO_3$ single crystals around morphotropic phase boundary determined by temperature-dependent transmittance spectra. *Acta Materialia*, 59, 17, 6684-6690

[25] Zhang, W. J.; Li, W. W.; Chen, X. G.; Hu, Z. G.; Liu, W.; Wang, G. S.; Dong, X.L. & Chu, J.H. (2011). Phonon mode and phase transition behaviors of $(1-x)PbSc_{1/2}Ta_{1/2}O_3$-$xPbHfO_3$ relaxor ferroelectric ceramics determined by temperature-dependent Raman spectra. *Appl. Phys. Lett.*, 99, 04, 041902(1-3)

[26] Zhu, J. J.; Li, W. W.; Xu, G.S.; Jiang, K.; Hu, Z. G.; Zhu, M. & Chu, J. H. (2011). Abnormal temperature dependence of interband electronic transitions in relaxor-based ferroelectric $(1-x)Pb(Mg_{1/3}Nb_{2/3})O_3$-$xPbTiO_3$ (x=0.24 and 0.31) single crystals. *Appl. Phys. Lett.*, 98, 09, 091913(1-3)

[27] Chen, X.; Jiang, K.; Hu, Z. G.; Chen, X. F.; Wang, G. S.; Dong, X. L. & Chu, J. H. (2012). Abnormal electronic transition variations of lanthanum-modified lead zironate stannate titanate ceramics near morphotropic phase boundary: A spectroscopic evidence. *Appl. Phys. Lett.*, 101, 01, 011914(1-5)

[28] Liao, Y. Y.; Li, Y. W.; Hu, Z. G. & Chu, J. H. (2012). Temperature dependent phonon Raman scattering of highly a-axis oriented $CoFe_2O_4$ inverse spinel ferromagnetic films grown by pulsed laser deposition. *Appl. Phys. Lett.*, 100, 07, 071905(1-4)

[29] Li, W. W.; Yu, Q.; Liang, J. R.; Jiang, K.; Hu, Z. G.; Liu, J.; Chen, H. D.; & Chu, J. H. (2011). Intrinsic evolutions of optical functions, band gap, and higher-energy electronic transitions in VO_2 film near the metal-insulator transition region. *Appl. Phys. Lett.*, 99, 24, 241903(1-3)

[30] Han, M. J.; Jiang, K.; Zhang, J. Z.; Li, Y. W.; Hu, Z. G. & Chu, J. H. (2011). Temperature dependent phonon evolutions and optical properties of highly c-axis oriented $CuGaO_2$ semiconductor films grown by the sol-gel method. *Appl. Phys. Lett.*, 99, 13, 131104(1-3)

[31] Yu, W. L.; Jiang, K.; Wu, J. D.; Gan, J.; Zhu, M.; Hu, Z. G. & Chu, J. H. (2011). Electronic structures and excitonic transitions in nanocrystalline iron-doped tin dioxide diluted magnetic semiconductor films: an optical spectroscopic study. *Phys. Chem. Chem. Phys.*, 13, 13, 6211-6222

[32] Wang, J.; Neaton, J. B.; Zheng, H.; Nagarajan, V.; Ogale, S. B.; Liu, B.; Viehland, D.; Vaithyanathan, V.; Schlom, D. G.; Waghmare, U. V.; Spaldin, N. A.; Rabe, K. M.; Wuttig, M. & Ramesh, R. (2003). Epitaxial $BiFeO_3$ multiferroic thin film heterostructures. *Science*, 299, 1719-1722

[33] Singh, M. K. & Katiyar, R. S. (2011). Phonon anomalies near the magnetic phase transitions in $BiFeO_3$ thin films with rhombohedral R_{3C} symmetry. *J. Appl. Phys.*, 109, 07, 07D916(1-3)

[34] Choi, S. G.; Yi, H. T.; Cheong, S. W.; Hilfiker, J. N.; France, R. & Norman, A. G. (2011). Optical anisotropy and charge-transfer transition energies in $BiFeO_3$ from 1.0 to 5.5 eV. *Phys. Rev. B*, 83, 10, 100101(R)(1-4)

[35] Basu, S. R.; Martin, L. W.; Chu, Y. H.; Gajek, M.; Ramesh, R.; Rai, R. C.; Xu, X. & Musfeldt, J. L. (2008). Influence of sample processing parameters on thermal boundary conductance value in an Al/AlN system. *Appl. Phys. Lett.* 92, 09, 091905(1-3)

[36] Balke, N.; Choudhury, S.; Jesse, S.; Huijben, M.; Chu, Y. H.; Baddorf, A. P.; Chen, L. Q.; Ramesh, R. & Kalinin, S. V. (2009). Deterministic control of ferroelastic switching in multiferroic materials. *Nature Nanotechnology*, 4, 12, 868-875

[37] Yang, S. Y.; Martin, L. W.; Byrnes, S. J.; Conry, T. E.; Basu, S. R.; Paran, D.; Reichertz, L.; Ihlefeld, J.; Adamo, C.; Melville, A.; Chu, Y. H.; Yang, C. H.; Musfeldt, J. L.; Schlom, D. G.; Ager, I. J. W. & Ramesh, R. (2009). Photovoltaic effects in $BiFeO_3$. *Appl. Phys. Lett.*, 95, 06, 062909(1-3)

[38] Choi, T.; Lee, S.; Choi, Y. J.; Kiryukhin, V. & Cheong, S.-W. (2009). Switchable ferroelectric diode and photovoltaic effect in $BiFeO_3$. *Science*, 324, 5923, 63-66.

[39] Jiang, K.; Zhang, J. Z.; Yu, W. L.; Hu, Z. G. & Chu, J. H. (2012). Manganese doping effects on interband electronic transitions, lattice vibrations, and dielectric functions of perovskite-type $Ba_{0.4}Sr_{0.6}TiO_3$ ferroelectric ceramics. *Appl. Phys. A-Mater. Sci. Process.*, 106, 04, 877-884

[40] Zhu, M.; Sun, L.; Li, W. W.; Yu, W. L.; Li, Y. W.; Hu, Z. G. & Chu, J. H. (2010). Lattice vibrations and dielectric functions of ferroelectric $SrBi_{2-x}Nd_xNb_2O_9$ bismuth

layer-structured ceramics determined by infrared reflectance spectra. *Mater. Res. Bull.,* 45, 11, 1654-1658

[41] Yu, W. L.; Li, W. W.; Wu, J. D.; Sun, J.; Zhu, J. J.; Zhu, M.; Hu, Z. G. & Chu, J. H. (2010). Far-infrared-ultraviolet dielectric function, lattice vibration, and photoluminescence properties of diluted magnetic semiconductor $Sn_{1-x}Mn_xO_2$/c-sapphire nanocrystalline films. *J. Phys. Chem. C*, 114 , 18, 8593-8600

[42] Kohiki, S.; Arai, M.; Yoshikawa, H.; Fukushima, S.; Oku, M. & Waseda, Y. (2000). Energy-loss structure in core-level photoemission satellites of $SrTiO_3$, $SrTiO_3$:La, and $SrTiO_3$:Nb. *Phys. Rev. B*, 62, 12, 7964-7969

[43] Warren, W. L.; Vanheusden, K.; Dimos, D.; Pike, G. E. & Tuttle, B. A. (1996). Oxygen vacancy motion in perovskite oxides. *J. Am. Ceram. Soc.*, 79, 02, 536-538

[44] Burstein, E. (1954). Anomalous optical absorption limit in InSb. *Phys. Rev.*, 93, 03, 632-633

[45] Graves, P. R.; Hua, G.; Myhra, S. & Thompson, J. G. (1995). The Raman modes of the Aurivillius phases: Temperature and polarization dependence. *J. Solid State Chem.*, 114, 01, 112-122

[46] Jiang, K.; Chen, X. G.; Li, W. W.; Zhan, Z. N.; Sun, L.; Hu, Z. G. & Chu, J. H. (2012). Doping effect on the phase transition temperature in ferroelectric $SrBi_{2-x}Nd_xNb_2O_9$ layer-structured ceramics: A micro-Raman scattering study. *J. Raman Spectrosc.*, 43, 04, 583-587

[47] Liang, K.; Qi, Y. J. & Lu, C. J. (2009). Temperature-dependent Raman scattering in ferroelectric $Bi_{4-x}Nd_xTi_3O_{12}$ (x=0, 0.5, 0.85) single crystals. *J. Raman Spectrosc.*, 40, 12, 2088-2091

[48] Dobal, P. S. & Katiyar, R. S. (2002). Studies on ferroelectric perovskites and Bi-layered compounds using micro-Raman spectroscopy. *J. Raman Spectrosc.*, 33, 06, 405-423

[49] Kitaev, Y. E.; Aroyo, M. I. & Mato, J. M. P. (2007). Site symmetry approach to phase transitions in perovskite-related ferroelectric compounds. *Phys. Rev. B*, 75, 06, 064110(1-11)

[50] Jellison, G. E. Jr. & Modine, F. A. (1996). Parameterization of the optical functions of amorphous materials in the interband region. *Appl. Phys. Lett.*, 69, 03, 371-373. Erratum: "Parameterization of the optical functions of amorphous materials in the interband region". 69, 03, 2137

[51] Basu, S. R.; Martin, L. W.; Chu, Y. H.; Gajek, M.; Ramesh, R.; Rai, R. C.; Xu, X. & Musfeldt, J. L. (2008). Photoconductivity in $BiFeO_3$ thin films. *Appl. Phys. Lett.*, 92, 09, 091905(1-3)

[52] Chen, P.; Podraza, N. J.; Xu, X. S.; Melville, A.; Vlahos, E.; Gopalan, V.; Ramesh, R.; Schlom, D. G. & Musfeldt, J. L. (2010). Optical properties of quasi-tetragonal $BiFeO_3$ thin films. *Appl. Phys. Lett.*, 96, 13, 131907(1-3)

[53] Xu, X. S.; Brinzari, T. V.; Lee, S.; Chu, Y. H.; Martin, L. W.; Kumar, A.; McGill, S.; Rai, R. C.; Ramesh, R.; Gopalan, V.; Cheong, S. W. & Musfeldt, J. L. (2009). Optical properties and magnetochromism in multiferroic BiFeO$_3$. *Phys. Rev. B*, 79, 13, 134425(1-4)

[54] Ramirez, M. O.; Kumar, A.; Denev, S. A.; Podraza, N. J.; Xu, X. S.; Rai, R. C.; Chu, Y. H.; Seidel, J.; Martin, L. W.; Yang, S. -Y.; Saiz, E.; Ihlefeld, J. F.; Lee, S.; Klug, J.; Cheong, S. W.; Bedzyk, M. J.; Auciello, O.; Schlom, D. G.; Ramesh, R.; Orenstein, J.; Musfeldt, J. L. & Gopalan, V. (2009). Magnon sidebands and spin-charge coupling in bismuth ferrite probed by nonlinear optical spectroscopy. *Phys. Rev. B*, 79, 22, 224106(1-9)

[55] Ramirez, M. O.; Kumar, A.; Denev, S. A.; Chu, Y. H.; Seidel, J.; Martin, L. W.; Yang, S. -Y.; Rai, R. C.; Xue, X. S.; Ihlefeld, J. F.; Podraza, N. J.; Saiz, E.; Lee, S.; Klug, J.; Cheong, S. W.; Bedzyk, M. J.; Auciello, O.; Schlom, D. G.; Orenstein, J.; Ramesh, R.; Musfeldt, J. L.; Litvinchuk, A. P. & Gopalan, V. (2009). Spin-charge-lattice coupling through resonant multimagnon excitations in multiferroic BiFeO$_3$. *Appl. Phys. Lett.*, 94, 16, 161905(1-3)

Chapter 6

Relaxor Behaviour in Ferroelectric Ceramics

A. Peláiz-Barranco, F. Calderón-Piñar,
O. García-Zaldívar and Y. González-Abreu

Additional information is available at the end of the chapter

1. Introduction

Ferroelectric materials are commonly characterized by high dielectric permittivity values [1]. Usually, for the well-known 'normal' ferroelectrics the temperature of the maximum real dielectric permittivity (T_m) corresponds to the ferroelectric-paraelectric (FE-PE) phase transition temperature (T_C) [2]. On the other hand, there are some kinds of ferroelectrics, so-called relaxor ferroelectrics, which have received special attention in the last years because of the observed intriguing and extraordinary dielectric properties [3-25], which remain not clearly understood nowadays. For instance, some remarkable characteristics of the dielectric response of relaxor materials can be summarized as follows: i- they are characterized by wide peaks in the temperature dependence of the dielectric permittivity, ii- the temperature of the corresponding maximum for the real (ε') and imaginary (ε'') component of the dielectric permittivity (T_m and $T_{\varepsilon''max}$, respectively) appears at different values, showing a frequency dependent behaviour, and iii- the Curie-Weiss law is not fulfilled for temperatures around T_m. So that, the temperature of the maximum real dielectric permittivity, which depends on the measurement frequency, cannot be associated with a FE-PE phase transition.

Lead zirconatetitanate (PZT) system is a typical ferroelectric perovskite showing 'normal' FE-PE phase transition [1]. Nevertheless, the partial substitution by different elements, such as lanthanum, contributes to enhance such relaxor characteristics [11-15]. In fact, for some lanthanum concentrations, the distortion of the crystalline lattice in the PZT system due to ions displacement could promote the formation of the so-called polar nanoregions (PNRs). Another interesting relaxor ferroelectric perovskite is known as PZN-PT-BT [16-19]. The $Pb(Zn_{1/3}Nb_{2/3})O_3$ ferroelectric material (PZN) belongs to the relaxor ferroelectrics family, receiving special attention for its technical importance [20]. However, its preparation usually needs the addition of $BaTiO_3$ (BT) and $PbTiO_3$ (PT) to obtain pure phases [16,21].

On the other hand, a considerable number of compositions from the Aurivillius family exhibit a relaxor ferroelectric behaviour [22-26]. The Aurivillius compounds are layered bismuth $[Bi_2O_2]^{2+}[A_{n-1}B_nO_{3n+1}]^{2-}$, where the sites A and B can be occupied for different elements. These are formed by the regular stacking of Bi_2O_2 slabs and perovskite-like blocks $A_{n-1}B_nO_{3n+1}$. These materials have received great attention due to their large remanent polarization, lead-free nature, relatively low processing temperatures, high Curie temperatures and excellent piezo-electric properties, which made them good candidates for high-temperature piezoelectric applications and memory storage [27]. The origin of the relaxor behaviour for these materials have been associated to a positional disorder of cations on A or B sites of the perovskite blocks that delay the evolution of long-rage polar ordering [28].

Several models have been proposed to explain the dielectric behaviour of relaxor ferroelec-trics. The basic ideas have been related to the dynamics and formation of the polar nanore-gions (PNRs). In this way, Smolenskii proposed the existence of compositional fluctuations on the nanometer scale taking into account a statistical distribution for the phase transition temperature [29]. On the other hand, Cross extended the Smolenskii's theory to a superpara-electric model associating the relaxor behaviour to a thermally activated ensemble of super-paraelectric clusters [30]. Viehland et al. have showed that cooperative interactions among these superparaelectric clusters could produce a glass-like freezing behaviour, commonly exhibited in spin-glass systems [31]. Later, Qian and Bursill [32] analyzed the possible influ-ence of random electric fields on the formation and dynamics of the polar clusters, which can be originated from nano-scaled chemical defects. They have also proposed that the re-laxor behaviour can be associated to a dipolar moment in an anisotropic double-well poten-tial, taking into account only two characteristic relaxation times. According to this model, the dispersive behaviour is produced by changes in the clusters size and the correlation length (defined as the distance, above which such PNRs become non-interactive regions) as a function of the temperature, which provides a distribution function for the activation ener-gy. However, despite their very attractive physical properties, the identification of the na-ture of the dielectric response in relaxors systems still remains open and requires additional theoretical and experimental information, which can be very interesting to contribute to the explanation of the origin of the observed anomalies.

The present chapter shows the studies carried out on several relaxor ferroelectric ceramic materials, which have been developed by the present authors. Lanthanum modified lead zirconatetitanate ceramics will be evaluated considering different La^{3+} concentrations and Zr/Ti ratios. It will be discussed the dynamical behaviour of the PNRs taking into account a relaxation model, which considers a distribution function for the relaxation times. The influ-ence of A or B vacancies on the relaxor behaviour will be also analyzed considering the de-coupling effects of these defects in the Pb-O-Ti/Zr bounding. For the PZN-PT-BT system, the relaxor behaviour will be explained considering the presence of local compositional fluctua-tion on a macroscopic scale. Finally, the relaxor behaviour of $Sr_{0.50}Ba_{0.50}Bi_2Nb_2O_9$ ferroelectric ceramics, which belong to the Aurivillius family, will be discussed considering a positional disorder of cations on A or B sites of the perovskite blocks.

2. Lanthanum modified Lead Zirconate Titanate (PLZT)

2.1. Relaxor behaviour and coexistence of AFE and FE phases

Several researches concerning PZT properties have shown that very good properties can be obtained by suitable selection of the Zr/Ti ratio and the substitution of a small amount of isovalent or heterovalent elements for the Pb or (Zr,Ti) sublattices [1]. Especial attention receives the Zr-rich PZT-type ceramics, which exhibit interesting dielectric and pyroelectric characteristics, suitable for pyroelectric detectors, energy converters, imaging systems, etc [1,33-34]. They show orthorhombic antiferroelectric (AFE-O), low temperature rhombohedral ferroelectric (FR-LT), high temperature rhombohedral ferroelectric (FR-HT) and cubic paraelectric (PC) phase sequence [1].

One of the most common additives to the PZT ceramics is the lanthanum cation, which substitute the Pb-ions in the A-site of the perovskite structure [1]. The lanthanum-modified lead zirconatetitanate $Pb_{1-x}La_x(Zr_{1-y}Ti_y)_{1-x/4}O_3$ (PLZT) ferroelectric ceramics show excellent properties to be considered for practical applications, especially PLZT x/65/35, x/70/30 and x/80/20 compositions due to the relaxor ferroelectric behaviour [35-36]. For Zr-rich PZT ceramics, which are close to the antiferroelectric-ferroelectric (AFE-FE) phase boundary, the FE phase is only marginally stable over the AFE [33]. In this way, as the FE state is disrupted by lanthanum modification, the AFE state is stabilized [33].

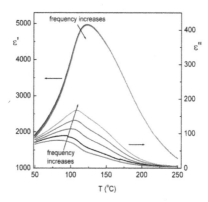

Figure 1. Temperature dependence of the real (ε') and imaginary (ε'') parts of the dielectric permittivity at several frequencies for PLZT 6/80/20 ferroelectric ceramic system.

Figure 1 shows the temperature dependence of the real (ε') and imaginary (ε'') parts of the dielectric permittivity, at several frequencies, for PLZT 6/80/20 ferroelectric ceramic system. Typical characteristics of relaxor ferroelectric behaviour are observed. The frequency dependence of ε' is very weak but the frequency dependence of ε'' clearly reflects the relaxor nature of the system. The asymmetrical shape of the curves could suggest the existence of more than one contribution. X-ray diffraction analysis for the PLZT 6/80/20 ceramic samples have previously shown a mixture of FE and AFE phases at room temperature [13]. Polarized light microscopy studies were performed on the ceramic system in a wide temperature range. Aggregates of individual regions were observed, which were associated to FE and AFE phases. As the temperature increased, a gradual change was observed, suggesting a transition to a single phase [13].

Figure 2 shows the hysteresis loops obtained for the PLZT 6/80/20 ceramics for two temperatures, as example of the obtained behaviour in a wide temperature range. Around 70°C a double-loop-like behaviour was observed suggesting an induced FE-AFE transformation.The FE state is disrupted by the lanthanum modification and the AFE state is stabilized, which may be a consequence of the short-range nature of the interaction between the AFE sublattices. The double-loop remains for higher temperature and around 120°C the loop disappears suggesting a transition to a PE phase. Then, it is possible to consider the coexistence of two phase transitions: a FE-AFE phase transition observed around 70°C and an AFE-PE phase transition around 120°C. Note, that the maximum of ε' is observed around 120°C. Negative values for T_C were found, which has confirmed that around 120°C an AFE-PE transition take places [13]. Thus, the relaxor ferroelectric behaviour for this system could be associated to the coexistence of FE and AFE phases in the studied material.

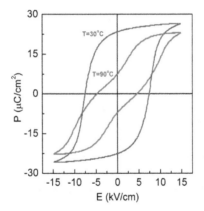

Figure 2. Hysteresis loops for PLZT 6/80/20 ferroelectric ceramic system.

2.2. A relaxation model

The nature of the relaxor behaviour is determined by the existence of PNRs, which possess different relaxation times [32,37]. The relaxation time (τ) represents the time response of such PNRs or polarization mechanisms to change with the applied electric field. However, this process does not occur instantaneously. Indeed, there exists certain inertia, which is the cause of the pronounced dielectric relaxation in relaxor ferroelectrics. The polar nanoregions appear below a certain temperature, the so-called the Burns' temperature (T_B) [38], which i typically hundreds degrees above the temperature of the maximum real dielectric permittivity (T_m). On cooling, the number, size (some of them) and the interaction of the PNRs increase [39]. The increase of such interactions promotes the freezing of some regions around certain temperatures below T_m, known as the freezing temperature (T_F). Two fundamental polarization mechanisms have been reported, which have been associated with the dynamics of the PNRs; (i) dipole reorientation [37,40] and (ii) domain wall vibrations [32]. Both mechanisms have a characteristic time response, which depends on the temperature and size of the PNRs. If the contributions of an ensemble of these regions are considered, the macroscopic polarization function, at fixed temperature, can be expressed by [41]:

$$P(t) = P_{01}e^{-\frac{t}{\tau_1}} + P_{02}e^{-\frac{t}{\tau_2}} + P_{03}e^{-\frac{t}{\tau_3}} + ... = \sum_i P_{0i}e^{-\frac{t}{\tau_i}} \tag{1}$$

where τ_i is the relaxation time of the ith PNR and P_{0i} takes the form of equation (2), in analogy with the Debye single relaxation time model, which takes into account a distribution function of relaxation times [41]:

$$P_{0i} = \frac{\varepsilon_s - \varepsilon_\infty}{\tau_i} g(\tau_i) \tag{2}$$

$$g(\tau_i) = \frac{2\sigma}{\pi} \frac{1}{4(\ln\frac{\tau_i}{\tau_0})^2 + \sigma^2} \tag{3}$$

In these relations $g(\tau_i)$ is a distribution function for the logarithms of the relaxation times, which has been assumed to be a Lorentz rather than a Gaussian distribution function; τ_0 and σ are the mean relaxation time and the Standard Deviation, respectively. According to the Debye model, the frequency dependence of the complex dielectric permittivity can be expressed as:

$$\varepsilon * (\omega) = \varepsilon' - i\varepsilon'' = \varepsilon_\infty + \int P(t)e^{-i\omega t}dt \tag{4}$$

where ε_s and ε_∞ are the low (static) and high (optical) dielectric permittivity, respectively, ω the measurement frequency, and P(t) the decay polarization function. Substituting equations (1), (2) and (3) into equation (4), the real and imaginary component of the dielectric permittivity for the multi-relaxation times approximation can be obtained and expressed as in equations (5) and (6). It is important to point out that equations (5) and (6) have been derived from the discrete expression (1), taking into account the values of the relaxation times (τ_i) close to each other [41].

$$\varepsilon'(\omega,T) = \varepsilon_\infty + (\varepsilon_s - \varepsilon_\infty) \frac{2\sigma}{\pi} \int_{-\infty}^{+\infty} \frac{1}{(4z^2 + \sigma^2)(1 + \omega^2 \tau_0^2 \exp(2z))} dz \tag{5}$$

$$\varepsilon''(\omega,T) = (\varepsilon_s - \varepsilon_\infty) \frac{2\sigma}{\pi} \int_{-\infty}^{+\infty} \frac{\omega \tau_0 \exp(z)}{(4z^2 + \sigma^2)(1 + \omega^2 \tau_0^2 \exp(2z)} dz \tag{6}$$

It has been considered in equations (5) and (6) that $z = \ln\tau/\tau_0$. By using the experimental results of $\varepsilon'(\omega,T)$ and $\varepsilon''(\omega,T)$ for two frequencies, the temperature dependence of dielectric parameters, such as, τ_0, ε_s and σ can be obtained as a solution of the equations system. After that, by using the theoretical results of $\tau_0(T)$, $\varepsilon_s(T)$ and $\sigma(T)$, the theoretical dependences of $\varepsilon'(\omega,T)$ and $\varepsilon''(\omega,T)$ can be obtained for the studied frequency and temperature ranges. The parameter ε_∞ has been considered negligible because of the high values of the dielectric permittivity in ferroelectric systems [1].

Figures 3 and 4 show the temperature dependence of the real (ε') and imaginary (ε'') components of the dielectric permittivity (symbols), at several frequencies, for the studied PLZT 10/80/20 composition. A relaxor characteristic behaviour can be observed. The maximum real dielectric permittivity decreases, while its corresponding temperature (T_m) increases, with the increase of the measurement frequency. There is not any peak for ε'' due to the low temperature range where the maximum of the real part of the dielectric permittivity appears, i.e. the maximum for ε'' should appear below the room temperature.

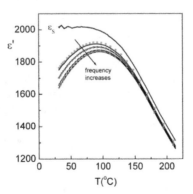

Figure 3. Temperature dependence of the real part (ε') of the dielectric permittivity at several frequencies for PLZT 10/80/20 ferroelectric ceramic system. The experimental values are represented by solid points and the theoretical results by solid lines. It has been included the theoretical temperature dependence of the static dielectric permittivity (ε_s)

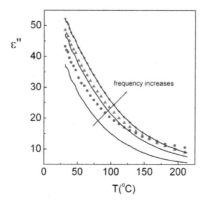

Figure 4. Temperature dependence of the imaginary part (ε'') of the dielectric permittivity at several frequencies for PLZT 10/80/20 ferroelectric ceramic system. The experimental values are represented by solid points and the theoretical results by solid lines.

The number of PNRs able to follow the applied external electric field switching decreases with the increase of the frequency, so that only such inertial-less regions contribute to the dielectric permittivity (those regions whose activation energy is close to thermal energy $k_B T$, being k_B the Boltzmann constant). Therefore, due to the cooperative nature, the real component of the dielectric permittivity should decrease with the increase of the measurement frequency.

The maximum for ε' in relaxor systems is not related to a crystallographic transition. Indeed, such a maximum corresponds to rapid changes of the fraction of the frozen Polar Regions [39]. Hence, the shift up to higher temperatures of the maximum for ε' with the increase of the frequency is a direct consequence of the delay in dielectric response of the frozen regions.

By using equations (5) and (6) and the experimental data of Figures 3 and 4, the temperature dependences of ε_s, σ and τ_0 were obtained. The results were obtained by numerical methods because there was no analytical solution for the equation system. The theoretical curves for ε' and ε'' were obtained for all the studied frequency range, and shown in the same Figures 3 and 4 as solid lines, for the studied system. The temperature dependence of the static dielectric permittivity (ε_s) is also shown in Figure 3. A good agreement between experimental and theoretical results can be observed. It is important to point out that the deviation between the experimental and theoretical results, observed at low frequencies for the temperature dependence of ε'', can be associated with the contribution of the electric conductivity to the dielectric response, which has not been considered in the proposed model.

Figure 5 shows the temperature dependence of $\ln\tau_0$ and σ for the studied composition. As can be seen, $\ln\tau_0$ increases with the increase of the temperature, passes through a maximum and then decreases for higher temperatures. Similar results have been previously reported by Lin et al [41]. According to the model, the logarithmic mean relaxation time ($\ln\tau_0$) should have has approximately the same tendency as that of $\sigma(T)$. Both parameters increases when the temperature decreases which can be understood as a consequence of the increased inertia of the dipolar clusters as they become more correlated with each other upon cooling. This observation also could reflect a dynamic change in the well potential for the shifting ions. However, the reduction in $\ln\tau_0(T)$ at lower temperatures possibly could reflect a freezing phenomenon of some clusters that are saturated in correlation, which leads to the frustration of cooperative interactions and, hence, leaves only relatively unstrained or smaller clusters available to couple with the electric field. Also, one could speculate that such clusters are the regions within the domain boundaries between the frozen regions.

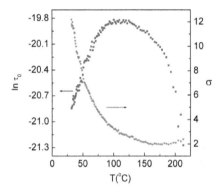

Figure 5. Temperature dependence of the main relaxation time (logarithmic representation, $\ln\tau_0$) and its standard deviation (σ), for PLZT 10/80/20 ferroelectric ceramic system.

In this way, the size and the interaction between PNRs increase on cooling from the high temperature region and its contribution to the dielectric permittivity becomes negligible below the freezing temperature (T_F) [39]. In this temperature range ($T < T_F$), the applied electric field is not strong enough to break such interactions and only the smallest regions can switch with the electric field, i.e. that is to say, there is a frustration of the cooperative effect. For temperatures above T_F, there are fluctuations between equivalent polarization states [30], leading to a decrease of the macroscopic polarization upon heating [42]. The applied electric field cannot reorient the ferroelectric dipoles because of the thermal fluctuation. The potential barrier between equivalent polarization states decreases with the increase of the temperature and the thermal energy promotes the spontaneous switching of the dipoles, even when an electric field is applied. The contribution to the dielectric permittivity is due to those PNRs, which can switch with the applied electric field. Thus, for temperatures slightly above T_F all the PNRs could contribute to the dielectric permittivity and a maximum value of the mean relaxation time could be expected. Therefore, the temperature corresponding to the maximum of $\ln\tau_0$ can be related to the freezing temperature. The standard deviation (σ), which can be interpreted as the correlation between the PNRs [32,41], decreases upon heating and it is relatively small at high temperatures. When the temperature increases the thermal energy is high enough to break down the interaction between the PNRs.

So that, a decrement of the correlation between the PNRs is promoted by an increase of the temperature and the Standard deviation decreases up to a relatively constant value around the Burns' temperature (T_B) [38,43], where the PNRs completely disappear.

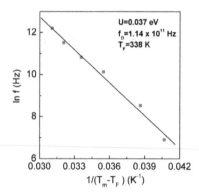

Figure 6. Temperature dependence of the frequency (lnf versus $1/(T_m–T_F)$ curve) for PLZT 10/80/20 ferroelectric ceramic system. Solid lines represent the fitting by using the Vogel–Fulcher relation.

The freezing temperature (T_F) was determined by using the Vogel–Fulcher relation [44], as expressed by equation (7):

$$f = f_D e^{-\frac{U}{k_B(T_m-T_F)}} \tag{7}$$

where f_D is the Debye frequency and U is the activation energy. Figure 6 show the temperature dependence of the frequency (lnf versus $1/(T_m–T_F)$ curve) and the fitting by using equation (7). In order to maintain the standard representation, the temperature in Figure 6 was expressed in Kelvin. The fitting parameters U, f_D and T_F have been included in the figure. The results are in agreement with previously reported results in the literature [42]. The disagreement between the T_F value, which has been determined by using the Vogel-Fulcherrelation, and the temperature corresponding to the maximum value of $ln\tau_0$ could be associated to the diffusivity of the phase transition in the studied system.

2.3. Relaxor behaviour and vacancies.

Relaxor ferroelectrics are formed by temperature dependent Polar Nanometric Regions (PNRs), which possess different volumes [45] and orientations for the polarization [30,37,46-47]. The PNRs appear at elevated temperatures (in the paraelectric state), at the so-called Burns' temperature (T_B) [38,48-49], due to short-range interactions, which establish a local polarization thermally fluctuating between equivalents polar states.

The disorder in the arrangement of different ions on the crystallographic equivalent sites is the common feature of relaxors [50]. This is associated, in general, with a complex perovskite of the type $A(B'B'')O_3$, where the B sites of the structure are occupied by different cations. Nevertheless, in lead zirconate titanate based ceramics (PZT) the disorder due to the arrangement of the isovalent cations Zr^{4+} and Ti^{4+}, in the B site of the structure, does not lead to relaxor behaviour for any Zr/Ti ratio. The lanthanum modification on this system, above certain La^{3+} concentrations, provides a relaxor ferroelectric state [11-12]. Electrical neutrality can be achieved, stoichiometrically, considering the vacancies formation either in the A-site (Pb^{2+}) or in the B-site (Zr^{4+}, Ti^{4+}), or on both. The distortion of the crystalline lattice in the PZT system could promote the formation of the PNRs. In this section, the results concerning the influence of the A or B vacancies defects in the relaxor ferroelectric behaviour of lanthanum modified PZT ceramic samples will be presented. The studied compositions will be $Pb_{0.85}La_{0.10}(Zr_{0.60}Ti_{0.40})O_3$ and $Pb_{0.90}La_{0.10}(Zr_{0.60}Ti_{0.40})_{0.975}O_3$, labelled as PLZT10-VA and PLZT10-VB, respectively.

The temperature dependence of the real dielectric permittivity (ε') and the dielectric losses (tan δ) are shown in the Figures 7 and 8, respectively, for several frequencies. Typical characteristics of relaxor ferroelectrics are exhibited for both samples, i.e. wide peaks for ε', the maximum real dielectric permittivity shifts to higher temperatures with the increase of the frequency, and the maximum dielectric losses temperature appears at temperatures below T_m. On the other hand, it can be observed that the dielectric response is highly affected by the type of compensation. The maximum values of the real dielectric permittivity in the PLZT10-VA sample are lower than those obtained for the PLZT-10VB sample, and appear at lower temperatures. The PLZT10-VA sample also exhibits a larger temperature shift of T_m (ΔT=15°C), from 1 kHz to 1 MHz, than that of the PLZT10-VB sample (ΔT=9°C).

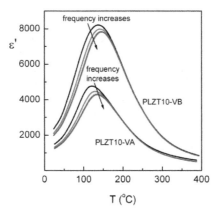

Figure 7. Temperature dependence of the real (ε') component of the dielectric permittivity at several frequencies, for PLZT10-VA and PLZT10-VB ceramic samples.

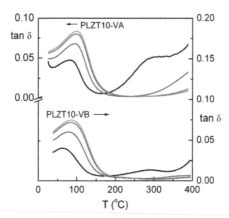

Figure 8. Temperature dependence of the dielectric losses (tan δ) at several frequencies for PLZT10-VA and PLZT10-VB ceramic systems.

It has been previously commented that, in relaxors, the maximum real dielectric permittivity is not a consequence of crystallographic changes but it is associated with a rapid change in the volume fraction of frozen polar nanoregions [39]. During cooling, the PNRs grow in size and number [39,43] as well as the interactions among them increases. Thus, a decrement of its mobility is expected; the thermal energy ($k_B T$) is not enough to switch the polar state of the PNRs (freezing phenomenon). While the regions grow, with the decreasing of the temperature, the thermal energy $k_B T$ decreases. The freezing process begins at high temperatures and finish at low temperatures, below T_m, in the so called freezing temperature (T_F), where all the PNRs are frozen.

The dielectric and ferroelectric response in relaxor ferroelectrics strongly depend of the temperature dependence of the freezing process (dynamic of the PNRs) as well as on the interactions between external electric field and the dipole moment of the PNRs. When a weak alternating electric field is applied, only the unfrozen regions and the frozen ones with activation energies close to $k_B T$ could switch with the applied alternating electric field. Therefore, only these regions contribute to the dielectric permittivity. However, not all these regions can switch at any excitation frequency; the number of PNRs able to follow the applied external electric field switching decreases with the increase of the frequency [14]. The response also depends on the relaxation time of each region [14, 41]. Thus, the dielectric response in relaxors depends of the number of regions that can contribute at each temperature and the distribution of the corresponding relaxation times.

The Pb^{2+} ions (A-sites of the perovskite structure) establish the long-range order in PZT based ceramics due to strong coupling of Pb-O-Ti/Zr bonding [51]. The coupling defines the height of the energy barrier between polar states via Ti/Zr-hopping [51]. The Pb^{2+} substitution by La^{3+} ions, above certain level of La^{3+} ions concentration, weakens this bonding resulting in small and broadly distributed energy barriers [51], which disrupts the long-range order and promotes the PNRs formations. An additional contribution, which plays a fundamental role in the relaxor behaviour, can be originated from the A- or B-sites vacancies formations. There are two important differences between the A- or B-site compensation, which can explain the different dielectric behaviour observed in the studied samples: i) for the same lanthanum concentration the number of lead vacancies is twice the Zr/Ti vacancies, if it is considered A- or B-site compensation, respectively, i.e. $2La^{3+} \rightarrow 1V_{Pb}$ and $4La^{3+} \rightarrow 1V_{Zr/Ti}$; ii) the A-site vacancies generate large inhomogeneous electric fields, which reduce the barriers between energy minima for different polarization directions [20], and promote the decoupling of the Pb-O-Ti/Zr bonding, making the corresponding Ti/Zr-hopping easier due to the lack of lead ions. For the case of the B-site vacancies, all the Ti/Zr ions are well coupled to the lead or the lanthanum ions.

From this point of view, it could be pointed out that for the PLZT10-VA sample the defects concentration, which could promote the PNRs formation, is higher than that for the PLZT10-VB sample. On the other hand, the thermal energy, which could change the nanoregions polarization state, is smaller for the PLZT10-VA sample as a consequence of the smaller hopping barrier. Considering these analyses, it could be explained the lower T_m values of the real part of the dielectric permittivity for the PLZT10-VA. The observed results are in agreement with previous reports for lead based systems with and without Pb^{2+} vacancies [52].

The larger temperature shift of T_m, which was analyzed for the PLZT10-VA, could be related to a higher temperature dependence of the freezing process around T_m, i.e. higher temperature dependence in the change of the volume fraction of frozen polar nanoregions around T_m. The number of PNRs, which can follow the electric field switching, decreases with the increase of the frequency. Thus, the maximum value of ε' decreases when the frequency increases, and also shifts to higher temperatures.

From the Figure 8, it is observed that the maximum values of the dielectric losses, for both samples, are observed in the same temperature range. No remarkable differences exist as can be expected from the difference observed in the T_m values between both systems. This could suggest similar dipolar dynamic in the temperature range for both ceramics.

Figure 9 shows the temperature dependence of the remanent polarization (P_R) for both samples. The PLZT10-VB sample exhibits higher values, which is in agreement with the previous discussion concerning the hopping barrier. The decoupling introduced by A-site vacancies affects the total dipolar moment of the system and, consequently, the macroscopic polarization is affected. Thus, the magnitude of the polarization decreases with the increase of the A-site vacancies concentration.

On the other hand, the remanent polarization for both samples shows an anomaly in the same temperature region (70–85°C), which could be associated with the 'offset' of the freezing process [14, 53]. This result suggests that the freezing temperature (T_F) could be the same for both samples or, at less the dynamic of PNRs is the same in both systems for that temperature range. Furthermore, the anomaly is observed in the same region where the maximum values of the dielectric losses have been observed, which could suggest a relation between the freezing phenomenon and the temperature evolution of dielectric losses.

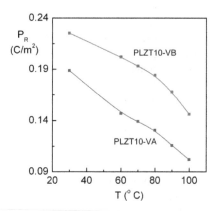

Figure 9. Temperature dependence of remnant polarization for the PLZT10-VA and PLZT10-VB samples. The lines between dots are only a guide to the eyes.

3. Other typical relaxor ferroelectric perovskites

Relaxor-based ferroelectric materials $PbMg_{1/3}Nb_{2/3}O_3-PbTiO_3$ (PMN–PT) and $PbZn_{1/3}Nb_{2/3}O_3-PbTiO_3$ (PZN–PT) have been extensively studied due to their high electromechanical coupling factor and piezoelectric coefficient [3,8,16-21,43-44,54-57].

It is known that the large dielectric permittivity of perovskite relaxor ferroelectrics, such as $(PbMg_{1/3}Nb_{2/3}O_3)_{1-x}(PbTiO_3)_x$, comes from ferroelectric like polar nanoregions [30]. Upon lowering the temperature these nanoregions grow slightly [43] but do not form long-range ferroelectric order. The temperature dependence of the dielectric permittivity shows a frequency-dependent peak as the polar nanoregion orientations undergo some sort of glassy kinetic freezing. Analyses of the phase and microstructure evolution for these materials have showed that the increase in the sintering temperature caused the weakening of the dielectric relaxor behaviour [54]. It has been related to compositional fluctuation and the release of the internal stresses, leading to the decrease in the short-range B-site order.

The PNRs freezing process in $PbMg_{1/3}Nb_{2/3}O_3$ has been analyzed considering both polarization and strain operating on subtly different timescales and length scales [57]. The strain fields are considered as relatively weak but longer ranging, while the dipole interactions tend to be short range and relatively strong. It has been discussed that the elastic shear strain fields is partly suppressed by cation disordering and screened by the presence of antiphase boundaries with their own distinctive strain and elastic properties, which provide a mechanism for suppressing longer ranging correlations of strain fields.

On the other hand, the barium modified PZN-PT system has shown excellent piezoelectric properties [58]. The preparation of PZN system usually needs the addition of $BaTiO_3$ (BT) to obtain pure phases. Therefore, the ternary system PZN-PT-BT has received wider attention, showing high values of the dielectric permittivity and the pyroelectric coefficient [21,59]. This material shows lower diffuseness of the phase transition and weaker frequency dispersion of the dielectric response than that of the PZN-BT system [16]. It has been shown that it could be interesting try to decrease the transition temperature and to grow the value of the figure of merit in order to obtain better materials for practical applications [17-18,21,60-61].

The present authors have developed several researches concerning $(Pb_{0.8}Ba_{0.2})$ $[(Zn_{1/3}Nb_{2/3})_{0.7}Ti_{0.3}]O_3$ (PZN-PT-BT) system [19,62-63]. A Positive Temperature Coefficient of Resistivity (PTCR) effect has been studied. This effect has found extensive applications as thermal fuses, thermistors, safety circuits and other overload protection devices [64]. It has been discussed that the PTCR behaviour primarily arise from the Schottky barrier formed at grain boundary regions, which act as effective electron traps of the available electrons from the oxygen vacancies in the ceramic, increasing the Schottky barrier height of the material. Excellent properties to be used in dielectric and pyroelectric applications have been also reported [63].

Figure 10 shows the temperature dependence of the real part (ε') of the dielectric permittivity for the $(Pb_{0.8}Ba_{0.2})[(Zn_{1/3}Nb_{2/3})_{0.7}Ti_{0.3}]O_3$ ferroelectric ceramic system at several frequencies. Typical characteristics of relaxor ferroelectrics are observed. There is a strong dispersion of the maximum of ε' and its corresponding temperature shift towards higher temperatures when the frequency increases. On the other hand, the system did not follow a Curie–Weiss–like behaviour above T_m. For relaxor ferroelectrics, some of the dipoles are frozen during the time scale of measurement. The fraction of frozen dipoles in itself is a function of the temperature, so Curie-Weiss's law is no longer valid.

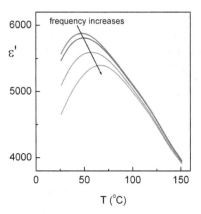

Figure 10. Temperature dependence of the real (ε') part of the dielectric permittivity at several frequencies for $(Pb_{0.8}Ba_{0.2})[(Zn_{1/3}Nb_{2/3})_{0.7}Ti_{0.3}]O_3$ ferroelectric ceramic system.

For $Pb(B'_{1/3}B''_{2/3})O_3$ type pervoskites, it has been reported a particular microstructure where the B-site 1:1 short-range-ordered nanodomains (rich in B' ions) are embedded in a matrix rich in B'' ions, which promotes a compositional inhomogeneity [65]. The nonstoichiometric ordering induces strong charge effects; i.e., the ordered domains have a negative charge with respect to the disordered matrix. Thus, the charge imbalance inhibits the growth of domains and nanodomains are obtained in a disordered matrix with polar micro-regions.

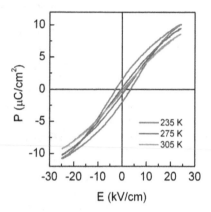

Figure 11. Hysteresis loops for $(Pb_{0.8}Ba_{0.2})[(Zn_{1/3}Nb_{2/3})_{0.7}Ti_{0.3}]O_3$ ferroelectric ceramic system.

Figure 11 shows the hysteresis loops at several temperatures for the studied system. Slim P-E loops and small remanent polarization values are observed. The slim-loop nature suggests that most of the aligned dipole moments switch back to a randomly oriented state upon removal of the field. It could be interpreted in terms of correlated polar nanodomains embedded in a paraelectric matrix [44]. For relaxor ferroelectric materials, there is a micro- to macro-domain transition [66]. In the absence of any external field, the domain structure of relaxor ferroelectrics contains randomly oriented micropolar regions. When an electric field is applied, the micro-domains are oriented along the field direction and the macro-domains appear. The micro- to macro-domain transition has been confirmed by 'in-situ' switching by means of an electron beam inducing local stresses to align the domains [66].

4. Lead free Aurivillius relaxor ferroelectrics

The temperature dependence of the real (ε') and imaginary (ε'') parts of the dielectric permittivity, at several frequencies, for the $Sr_{0.5}Ba_{0.5}Bi_2Nb_2O_9$ sample are showed in the Figure 12. The maximum of ε' decreases with the increase of the frequency; its corresponding temperature (T_m) shifts with the frequency, showing a high frequency dispersion. For the imaginary part (ε'') the maximum values are observed at lower temperatures than those the observed for ε' and the corresponding temperature again shows a significant frequency dispersion. These characteristics are typical of relaxor ferroelectric materials. On the other hand, an abrupt increase of ε'' is observed in the higher temperature zone, which is more clear in the low frequency range. This behaviour could be associated to the conductivity losses.

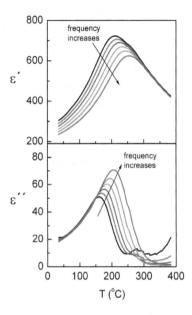

Figure 12. Temperature dependence of the real (ε') and imaginary (ε'') parts of the dielectric permittivity, at several frequencies, for the $Sr_{0.5}Ba_{0.5}Bi_2Nb_2O_9$.

The origin of the relaxor behaviour for this material can be explained from a positional disorder of cations on A or B sites of the perovskite blocks that delay the evolution of long-range polar ordering [28]. Previous studies in the $SrBi_2Nb_2O_9$ system have showed that the incorporation of barium to this system conduces to a relaxor behaviour in the ferroelectric-paraelectric phase transition [28,67] and the increases of barium content enhance the degree of the frequency dispersion of the dielectric parameters [67]. This behaviour can be explained considering that the barium ions can substitute the strontium ions in the A site of the perovskite block but enter in the $Bi_2O_2^{2+}$ layers, conducting to an inhomogeneous distribution of barium and local charge imbalance in the layered structure. The incorporation of barium ions to bismuth sites takes care to reduce the constrain existent between the perovskite blocks and the layered structure [68].

The relaxor behaviour is usually characterized by using the variation of T_m with frequency (ΔT), degree of the frequency dispersion, and the critical exponent (δ) obtained from the following law [67]:

$$\frac{1}{\varepsilon'} - \frac{1}{\varepsilon'_{max}} = \frac{\left(T - T_m\right)^{\delta}}{C} \qquad (8)$$

whereε'_{max} is the maximum value for the real part of the dielectric permittivity, and δ and C are constants. For an ideal relaxor ferroelectric $\delta = 2$, while for a normal ferroelectric $\delta = 1$ and the system follows the Curie-Weiss law.

For the studied material, $\Delta T = 50°C$ between 1 kHz and 1 MHz, which is higher than that of the obtained values for the PMN and PLZT 8/65/35 systems [67]. Thus, the studied material shows higher frequency dispersion. However, ΔT is lower than that of the $BaBi_2Nb_2O_9$ system (BBN), which is in agreement with previous reports where ΔT decreases with the decrease of the barium concentration [67].

The Figure 13 shows the dependency of the log $(1/\varepsilon'-1/\varepsilon'_{max})$ vs. log $(T-T_m)$ at 1 MHz for the study ceramic sample. The solid points represent the experimental data and the line the fitting, which was carried out by using the equation 8. The value obtained for δ parameter is 1.71, which is in agreement with other reports for the studied system [67].

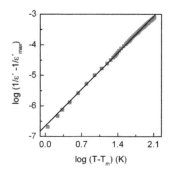

Figure 13. Dependence of the log $(1/\varepsilon'-1/\varepsilon''_{max})$ vs. log $(T-T_m)$ at 1 MHz for the $Sr_{0.5}Ba_{0.5}Bi_2Nb_2O_9$.

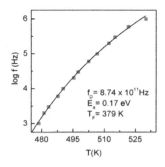

Figure 14. Dependence of the log f with T_m for the $Sr_{0.5}Ba_{0.5}Bi_2Nb_2O_9$ considering a frequency range from 1 kHz to 1 MHz. The fitting was carried out by using the Vogel-Fulcher law.

Figure 14 shows the dependence of log f with T_m for the studied material. The experimental data was fitted by using the Vogel-Fulcher law (equation 7). The fitting parameters have been included in the figure. The difference between T_m at 1 kHz and T_F is about 100 K, a higher value considering the studies carried out in PMN type relaxors [69]. On the other hand, the BBN system has showed a difference higher than 300 K, which is agreement with its strong shift of T_m [28, 69].

5. Conclusions

The relaxor behaviour has been discussed in ferroelectric perovskites-related structures. The influence of the coexistence of ferroelectric and antiferroelectric phases on the relaxor behaviour of lanthanum modified lead zirconate titanate ceramics has been analyzed. A relaxation model to evaluate the dynamical behaviour of the polar nanoregions in a relaxor ferroelectric PLZT system has been presented and the results have been discussed considering the correlation between the polar nanoregions and the freezing temperature. On the other hand, the influence of A or B vacancies on the relaxor behaviour for PLZT materials have been studied considering the decoupling effects of these defects in the Pb-O-Ti/Zr bounding and in terms of the dynamic of PNRs. Other typical relaxor ferroelectrics were analyzed too. For the PZN-PT-BT system, the relaxor behaviour was discussed considering the presence of local compositional fluctuation on a macroscopic scale. The $Sr_{0.50}Ba_{0.50}Bi_2Nb_2O_9$ ferroelectric ceramic, which is a typical relaxor from the Aurivillius family, was analyzed considering a positional disorder of cations on A or B sites of the perovskite blocks, which delay the evolution of long-rage polar ordering.

Acknowledgement

The authors wish to thank to TWAS for financial support (RG/PHYS/LA No. 99-050, 02-225 and 05-043) and to ICTP for financial support of Latin-American Network of Ferroelectric Materials (NET-43). Thanks to FAPESP, CNPq and FAPEMIG Brazilian agencies. Thanks to CONACyT, Mexico. Dr. A. Peláiz-Barranco wishes to thank the Royal Society. M.Sc. O. García-Zaldívar wishes to thanks to Red de Macrouniversidades / 2007 and to CLAF/ICTP fellowships NS 32/08.

Author details

A. Peláiz-Barranco, F. Calderón-Piñar, O. García-Zaldívar and Y. González-Abreu

*Address all correspondence to: pelaiz@fisica.uh.cu

Physics Faculty–Institute of Science and Technology of Materials, Havana University, San Lázaro y L, Vedado, La Habana, Cuba

References

[1] Xu Y. Ferroelectric Materials and Their Applications. Netherland: Elsevier Science Publishers; 1991.

[2] Strukov BA., Levanyuk AP. Ferroelectric Phenomena in Crystals. Berlin: Springer-Verlag; 1998.

[3] Tai CW., Baba-Kishi KZ. Relationship between dielectric properties and structural long-range order in $(x)Pb(In_{1/2}Nb_{1/2})O_3$:$(1 - x)Pb(Mg_{1/3}Nb_{2/3})O_3$ relaxor ceramics. Acta Materialia 2006;54; 5631-5640.

[4] Huang S., Sun L., Feng Ch., Chen L. Relaxor behaviour of layer structured $SrBi_{1.65}La_{0.35}Nb_2O_9$. J. Appl. Phys. 2006;99; 076104.

[5] Ko JH., Kim DH., Kojima S. Correlation between the dynamics of polar nanoregions and temperature evolution of central peaks in $Pb[(Zn_{1/3}Nb_{2/3})_{0.91}Ti_{0.09}]O_3$ ferroelectric relaxors. Appl. Phys. Lett. 2007;90; 112904.

[6] Kumar P., Singh S., Juneja JK., Prakash Ch., Raina KK. Dielectric behaviour of La substituted BPZT ceramics. Physica B 2009;404(16) 2126–2129.

[7] Craciun F. Strong variation of electrostrictive coupling near an intermediate temperature of relaxor ferroelectrics. Physical Review B 2010;81; 184111.

[8] Ko JH., Kim DH., Kojima S. Low-frequency central peaks in relaxor ferroelectric PZN-9%PT single crystals. Ferroelectrics 2007; 347(1) 25-29.

[9] Samara GA. The relaxational properties of compositionally disordered ABO_3 perovskites. J. Phys.: Condens. Matter 2003;15(9) R367.

[10] Shvartsman VV., Lupascu DC. Lead-free relaxor ferroelectrics. J. Am. Ceram. Soc. 2012;95(1) 1–26.

[11] Dai X., Xu Z., Li JF., Viehland D. Effects of lanthanum modification on rombohedral $Pb(Zr_{1-x}Ti_x)O_3$ ceramics: Part I. Transformation from normal to relaxor ferroelectric behaviours. J. Mat. Res. 1996;11(3) 618-625.

[12] Dai X., Xu Z., Li JF., Viehland D. Effects of lanthanum modification on rombohedral $Pb(Zr_{1-x}Ti_x)O_3$ ceramics: Part II. Relaxor behaviour versus enhanced antiferroelectric stability. J. Mat. Res. 1996;11(3) 626-638.

[13] Peláiz-Barranco A., Mendoza ME., Calderón-Piñar F., García-Zaldívar O., López-Noda R., de los Santos-Guerra J., Eiras JA., Features on phase transitions in lanthanum-modified lead zirconate titanate ferroelectric ceramics. Sol. Sta. Comm. 2007;144(10-11) 425-428.

[14] García-Zaldívar O., Peláiz-Barranco A., Calderón-Piñar F., Fundora-Cruz A., Guerra JDS., Hall DA., Mendoza ME.. Modelling the dielectric response of lanthanum modi-

fied lead zirconate titanate ferroelectric ceramics - an approach to the phase transitions in relaxor ferroelectrics. J. Phys.: Cond. Matt. 2008;20; 445230.

[15] Peláiz Barranco A., Calderón Piñar F., Pérez Martínez O. Normal-Diffuse Transition in $(Pb,La)Zr_{0.53}Ti_{0.47}O_3$ Ferroelectric Ceramics. Phys. Sta. Sol. (b) 2000;220(1) 591-595.

[16] Fan HQ., Kong LB., Zhang LY., Yao X. Structure-property relationships in lead zinc niobate based ferroelectric ceramics. J. Appl. Phys. 1998;83; 1625-1630.

[17] Zhu WZ., Yan M., Effect of Mn-doping on the morphotropic phase boundary of PZN-BT-PT system. J. Mater. Sci. Lett. 2001;20(16) 1527-1529.

[18] Zhu WZ., Yan M., Kholkin AL., Mantas PQ., Baptista JL., Effect of tungsten doping on the dielectric response of PZN–PT–BT ceramics with the morphotropic phase boundary composition. J. Eur. Ceram. Soc. 2002;22(3) 375-381.

[19] Peláiz-Barranco A., González-Carmenate I., Calderón-Piñar F., Relaxor behaviour in PZN-PT-BT ferroelectric ceramics. Sol. Sta. Comm. 2005;134; 519-522.

[20] Haertling HG. Ferroelectric Ceramics: History and Technology. J. Am. Ceram. Soc. 1999;82(4) 797-818.

[21] Halliyal A., Kumar U., Newnham RE., Cross LE. Stabilization of the perovskite phase and dielectric properties of ceramics in the $Pb(Zn_{1/3}Nb_{2/3})O_3$-$BaTiO_3$ system. Am. Ceram. Soc. Bull. 1987;66(4) 671-676.

[22] Jennet DB., Marchet P., El Maaoui M., Mercurio JP., From ferroelectric to relaxor behaviour in the Aurivillius-type $Bi_{4-x}Ba_xTi_{3-x}Nb_xO_{12}$ (0<x<1.4) solid solutions. Mat. Lett. 2005;59; 376-382.

[23] Karthik C., Ravishankar N., Maglione M., Vondermuhll R., Etourneau J., Varma KBR. Relaxor behaviour of $K_{0.5}La_{0.5}Bi_2Ta_2O_9$ ceramics. Sol. Sta. Comm. 2006;139; 268–272.

[24] González-Abreu Y., Peláiz-Barranco A., Araújo EB., Franco Júnior A. Dielectric relaxation and relaxor behaviour in bi-layered perovskite. Appl. Phys. Lett. 2009;94(26) 262903.

[25] Zhang H., Yan H., Reece MJ. High temperature lead-free relaxor ferroelectric: intergrowth Aurivillius phase $BaBi_2Nb_2O_9$-$Bi_4Ti_3O_{12}$ ceramics. J. Appl. Phys. 2010;107; 104111.

[26] Chakrabarti A., Bera J. Effect of La-substitution on the structure and dielectric properties of $BaBi_4Ti_4O_{15}$ ceramics. J. Alloys Comp. 2010;505; 668–674.

[27] Rabe KM., Ahn CA. Triscone JM., editors. Physics of Ferroelectrics. Berlin Heidelberg, New York: Springer; 2007.

[28] Miranda C., Costa MEV., Avdeev M., Kholkin AL., Baptista JL., Relaxor properties of Ba-based layered perovskites. J. Eur. Ceram. Soc. 2001;21; 1303-1306.

[29] Smolienskii GA. Physical phenomena in ferroelectric with diffused phase transition. J. Phys. Soc. Japan 1970;28,Supplement; 26-30.

[30] Cross LE. Relaxor ferroelectrics. Ferroelectrics 1987;76; 241-267.

[31] Viehland D., Jang SJ., Cross LE., Wutting M. Freezing of the polarization fluctuations in lead magnesium niobate relaxors. J. Appl. Phys. 1990;68; 2916-2921.

[32] Quian H., Bursill LA., Phenomenological theory of the dielectric response of lead magnesium niobate and lead scandium tantalate. Int. J. Mod Phys. B 1996;10; 2007-2025.

[33] Ishchuk VM., Baumer VN., Sobolev VL., The influence of the coexistence of ferroelectric and antiferroelectric states on the lead lanthanum zirconate titanate crystal structure. J. Phys. Condens. Matter 2005;17(19) L177-L182.

[34] Peláiz-Barranco A., Hall DA. Influence of composition and pressure on the electric field-induced antiferroelectric to ferroelectric phase transformations in lanthanum modified lead zirconate titanate ceramics. IEEE Trans. Ultras. Ferroel. Freq. Cont. 2009;56(9) 1785-1791.

[35] Jullian Ch., Li JF., Viehland D. Comparisons of polarization switching in "hard," "soft," and relaxor ferroelectrics. J. Appl. Phys. 2005;95(8) 4316-4318.

[36] Peláiz-Barranco A., García-Zaldívar O., Calderón-Piñar F., López-Noda R., Fuentes-Betancourt J. A multi-Debye relaxation model for relaxor ferroelectrics showing diffuse phase transition. Phys. Sta. Sol. (b) 2005;242(9) 1864-1867.

[37] Bovtun V., Petzelt J., Porokhonskyy V., Kamba S., Yakimenko Y. Structure of the dielectric spectrum of relaxor ferroelectrics. J. Eur. Ceram. Soc. 2001;21(10-11) 1307-1311.

[38] Burns G., Dacol FH., Crystalline ferroelectrics with glassy polarization behaviour. Phys. Rev. B 1983;28; 2527-2530.

[39] Cheng ZY., Katiyar RS., Yao X., Bhalla AS. Temperature dependence of the dielectric constant of relaxor ferroelectrics. Phys. Rev. B 1998;57; 8166-8177.

[40] Bokov AA. Recent advances in diffuse ferroelectric phase transitions. Ferroelectrics 1992;131; 49-55.

[41] Lin HT., Van Aken DC., Huebner W. Modelling the dielectric response and relaxation spectra of relaxor ferroelectrics. J. Am. Ceram. Soc. 1999;82(10) 2698-2704.

[42] Dal-Young K., Jong-Jin Ch., Hyoun-Ee K. Birefringence study of the freezing mechanism of lanthanum-modified lead zirconate titanate relaxor ferroelectrics. J. Appl. Phys. 2003;93; 1176-1179.

[43] Xu G., Shirane G., Copley JRD., Gehring PM. Neutron elastic diffuse scattering study of $Pb(Mg_{1/3}Nb_{2/3})O_3$. Phys. Rev. B 2004;69; 064112.

[44] Viehland D., Li JF., Jang SJ., Cross LE., Wuttig M. Dipolar-glass model for lead magnesium niobate. Phys. Rev. B 1991;43; 8316-8320.

[45] Xu G., Gehring PM., Shirane G. Coexistence and competition of local- and long-range polar orders in a ferroelectric relaxor. Phys. Rev B 2006;74; 104110.

[46] Kirillov VV., Isupov VA. Relaxation polarization of $PbMg_{1/3}Nb_{2/3}O_3$ (PMN) a ferroelectric with a diffused phase transition. Ferroelectrics 1973;5; 3-9.

[47] Isupov VA. Nature of physical phenomena in ferroelectric relaxors. Phys. Sol. Sta. 2003;45(6) 1107-1111.

[48] Pirc R., Blinc R. Vogel-Fulcher freezing in relaxor ferroelectrics. Phys. Rev. B 2007;76; 020101R.

[49] Jiménez R., Jiménez B., Carreaud J., Kiat JM., Dkhil B., Holc J., Kosec M., Algueró M. Transition between the ferroelectric and relaxor states in $0.8Pb(Mg_{1/3}Nb_{2/3})O_3$-$0.2PbTiO_3$ ceramics. Phys. Rev. B 2006;74; 184106.

[50] Bokov AA., Ye Z-G. Recent progress in relaxor ferroelectrics with perovskite structure. J. Mat. Sci. 2006;4(1) 31-52.

[51] Chen IW., Li P., Wang Y. Structural origin of relaxor perovskites. J. Phys. Chem. Sol. 1996;57(10); 1525-1536.

[52] Bellaiche L., Íñiguez J., Cockayne E., Burton BP. Effects of vacancies on the properties of disordered ferroelectrics: A first-principles study. Phys. Rev. B 2007;75; 014111.

[53] Shen M., Han J., Cao W. Electric-field-induced dielectric anomalies in C-oriented $0.955Pb\ Zn_{1/3}Nb_{2/3}O_3$-$0.045PbTiO_3$ single crystals. Appl. Phys. Lett. 2003;83(4) 731-733.

[54] Wu N-N., Hou Y-D, Wang Ch., Zhu M-K, Song X-M, Hui Y. Effect of sintering temperature on dielectric relaxation and Raman scattering of $0.65Pb(Mg_{1/3}Nb_{2/3})O_3$-$0.35PbTiO_3$ system. J. Appl. Phys. 2009;105; 084107.

[55] Park SE., Shrout TR. Ultrahigh strain and piezoelectric behaviour in relaxor based ferroelectric single crystals. J. Appl. Phys. 1997;82(4) 1804-1811.

[56] Park SE., Shrout TR. Characteristics of relaxor-based piezoelectric single crystals for ultrasonic transducers. IEEE Trans. Ultrason. Ferroelectr. Freq. Control 1997;44(5) 1140-1147.

[57] Carpenter MA., Bryson JFJ., Catalan G., Zhang SJ., Donnelly NJ. Elastic and inelastic relaxations in the relaxor ferroelectric $Pb(Mg_{1/3}Nb_{2/3})O_3$: II. Strain–order parameter coupling and dynamic softening mechanisms. J. Phys.: Condens. Matter 2012;24; 045902.

[58] Uchino K. High electromechanical coupling piezoelectrics: relaxor and normal ferroelectric solid solutions. Sol. Sta. Ion. 1998;108(1-4) 43-52.

[59] Deb KK., Hill MD., Roth RS., Kelly JF. Dielectric and pyroelectric properties of doped lead zinc niobate (PZN) ceramic materials. Am. Cer. Bull. 1992;71; 349-352.

[60] Krumin A. Specific solid state features of transparent ferroelectric ceramics. Ferroelectrics 1986;69(1) 1-16.

[61] Xia F., Yao X. Piezoelectric and dielectric properties of PZN-BT-PZT solid solutions. J. Mat. Sci. 1999;34; 3341-3343.

[62] Peláiz Barranco A., González Carmenate I., Calderón Piñar F., Torres García E. AC behaviour and PTCR effect in PZN-PT-BT ferroelectric ceramics. Sol. Sta. Comm. 2004;132; 431-435.

[63] Peláiz Barranco A., Gonzalez Carmenate I., Calderón Piñar F. Dielectric and pyro-electric behaviour of $(Pb_{0.8}Ba_{0.2})[(Zn_{1/3}Nb_{2/3})_{0.7}Ti_{0.3}]O_3$ ferroelectric ceramics. Rev. Cub. Fis. 2009;26(2B) 238-241.

[64] Chang H-Yi., Liu K-S., Lin IN. Modification of PTCR behaviour of $(Sr_{0.2}Ba_{0.8})TiO_3$ materials by post-heat treatment after microwave sintering. J. Eur. Ceram. Soc. 1996;16(1) 63-70.

[65] Chen J., Chan H., Harmer MP. Ordering structure and dielectric properties of undoped and La/Na-doped $Pb(Mg_{1/3}Nb_{2/3})O_3$. J. Am. Ceram. Soc. 1989;72(4) 593-598.

[66] Randall CA., Barber DJ., Whatmore RW., Groves P. Short-range order phenomena in lead-based perovskites. Ferroelectrics 1987;76(1) 277-282.

[67] Huang S., Feng Ch., Chen L., Wang Q. Relaxor Behaviour of $Sr_{1-x}Ba_xBi_2Nb_2O_9$ Ceramics. J. Am. Ceram. Soc. 2006;89(1) 328-331.

[68] Haluska MS., Misture ST. Crystal structure refinements of the three-layer Aurivillius ceramics $Bi_2Sr_{2-x}A_xNb_2TiO_{12}$ (A=Ca,Ba, x=0,0.5,1) using combined X-ray and neutron powder diffraction. J. Solid State Chem. 2004;177; 1965-1975.

[69] Kholkin AL., Avdeev M., Costa MEV., Baptista JL. Dielectric relaxation in Ba-based layered perovskites. Appl. Phys. Letters 2001;79(5) 662-664.

Phase Diagram of The Ternary BaO-Bi$_2$O$_3$-B$_2$O$_3$ System: New Compounds and Glass Ceramics Characterisation

Martun Hovhannisyan

Additional information is available at the end of the chapter

1. Introduction

Restriction and an interdiction on use of toxic materials in electronics products since 2006 have promoted an intensification of development new ecologically friendly materials (glasses, glass ceramics, ceramics) with attractive properties. It has stimulated new lead (cadmium) free systems with good glass forming abilities investigations and new stoiciometric and eutectic points revealing and characterization. Alkaline-earth bismuth borate ternary systems were a good candidate for this purpose, because the binary Bi$_2$O$_3$-B$_2$O$_3$ system have propensity for glass formation and set of binary compounds and eutectics [1 - 3]. Furthermore, bismuth borate single crystals and glass ceramics have nonlinear optical (NLO) properties and other attractive properties [4 - 7]. Both these factors are reasons for further study of binary and ternary bismuth borate systems, and the glasses which they form.

The phase diagram of the Bi$_2$O$_3$–B$_2$O$_3$ system was first determined by Levin &Daniel in 1962 [2] and five crystalline compounds, Bi$_{24}$B$_{12}$O$_{39}$, Bi$_4$B$_2$O$_9$, Bi$_3$B$_5$O$_{12}$, BiB$_3$O$_6$ and Bi$_2$B$_8$O$_{15}$, were identified. Later Pottier revealed a sixth compound, BiBO$_3$ (bismuth orthoborate) [8], which was missing in the original phase diagram [2]. There are no doubts about the existence of BiBO$_3$ now: Becker with co-workers have confirmed existence of bismuth orthoborate [5, 9, 10] and its transparent colourless single crystals of BiBO$_3$ have recently been grown from the melt and characterized by Becker & Froehlich [10]. Monophase samples of both crystalline BiBO$_3$ modifications were obtained by crystallisation below 550°C of bismuth borate glasses with 50-57 mol% B$_2$O$_3$ [10]. However, these authors did not correct the phase diagram, and did not determine the melting point of BiBO$_3$ or the eutectic composition between BiBO$_3$ and Bi$_3$B$_5$O$_{12}$. The compound BiBO$_3$ and this eutectic point are clearly given on the Zargarova & Kasumova's version of the B$_2$O$_3$–Bi$_2$O$_3$ phase diagram, without indication of their melting points and the eutectic composition [11].

Kargin with co-authors [12] by DTA and X-ray analysis have studied conditions of metastable phases formation at system Bi_2O_3-B_2O_3 melts crystallization. They have confirmed existence of metastable $BiBO_3$ compound and for the first time have specified on congruent
character of its melting. Authors also establish formation of a metastable phase of
$5Bi_2O_3 \bullet 3B_2O_3$ composition. Both compounds together with initial Bi_2O_3 and B_2O_3 are present
on the metastable state diagram of the Bi_2O_3-B_2O_3 system constructed by them.

Presence of five compounds on the known Bi_2O_3-B_2O_3 phase diagram has naturally led to
formation of five eutectics compositions containing (mol % B_2O_3): 19.14 (622°C), 44.4 (646 °C),
73.5 (698°C), 76.6 (695°C) and 81.04 (709°C). There is an area of phase separation traditional
for borate systems, observed for compositions containing 81-100 mol % B_2O_3 [2]. Though
according to [12], the area of stable phase separation is stretched to 58-95 mol % B_2O_3.

Interest to ternary alkali free bismuth borate systems M_xO_y-Bi_2O_3-B_2O_3 (M=Zn,Sr,Ca,Ba)
studies has amplified recently. Various research groups (Russian, Canadian, Armenian)
worked in this area during 1990-2009 and revealed a number of ternary compounds, determined their structure, optical and nonlinear optical properties. Thus, three ternary zinc bismuth borate compounds have been revealed in the ZnO-Bi_2O_3-B_2O_3 system. At first
Zargarova& Kasumova have revealed $ZnBi_4B_2O_{10}$ and $ZnBiBO_4$ compounds [11]. Later Barbier with co-authors by solid-state reaction have synthesized third melilite type $ZnBi_2B_2O_7$
compound with large SHG (four time higher as KDP) [13].

Barbier & Cranswick at first two novel noncentrosymmetric $MBi_2B_2O_7$ or $MBi_2O(BO_3)_2$
(M=Ca, Sr) compounds have synthesized by solid-state reactions in air at temperatures in
the 600–700°C range [14]. Their crystal structures have been determined and refined using
powder neutron diffraction data. $CaBi_2B_2O_7$ compound has SHG response two time higher
as KDP [14]. However, authors didn't pay attention for both compounds melting behavior.

Egorisheva with co-authors have studied phase relation in the CaO-Bi_2O_3-B_2O_3 system and
constructs the 600 °C (subsolidus) section of its phase diagram [15]. A new ternary compound of composition $CaBi_2B_4O_{10}$ was identified and the existence of $CaBi_2B_2O_7$ ternary compound was comfirmed. Both compounds had incongruent melting at 700 and 783 °C
respectively and liquidus temperature about 900-930 °C.

Kargin with co-workers have studied phase relation in the SrO-Bi_2O_3-B_2O_3 system in subsolidus at 600 °C [16]. Two new ternary compound of $Sr_7Bi_8B_{18}O_{46}$ and $SrBiBO_4$ compositions were
identified. Both compounds had incongruent melting at 760 and 820 °C without indication
liquidus temperature. However, later Barbier et el. have discribe new novel centrosymmetric borate $SrBi_2OB_4O_9$ ($SrBi_2B_4O_{10}$) forming in the SrO–Bi_2O_3–B_2O_3 system [17], thereby having substituted under doubt existence of previously reported $Sr_7Bi_8B_{18}O_{46}$ compound [16].

The uniqueness of the BaO-Bi_2O_3-B_2O_3 system is shown by the available sets of compounds
and eutectics both in the binary Bi_2O_3-B_2O_3 and BaO–B_2O_3 systems. Seven compounds are
known in the BaO-B_2O_3 system. Four congruent melting binary compounds $Ba_3B_2O_6$, BaB_2O_4,
BaB_4O_7, BaB_8O_{13} with melting points(m.p.) 1383, 1105, 910, 889°C accordingly were found by
Levin & McMurdie [18, 19]. Further, Green and Wahler have found out new congruent melted at 890°C $Ba_2B_5O_{17}$ compound at the ternary BaO-B_2O_3-Al_2O_3 system investigation [20].

Hubner confirmed an existence of the congruent melted Ba$_2$B$_5$O$_{17}$ compound with m.p. 890 °C, and revealed two new compounds Ba$_4$B$_2$O$_7$, Ba$_2$B$_2$O$_5$ [21]. However, all scientists and researches have used the melting diagram of the BaO-B$_2$O$_3$ system created by Levin & McMurdie up to now, without the indication in it the specific areas of existence of new compounds and eutectic points among them [18].

Both these factors were the reason of the BaO-B$_2$O$_3$ system phase diagram correction made by Hovhannisyan R.M. [22]. Author has revealed fields of Ba$_2$B$_5$O$_{17}$ and BaB$_4$O$_7$ compounds crystallisation and new eutectic points which are absent on the diagram constructed by Levin & McMurdie [18]. Six binary eutectic compositions containing 31.5, 37.5, 63.5, 68.5, 76.0, 83.4 mol % B$_2$O$_3$ with melting points 1025, 915, 905, 895, 869 and 878°C accordingly were on the diagram after correction.

Area of two immiscible liquids established by Levin & McMurdie [35] in the BaO-B$_2$O$_3$ in an interval of 1.5 to 30wt. % BaO content, it has been confirmed in the subsequent by other authors. However, the temperature of the liquation couple, which are 1150, 1180, 1256 °C according to [23] and 1539°C according to [24] is discussed till now.

There are no full version of the phase diagram of the BaO-Bi$_2$O$_3$ system till now [1]. It is very complex system, which is very critical to atmosphere and pressure at experiment carrying out [25 - 28]. Two low melted eutectic areas (740-790°C) clear observed on phase diagram studied in air or oxigen in high bismuth content region around 5-7 mol%BaO and 25-30 mol %BaO [26 - 28].

All research groups payd special attention to BaO-Bi$_2$O$_3$-B$_2$O$_3$ system studies and new ternary compounds revealing and characterisation. Barbier *et al.* have studied seven compositions in the ternary BaO-B$_2$O$_3$-Bi$_2$O$_3$ system by solid state synthesis at temperatures below 650°C and BaBiBO$_4$, or BaBi(BO$_3$)O, a novel borate compound, has been made and chracterisised [29]. Above 650°C it decays with bismuth borate glass formation. A powder sample of BaBiBO$_4$ had a second harmonic signal with a NLO efficiency equal to five times that of KDP.

Practically in parallel, Egorysheva with co-workers have been investigated phase equilibrium in the Bi$_2$O$_3$-BaB$_2$O$_4$-B$_2$O$_3$ system by X-ray analysis and DTA [30, 31]. Studies were spent by the samples solid state synthesis in closed Pt crucibles in muffle furnaces at the temperature range 500-750 ^{0}C, that corresponds to sub-solidus area. The synthesis duration (with intermediate cakes regrinding) were 6-16 days. They confirmed presence of BaBiBO$_4$ and have revealed three new compounds: BaBiB$_{11}$O$_{19}$, BaBi$_2$B$_4$O$_{10}$, Ba$_3$BiB$_3$O$_9$. BaBiB$_{11}$O$_{19}$, BaBi$_2$B$_4$O$_{10}$ have congruent melting at 830 and 730 °C respectively and BaBiBO$_4$ melt incongruently at 780^{0}C. Ba$_3$BiB$_3$O$_9$ undergoes a phase transition at 850°C and exist up to 885°C, were decompose in the solid state [31].

Recently single crystals of BaBi$_2$B$_4$O$_{10}$ composition were grown by cooling of a melt with the stoichiometric composition with cooling rate 0.5 K/h [32]. They have once again confirmed existence of BaBi$_2$B$_4$O$_{10}$ stoichiometric compound earlier obtained by solid state synthesis.

In 1972 Elwell with co-workers investigated the $BaO-B_2O_3-Bi_2O_3$ system by hot stage microscopy and a new ternary eutectic composition, $23.4BaO\bullet62.4Bi_2O_3\bullet14.2B_2O_3$ (wt%), with a low liquidus temperature of 600°C, was revealed for ferrite spinel growth [33].

Using different melts cooling rates Hovhannisyan, M. with co-authors at first have determined large glass-forming field in the $BaO-Bi_2O_3-B_2O_3$ system, which includes all eutectics in the binary $Bi_2O_3-B_2O_3$, $BaO-B_2O_3$ and $BaO-Bi_2O_3$ systems and covers majority of the concentration triangles, reaching up to 90 mol% Bi_2O_3. [34, 35].

The methodology based on glass samples investigation was more effective at $BaO-Bi_2O_3$-B_2O_3 system phase diagram construction, than a traditional technique based on solid state sintered samples studies. Because DTA curves of glasses, to the contrary DTA curves of solid state sintered samples, indicates their all characteristics temperatures, includes exothermal effects of glass crystallizations and endothermic effects of formed crystalline phases melting. It has allowed us to reveal two new $BaBi_2B_2O_7$ and $BaBi_{10}B_6O_{25}$ congruent melted at 725 and 690°C respectively compounds in the $BaO-Bi_2O_3-B_2O_3$ system [34 - 36].

However, our further studies of glasses and glass ceramics in this system have shown necessity of glass forming diagram correction and phase diagram construction in the ternary $BaO-Bi_2O_3-B_2O_3$ system and present these data to scientific communitty. Another aim of this work is both known and novel stoichiometric ternary barium bismuth borates compounds characterisation in glassy, glass ceramic and ceramic states for further practical application.

2. Experimental

About three hundred samples of various binary and ternary compositions have been synthesized and tested in $BaO- Bi_2O_3-B_2O_3$ system. Compositions were prepared from "chemically pure" grade $BaCO_3$, H_3BO_3 and Bi_2O_3 at 2.5-5.0 mol % intervals. The most part of samples has been obtained as glasses by various cooling rates depending on melts glass forming abilities: as bulk glass plates with thickness 6,5 ÷7mm by casting on metallic plate (up to $10^K/s$), as monolithic glass plates with thickness up to 3mm by casting between two steel plates($\sim10^2$ $^K/s$), and glass tapes samples with thickness 30–400 μm through super cooling method ($10^3\div10^4$ $^K/s$). Glass formation was determined visually or by x-ray analysis. The glass melting was performed at 800–1200°C for 15–20 min with a 20–50 g batch in a 20–50 ml uncovered quartz glass or corundum crucible, using an air atmosphere and a "Superterm 17/08" electric furnace. Compositions in the $BaO-B_2O_3$ system were melted in a 25 or 50 ml uncovered Pt crucibles at 1400–1500°C for 30 min with a 20–50 g batch. The chemical composition of some glasses was determined by traditional chemical analysis, and the results indicate a good compatibility between the calculated and analytical amounts of B_2O_3, BaO and Bi_2O_3. SiO_2 contamination from quartz glass crucibles did not exceed 2 wt%, and alumina contamination did not exceed 0 5–1 wt%, according to the chemical analysis data.

Samples of compositions laying outside of a glass formation field or having high melting temperature, have been obtained by solid-phase synthesis. Mixes (15-20 g) were carefully frayed in an agate mortar, pressed as tablets, located on platinum plates and passed the thermal treatment in "Naber"firm electric muffles. After regrinding powders were tested by DTA and X-ray methods. The synthesized samples of binary barium borate system compositions containing 60 mol% and more of BaO and also compositions containing over 90mol % B_2O_3 had very low chemical resistance and were hydrolyzed on air at room temperature. In this connection the synthesized samples were kept in a dryer at 200°C.

DTA and X-ray diffraction data of glass and crystallized glass samples have been used for phase diagram construction in the ternary BaO- Bi_2O_3 -B_2O_3 system. The DTA analysis (pure Al_2O_3 crucible, powder samples weight ~600 mg, heating rates 10 K/min) on Q-1500 type derivatograph were carried out. Glass transition -T_g, crystallization peaks -T_{cr}, melting -T_m and liquidus -T_L temperatures have been determined from DTA curves. Reproducibility of temperatures effects on DTA curves from melting to melting was ±10K. The accuracy of temperature measurement is ±5 K.

Thermal expansion coefficient (TEC) and glass transition temperature (T_g) measurements were made on a DKV-4A type vertical quartz dilatometer with a heating rate of 3K/min. Glass samples in the size of 4×4×50 millimeters have been prepared for TEC measurement. The dilatometer was graduated by the quartz glass and sapphire standards. The TEC measurement accuracy is $±(3÷4)•10^{-7}K^{-1}$, T_g ±5 °C.

X-ray patterns were obtained on a DRON-3 type diffractometer (powder method, $CuK\alpha$–radiation, Ni-filter). Samples for glass crystallization were prepared with glass powder pressed in the form of tablets. Crystallization process was done in the electrical muffles of "Naber" firm by a single-stage heat treatment. This was done within 6-12 hours around a temperature at which the maximum exothermal effects on glasses by DTA were observed.

Crystalline phases of binary and ternary compounds formed both at glasses crystallization and at solid-phase synthesis have been identified by using JCPDS-ICDD PDF-2 release 2008 database [43].

Computerized methodic of ferroelectric hysteresis test and measurement of ferroelectric properties such as coercive field and remanent polarization at wide temperature (up to 250°C and frequency (10-5000Hz) ranges was used. Methodic based on the well known Sawyer – Tower's [44] modified scheme, which is allowing to compensate phase shifts concerned with dielectric losses and conductivity. The desired frequency signal from waveform generator is amplifying by high voltage amplifier and applying to sample. The signals, from the measuring circuit output, proportional to applied field and spontaneous polarization are passing throw high impedance conditioning amplifiers, converting by ADC and operating and analyzing in PC. The technique allows to perform tests of synthesized glass ceramics obtained by means of controlling crystallization of thin (above 30 micrometer thick) monolithic tape (film) specimens by applying up to 300kV/cm field to our thin samples (~50 micrometer thick) and obtain hysteresis loops for wide diversity of hard FE materials.

3. Results

3.1. Glass forming and phase diagrams of the BaO-Bi$_2$O$_3$-B$_2$O$_3$ system

The traditional method of phase diagram construction based on solid-phase sintered samples investigation takes long time and is not effective. The glass samples investigation technique is progressive, because the DTA curves have registered all processes taking place in glass samples, including the processes of glass crystallizations, quantity of crystal phases and temperature intervals of their formation and melting. However, inadequate amount of glass samples restrict their use during phase diagram construction. The super-cooling method promotes the mentioned problem solving and open new possibilities for phase diagrams constructions.

Hovhannisyan R.M. with co-workes successfully developed this direction last time and have constructed phase diagrams in binary and ternary alkiline-eath bismuth borate, barium boron titanate, barium aluminum boron titanate, barium gallium borate, yttrium aluminum borate, yttrium gallium borate, lantanum gallium borate, zinc tellurium molibdate and other systems [34 - 42].

3.1.1. Glass forming diagram of the BaO-Bi$_2$O$_3$-B$_2$O$_3$ system

Figure 1 shows the experimental data on glass formation in the BaO-Bi$_2$O$_3$-B$_2$O$_3$ system obtained by different authors from 1958 to 2007 [45 - 49]. For defining the glass forming ability of the pointed system, the authors of the mentioned works used different amounts of melt, glass melting crucibles, temperature–time melting regimes, and technological methods of melt cooling. Imaoka & Yamazaki studied glass formation by melts cooling on air Glasses were melted at temperatures below 1200 °C in gold-palladium or platinum-rhodium crucibles (Fig.1.1) [45]. Janakirama-Rao glass formation studied by melts cooling on air. Glasses melted in platinum crucibles at 600- 1400 °C with 0.5-1.0 h melts exposition and its cooling in air (Fig. 1.2) [46]. Izumitani [47] experiments spent in 10g crucibles at 1100-1350 °C with melts cooling on air (Fig. 1.3). Milyukov with co-authors glass formation studied by melts casting in steel mold. Glasses melted in platinum crucibles at 600-1400 °C with melts stirring by Pt stirrer for 1h (Fig. 1.4) [48]. Kawanaka & Matusita glass formation studied by silica rod stirred melts pouring into preheated to 250-300°C carbon mold (Fig. 1.5) [49].

Authors used different weights of glass forming melts, melting crucibles, temperature-time of melting regimes and technological methods of melts cooling. Obtained data are difficultly comparable and remote from two basic criteria promoting glass formation: liquidus temperature and speeds of melts cooling.

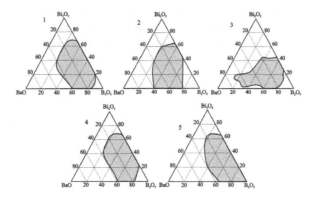

Figure 1. Glass forming regions in the BaO-Bi₂O₃-B₂O₃ system according to the data of the authors: 1- [45], 2-[46], 3-[47], 4-[48], 5-[49].

Figure 2 shows corrected glass formation diagram in the BaO-Bi₂O₃-B₂O₃ system based on phase diagrams of the BaO–B₂O₃, BaO–Bi₂O₃, and B₂O₃–Bi₂O₃ binary systems and controllable melt cooling rates. Using the term "diagram," but not the glass formation region, we take into account the interrelation between the phase diagram and the glass forming ability of the system.

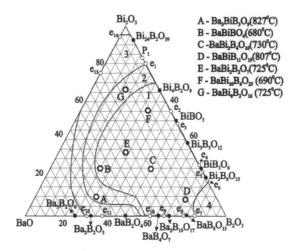

Figure 2. Glass forming diagram in the BaO-Bi₂O₃-B₂O₃ system depending of melts cooling rates: 1-up to 10 K/s; 2~10²K/s; 3-(10³-10⁴) K/s; 4- stable phase separation region.

Area of glass compositions with low crystallization ability and stable glass formation in the studied ternary BaO-Bi₂O₃ -B₂O₃ system have been determined at melts cooling rate ~ 10 K/s

(Fig.2-1). It included binary $Bi_4B_2O_9$, $BiBO_3$, $Bi_3B_5O_{12}$, BiB_3O_6, $Bi_2B_8O_{15}$, $Ba_2B_{10}O_{17}$, BaB_8O_{13} compounds in the $BaO-Bi_2O_3$ and $BaO-B_2O_3$ systems and five ternary $BaBiB_{11}O_{19}$(D), $Ba-Bi_2B_4O_{10}$(C), $BaBi_2B_2O_7$ (E), $BaBi_{10}B_6O_{25}$ (F), and $BaBiBO_4$ (B) compounds. However, we didn't comfirm presence of $Ba_3BiB_3O_9$ (A) compound in area of stable glasses at melts cooling rate ~ 10 K/s, which was reported earlier [34, 35].

Increasing of melts cooling speed up to ~10^2 K/s has led to glass formation area expansion (Fig.2-2). This cooling rate is enough for monolithic glass plates with thickness up to 3mm fabrication by melts casting between two steel plates (Fig.2-2). The glass plates of compositions correspondings to $Ba_3BiB_3O_9$ (A) and supposed $BaBi_8B_2O_{16}$ (G) compounds have been obtained by this way.

Super cooling technique constructed by our group allowed to expand the borders of glass formation in studied system under high melts cooling rates equal to (10^3-10^4) K/s (Fig.2-3). Determined glassforming area include compositions content: 80 – 95 mol% Bi_2O_3 in the binary B_2O_3–Bi_2O_3 system; 43-70 mol% BaO in the binary BaO–B_2O_3 system, including BaB_2O_4 composition. Area of glass formation from both these areas moves to 55-95 mol% content compositions in the binary BaO–Bi_2O_3 system (Fig.2-3).

Traditional for borate systems a stable phase separation region was also observed for high B_2O_3 content compositions contents more than 84 -87 mol%B_2O_3 (Fig.2-4).

3.1.2. Phase diagram of the $BaO-Bi_2O_3-B_2O_3$ system

Our investigation of the ternary $BaO-B_2O_3-Bi_2O_3$ system have purposefully been directed on construction of the phase diagram through first of all glass forming diagram construction and revealing both new compounds and eutectic compositions. Constructed by us glass forming diagram (Fig.2) practically occupies the most part of the $BaO-B_2O_3-Bi_2O_3$ concentration triangles. It has allowed to use synthesized glasses as initial compositions at phase diagram construction. It was basic difference of our methodology from technologies used by other authors. Phase equilibriums reached at isothermal sections construction do not allow to have a full picture of processes in cases of the solid state synthesized samples investigations. Whereas at glass samples studies we determine not only characteristic points of glasses(T_g and T_s) by DTA, but also quantity of crystal phases, temperatures of their crystallization and then temperatures of their melting. It has allowed us to reveal new stoichiometric compositions which have been lost by other research groups at isothermal sections construction by traditional methods. In some cases we also in parallel used samples obtained by solid state synthesis for comparison with their glassy analogues or in those cases, when their obtaining in the glassy form was impossible.

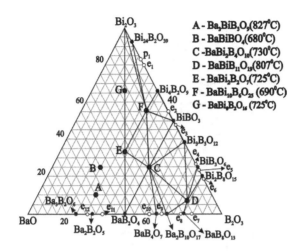

Figure 3. The BaO-B₂O₃-Bi₂O₃ system triangulation

Experimental data concerning phase diagrams of binary systems Bi_2O_3-B_2O_3, BaO-B_2O_3, BaO-Bi_2O_3 and pseudo-binary sections in the BaO-B_2O_3-Bi_2O_3 system have allowed us to estimate fields of primary crystallization of co-existing phases and divided all concentration triangle into elementary triangles, i.e. to make triangulation.

The triangulation scheme in the BaO-B_2O_3-Bi_2O_3 system is presented on Fig. 3. By means of a triangulation all concentration triangle is divided into following elementary triangles: $BiBO_3$-F-C, $BiBO_3$-F-$Bi_4B_2O_9$, $BiBO_3$-C-$Bi_3B_5O_{12}$, F-E-C, C-BaB_2O_4-E, C-$Bi_3B_5O_{12}$-D, $Bi_4B_2O_9$-F-Bi_2O_3, E-Bi_2O_3-F, C-BaB_2O_4-$Ba_2B_{10}O_{17}$, $Ba_2B_{10}O_{17}$-D-C, $Ba_2B_{10}O_{17}$-D-B_2O_3, $Bi_3B_5O_{12}$-D-$Bi_2B_8O_{15}$, D-$Bi_2B_8O_{15}$-B_2O_3.

3.1.2.1. Phase diagram of the binary Bi_2O_3–B_2O_3 system

First of all we have attempt to finished phase diagram construction in area of compositions around of $BiBO_3$ compound. Compositions containing 45–65 mol% B_2O_3 in the Bi_2O_3-B_2O_3 system were tested to determine the melting point of $BiBO_3$ and to determine the eutectic composition between $BiBO_3$ and $Bi_3B_5O_{12}$. The compositions used to correct the B_2O_3-Bi_2O_3 phase diagram were prepared by solid state synthesis at 520°C, with steps of 0.5–1.0 mol% B_2O_3 over the interval 45–55 mol% B_2O_3. As a result, the eutectic composition, 48.5Bi2O3•51.5B2O3 (mol%), between $BiBO_3$ and $Bi_3B_5O_{12}$, was determined, and its melting point was measured by DTA as 665±5°C (Fig. 4). It was also found that $BiBO_3$ melts congruently at 685±5°C.

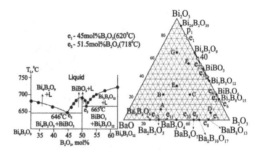

Figure 4. Corrected phase diagram of the Bi_2O_3-B_2O_3 system in the interval 30–65 mol% B_2O_3.

3.1.2.2. Phase diagram of the pseudo-binary BiBO₃–BaB₂O₄ system

$BaBi_2B_4O_{10}$ is a congruently melting compound, with a melting point of 730°C, and it occupies the central area of the $BiBO_3$–BaB_2O_4 pseudo-binary system (Fig. 5). This system forms two simple pseudo-binary eutectics, E_1 at 15 mol% BaB_2O_4, with a melting point of 620°C, and E_2 at 60 mol% BaB_2O_4, with a melting point of 718°C.

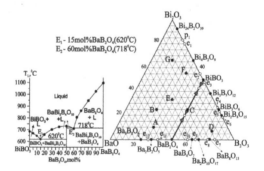

Figure 5. Phase diagram of the pseudo-binary system $BiBO_3$-BaB_2O_4

3.1.2.3. Phase diagram of the pseudo-binary Bi₄B₂O₉–BaBi₂B₂O₇ system

The introduction of 12 mol% $BaBi_2B_2O_7$ in the pseudo-binary system $Bi_4B_2O_9$-$BaBi_2B_2O_7$ reduced the melting point of initial $Bi_4B_2O_9$, and resulted in the formation of a simple pseudo-binary eutectic, E_3, with melting point 605°C (Fig. 6). A maximum of the liquidus with melting point of 690°C is seen at 33.33 mol% $BaBi_2B_2O_7$, which indicates the formation of the new congruently melting ternary compound $BaBi_{10}B_6O_{25}$ (11.11BaO•55.55Bi_2O_3•33.33B_2O_3). Further increase of the $BaBi_2B_2O_7$ content (49 mol%) leads to a second pseudo-binary eutectic, E_4, with melting point 660°C. Increasing of the liquidus temperature is observed in the post eutectic region of composition, with a maximum at 725°C. It corresponds to the formation of the new congruently melting ternary compound $BaBi_2B_2O_7$ (Fig. 6).

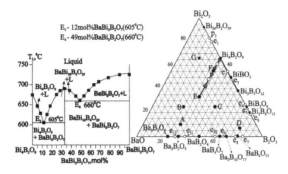

Figure 6. Phase diagram of the pseudo-binary system $Bi_4B_2O_9$-$BaBi_2B_2O_7$.

Two new crystalline ternary compounds, $BaBi_2B_2O_7$ and $BaBi_{10}B_6O_{25}$, were revealed by crystallisation at the same glass composition. Both compounds, $BaBi_2B_2O_7$ and $BaBi_{10}B_6O_{25}$, melt congruently at 725±5°C and 690±5°C, respectively. The X-ray characteristics of $BaBi_2B_2O_7$ and $BaBi_{10}B_6O_{25}$ were determined and are given in Tables 1 and 2.

No.	d_{exp}	I/I_0	hkl	No.	d_{exp}	I/I_0	hkl	No.	d_{exp}	I/I_0	hkl
1	6.23	9	101	20	2.15	25	140	39	1.349	7	125
2	5.02	9	111	21	2.12	5	232	40	1.33	3	543
3	4.80	5	201	22	2.06	24	123	41	1.28	8	035
4	4.29	5	020	23	2.01	6	523	42	1.24	6	560
5	4.11	14	120	24	1.977	25	241	43	1.217	5	263
6	3.88	6	300	25	1.84	15	142	44	1.21	9	843
7	3.67	4	021	26	1.826	7	133	45	1.206	13	271
8	3.59	26	301	27	1.786	6	004	46	1.19	14	145
9	3.56	50	121	28	1.729	52	114	47	1.173	12	245
10	3.52	23	220	29	1.679	23	250	48	1.17	4	1010
11	3.19	100	112	30	1.636	34	251	49	1.14	6	126
12	3.12	8	221	31	1.63	5	532	50	1.11	6	662
13	3.05	9	202	32	1.57	4	052	51	1.10	6	326
14	2.91	43	030	33	1.556	10	243	52	1.09	4	180
15	2.696	90	122	34	1.522	8	034	53	1.042	6	146
16	2.51	12	222	35	1.488	23	632	54	1.021	9	065
17	2.376	21	003	36	1.458	6	060	55	1.018	9	943
18	2.31	5	421	37	1.428	10	811	56	1.01	5	274
19	2.254	22	113	38	1.373	11	821				

Table 1. X-ray characteristics of the new ternary compound $BaBi_2B_2O_7$, synthesized at the same glass composition crystallization (640°C, 20 h).

Single crystals of $BaBi_{10}B_6O_{25}$ were obtained by cooling of a melt with the stoichiometric composition. Glass powder of composition $11.11BaO \bullet 55.55Bi_2O_3 \bullet 33.33B_2O_3$ (mol%) was heated in a quartz glass ampoule up to 750°C at a rate 10 K/min. After 2 h exposition at high temperature, the melt was cooled at a rate 0.5 K/h. Single crystals with sizes up to $1.66 \times 0.38 \times 0.19$ mm^3 were grown.

No.	d_{exp}	I/I_0	hkl	No.	d_{exp}	I/I_0	hkl	No.	d_{exp}	I/I_0	hkl
1	9.21	3.0	012	23	2.91	31	0110	45	2.01	5.2	321
2	6.26	3.0	101	24	2.8	6	221	46	1.98	23.7	323
3	6.02	3.0	005	25	2.7	75.9	206	47	1.92	3.0	1214
4	5.01	7.3	006	26	2.64	3.4	045	48	1.88	3.0	0314
5	4.89	10.8	104	27	2.57	3.0	2.07	49	1.86	3.0	162
6	4.63	6.5	024	28	2.53	4.3	046	50	1.84	15.1	254
7	4.19	4.3	025	29	2.52	5.2	144	51	1.83	3.0	1412
8	4.18	4.3	122	30	2.49	6.9	230	52	1.82	3.0	164
9	4.11	10	115	31	2.47	9.5	1210	53	1.81	3.0	255
10	3.92	6.0	030	32	2.45	4.7	232	54	1.79	3.4	165
11	3.80	3.4	032	33	2.38	16	0310	55	1.77	3.0	328
12	3.65	10.8	033	34	2.35	4.3	050	56	1.75	3.4	1413
13	3.56	39.7	107	35	2.34	5.2	051	57	1.73	55.2	341
14	3.51	9.5	125	36	2.33	3.9	1012	58	1.71	4.3	343
15	3.41	3.4	117	37	2.31	3.4	1212	59	1.69	3.0	069
16	3.38	7.3	130	38	2.25	17.2	228	60	1.68	16.4	070
17	3.33	8.6	131	39	2.21	7.8	150	61	1.65	3.0	259
18	3.27	9.1	132	40	2.19	3.0	055	62	1.64	31.0	074
19	3.18	100	133	41	2.15	21.2	237	63	1.61	2.9	400
20	3.07	14.7	203	42	2.09	3.4	312	64	1.6	6.0	402
21	3.04	9.2	212	43	2.06	21.6	304	65	1.55	6.5	421
22	2.94	14.2	040	44	2.03	3.0	314	66	1.52	9.5	177

Table 2. X-ray characteristics of the new ternary $BaBi_{10}B_6O_{25}$ single crystals.

The x-ray powder diffraction patterns of $BaBi_2B_2O_7$ and $BaBi_{10}B_6O_{25}$ could be indexed on an orthorhombic cell with lattice parameters as follows:

- for $BaBi_2B_2O_7$ $a=11.818$ Å, $b=8.753$ Å, $c=7.146$ Å, cell volume $V=739.203$ Å^3, $Z=4$;
- for $BaBi_{10}B_6O_{25}$ $a=6.434$ Å, $b=11.763$ Å, $c=29.998$ Å, cell volume $V=2270.34$ Å^3, $Z=8$.

3.1.2.4. Phase diagram of the pseudo-binary $BiBO_3$–$BaBi_{10}B_6O_{25}$ system

$BiBO_3$– $BaBi_{10}B_6O_{25}$ is a very important system (Fig.7). Initial $BiBO_3$ has a melting point of 685°C. The second maximum in the liquidus curve (Fig.7) of 690°C is connected with the formation of the new ternary compound $BaBi_{10}B_6O_{25}$. There is a simple pseudo-binary eutectic E_5 between these two compounds at 54 mol%$BaBi_{10}B_6O_{25}$, with a melting point of 595°C.

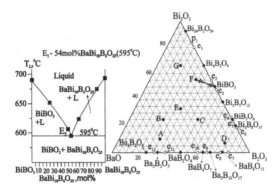

Figure 7. Phase diagram of the pseudo-binary system $BiBO_3$- $BaBi_{10}B_6O_{25}$.

3.1.2.5. Phase diagram of the pseudo-binary $BaBi_{10}B_6O_{25}$ - $BaBi_2B_4O_{10}$ system

The $BaBi_{10}B_6O_{25}$–$BaBi_2B_4O_{10}$ system confirms the presence of the new congruently melting ternary compound $BaBi_{10}B_6O_{25}$, with a melting point of 690°C (Fig.8). $BaBi_2B_4O_{10}$ melts congruently at 730°C. There is a simple pseudo-binary eutectic E_6 between these two compounds at 28 mol % $BaBi_2B_4O_{10}$, with a melting point of 660°C.

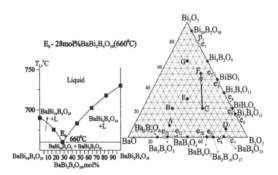

Figure 8. Phase diagrams of the pseudo-binary system $BaBi_{10}B_6O_{25}$ -$BaBi_2B_4O_{10}$.

3.1.2.6. Phase diagram of the pseudo-binary $BaBi_2B_4O_{10}$ – $50BaO•50Bi_2O_3$ section

This pseudo-binary section consists of two ternary compounds $BaBi_2B_4O_{10}$, $BaBi_2B_2O_7$ and two eutectics E_7, E_8 dividing fields primary crystallisations these compounds. Initial composition is $BaBi_2B_4O_{10}$ (Fig.9). The introduction of 20 mol% 50%$BaO•$50%Bi_2O_3 in the pseudo-binary system $BaBi_2B_4O_{10} - 50BaO \bullet 50Bi_2O_3$ reduced the melting point of initial $BaBi_2B_4O_{10}$, and resulted in the formation of a simple pseudo-binary eutectic, E_7, with melting point 680°C (Fig.9). A maximum of the liquidus with melting point of 725°C is seen at 33.33 mol% of 50%BaO • 50%Bi$_2$O$_3$, which indicates the formation of the new congruently melting ternary compound $BaBi_2B_2O_7$. Further increase of the 50%BaO•50%Bi$_2$O$_3$ content (52.5 mol%) leads to a second pseudo-binary eutectic, E_8, with melting point 700°C. Increasing of the liquidus temperature is observed in the post eutectic region of composition (more than 52.5mol% of 50BaO • 50Bi$_2$O$_3$). Un identified phase is in the post eutectic (E_8) region of composition (Fig.9).

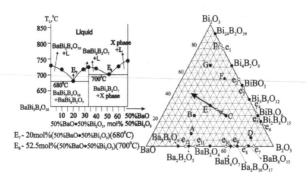

Figure 9. Phase diagram of the pseudo-binary section BaBi$_2$B$_4$O$_{10}$ – 50BaO•50Bi$_2$O$_3$.

3.1.2.7. Phase diagram of the pseudo-binary BaB₂O₄–Bi₂O₃ section

BaB_2O_4 - Bi_2O_3 section consist of two pseudo-binary BaB_2O_4- $BaBi_2B_2O_7$, $BaBi_2B_2O_7$- Bi_2O_3 systems (Fig.10). There are two eutectics: E_9 between BaB_2O_4 and $BaBi_2B_2O_7$, E_{10} between $BaBi_2B_2O_7$ and $BaBi_8B_2O_{16}$, and peritectic point P_1 between $BaBi_8B_2O_{16}$ and Bi_2O_3 (Fig.10). The introduction of 26 mol% Bi_2O_3 in the pseudo-binary system BaB_2O_4 – Bi_2O_3 sharp reduced the melting point of initial BaB_2O_4 on 445°C, and resulted in the formation of a simple eutectic, E_9, with melting point 685°C (Fig.10). A maximum of the liquidus with melting point of 725°C is seen at 33 33 mol% of Bi_2O_3, which indicates the formation of the new congruently melting ternary compound $BaBi_2B_2O_7$.

Further increase of the Bi_2O_3 content (42 mol%) leads to a second eutectic, E_{10}, formation with melting point 690°C. Increasing of the liquidus temperature is observed in the post eutectic region of composition (more than 42mol% of Bi_2O_3) and formation of new incongruent melted at 725 °C $BaBi_8B_2O_{16}$ ternary compound (Fig.10). It is very difficult determined of Ba-Bi$_8$B$_2$O$_{16}$ X-ray characteristics, because they very closed to Bi_2O_3 characteristics.

Constructed by us this section's diagram essentially differs from that constructed by Russian researches Egorisheva & Kargin [30] because they could not find out two new compounds revealed by us: congruent melted at 725°C $BaBi_2B_2O_7$ and incongruent melted at 725°C $BaBi_8B_2O_{16}$(Fig.10). At repeated, even more detailed studies they also could not find out these compounds [31].

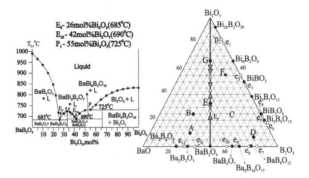

Figure 10. Phase diagram of the BaB₂O₄ – Bi₂O₃ section.

3.1.2.8. Phase diagram of the pseudo-binary $Bi_3B_5O_{12}$ – $BaBi_2B_4O_{10}$ system

It is very simple system with pseudo-binary eutectic E_{11} between two congruent melted $Bi_3B_5O_{12}$ and $BaBi_2B_4O_{10}$ compounds. Eutectic E_{11} content 38 mol% of $BaBi_2B_4O_{10}$ and has melting point 680°C (Fig.11).

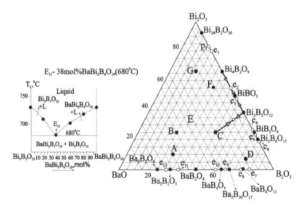

Figure 11. Phase diagram of the pseudo-binary system Bi₃B₅O₁₂ – BaBi₂B₄O₁₀

3.1.2.9. Phase diagram of the pseudo-binary $BaBi_2B_4O_{10}$ – $BaBiB_{11}O_{19}$ system

Initial $BaBi_2B_4O_{10}$ has a melting point of 730°C. The second maximum in the liquidus curve (Fig.12) of 807°C is connected with the formation of the ternary compound $BaBiB_{11}O_{19}$. There is a simple pseudo-binary eutectic, E_{12}, between these two compounds at 65.5 mol% Ba-Bi₂B₄O₁₀, with a melting point of 695°C.

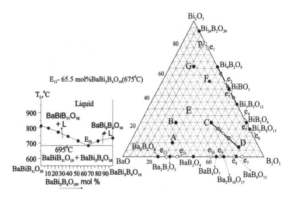

Figure 12. Phase diagram of the pseudo-binary system $BaBi_2B_4O_{10}$ – $BaBiB_{11}O_{19}$.

3.1.2.10. Phase diagram of the pseudo-binary $Bi_3B_5O_{12}$ – $BaBiB_{11}O_{19}$ system

Pseudo-binary system $Bi_3B_5O_{12}$ -$BaBiB_{11}O_{19}$ has simple eutectic E_{13} formed between two congruent melted $BaBiB_{11}O_{19}$ and $Bi_3B_5O_{12}$ compounds. According to DTA eutectic E_{13} has melting point 705°C and content 28 mol% $BaBiB_{11}O_{19}$ (Fig.13).

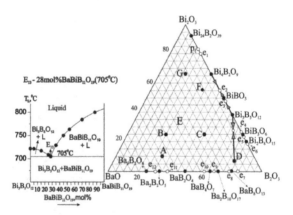

Figure 13. Phase diagram of the pseudo-binary system $Bi_3B_5O_{12}$ - $BaBiB_{11}O_{19}$.

3.1.2.11. Phase diagram of the pseudo-binary $BaBiB_{11}O_{19}$ – $Ba_2B_{10}O_{17}$ system

Pseudo-binary system $BaBiB_{11}O_{19}$ – $Ba_2B_{10}O_{17}$ has simple eutectic E_{14} formed between two congruent melted compounds $Ba_2B_{10}O_{17}$ and $BaBiB_{11}O_{19}$. According to DTA eutectic E_{14} has melting point 780°C and content 26 mol% $Ba_2B_{10}O_{17}$(Fig.14).

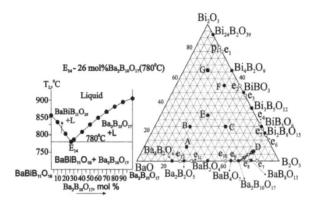

Figure 14. Phase diagram of the pseudo-binary system BaBiB₁₁O₁₉ – Ba₂B₁₀O₁₇.

3.1.2.12. Phase diagram of the pseudo-binary $BaBi_2B_4O_{10} – Ba_2B_{10}O_{17}$ system

It is very simple system with eutectic E_{15} formed between two congruent melted $Ba_2B_{10}O_{17}$ and $BaBi_2B_4O_{10}$ compounds. Pseudo-binary eutectic E_{15} content 24 mol% of $Ba_2B_{10}O_{17}$ and has melting point 710°C (Fig. 15).

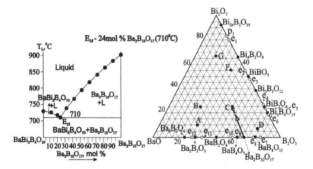

Figure 15. Phase diagram of the pseudo-binary system BaBi₂B₄O₁₀ – Ba₂B₁₀O₁₇.

3.1.2.13. Phase diagram of the pseudo-binary $BaBi_2B_4O_{10} – BaB_4O_7$ system

The same picture is observe for $BaBi_2B_4O_{10} – BaB_4O_7$ system: simple eutectic E_{16} is formed between two congruent melted compounds. Pseudo-binary eutectic E_{16} content 24 mol% of BaB_4O_7 and has melting point 715°C (Fig. 16).

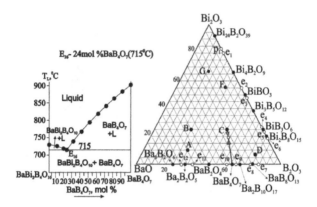

Figure 16. Phase diagram of the pseudo-binary system $BaBi_2B_4O_{10} - BaB_4O_7$.

3.1.2.14. Phase diagram of the pseudo-binary $BaBiB_{11}O_{19} - BaB_8O_{13}$ system

It is simple system with pseudo-binary eutectic E_{17} between two congruent melted $BaBiB_{11}O_{19}$ and BaB_8O_{13} compounds. Eutectic E_{17} content 38 mol% of BaB_8O_{13} and has melting point 770°C (Fig. 17).

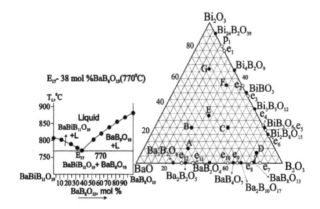

Figure 17. Phase diagram of the pseudo-binary system $BaBiB_{11}O_{19} - BaB_8O_{13}$.

3.1.2.15. Phase diagram of the $BaO-Bi_2O_3-B_2O_3$ ternary system

As result of huge work under project the phase diagram in the ternary $BaO-B_2O_3-Bi_2O_3$ system has been constructed for the first time and presented on Fig.18. Three new compounds $BaBi_2B_2O_7$, $BaBi_{10}B_6O_{25}$ and $BaBi_8B_2O_{16}$ have been revealed and characterized.

Fields of binary bismuth and barium borates as well as all ternary barium bismuth borates compounds crystallizations have been determined and outlined and sixteen ternary eutectic points E_1-E_{16} have been revealed as result of phase diagram construction (Fig. 18, table 3). The phase diagram evidently represents interaction of binary and ternary compounds taking place in the pseudo-ternary systems. The ternary eutectic E_1 with m.p 590°C has been determined among BiBO$_3$, F and Bi$_4$B$_2$O$_9$ compounds; ternary eutectic E_2 with m.p. 585°C has been formed among BiBO$_3$, F and C compounds; ternary eutectic E_3 with m.p. 640°C has been formed among F, E and C compounds; ternary eutectic E_4 with m.p. 622°C has been formed among C, BaB$_2$O$_4$ and E compounds; ternary eutectic E_5 with m.p. 610°C has been formed among BiBO$_3$, C and Bi$_3$B$_5$O$_{12}$ compounds; ternary eutectic E_6 with m.p. 675°C has been formed among C, Bi$_3$B$_5$O$_{12}$ and D compounds; ternary eutectic E_7 with m.p. 680°C has been formed among Bi$_3$B$_5$O$_{12}$, D and BiB$_3$O$_6$ compounds; ternary eutectic E_8 with m.p. 675°C has been formed among BiB$_3$O$_6$, D and Bi$_2$B$_8$O$_{15}$ compounds; ternary eutectic E_9 with m.p. 680°C has been formed among Bi$_2$B$_8$O$_{15}$, D and B$_2$O$_3$ compounds; ternary eutectic E_{10} with m.p. 730°C has been formed among BaB$_8$O$_{13}$, D and B$_2$O$_3$ compounds; ternary eutectic E_{11} with m.p. 750 °C has been formed among Ba$_2$B$_{10}$O$_{17}$-D- BaB$_8$O$_{13}$ compounds; ternary eutectic E_{12} with m.p. 680°C has been formed among C, Ba$_2$B$_{10}$O$_{17}$ and D compounds; ternary eutectic E_{13} with m.p. 690°C has been formed among Ba$_2$B$_{10}$O$_{17}$, C and BaB$_4$O$_7$ compounds; ternary eutectic E_{14} with m.p. 700°C has been formed among BaB$_4$O$_7$, C and BaB$_2$O$_4$ compounds; ternary eutectic E_{15} with m.p. 645 °C has been formed among G, F and C compounds; ternary eutectic E_{16} with m.p. 615°C has been formed among Bi$_{24}$B$_2$O$_{39}$, F and Bi$_4$B$_2$O$_9$ compounds (Fig18, Table 3).

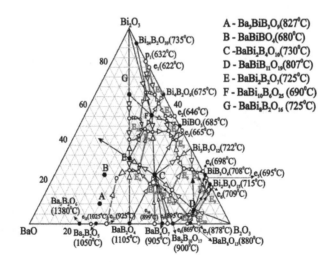

A - Ba$_3$BiB$_3$O$_9$(827°C)
B - BaBiBO$_4$(680°C)
C -BaBi$_2$B$_4$O$_{10}$(730°C)
D - BaBiB$_{11}$O$_{19}$(807°C)
E - BaBi$_2$B$_2$O$_7$(725°C)
F - BaBi$_{10}$B$_6$O$_{25}$ (690°C)
G - BaBi$_8$B$_2$O$_{16}$ (725°C)

Figure 18. Phase diagram of the BaO-Bi$_2$O$_3$-B$_2$O$_3$ system

Point	Composition, mol%			T_m, °C
	BaO	B_2O_3	Bi_2O_3	
E_1	4.5	40.5	55	590
E_2	7.3	45.1	47.6	585
E_3	15	38	47	640
E_4	32.4	41.3	26.3	622
E_5	5.4	52	42.6	610
E_6	9	63	28	675
E_7	3	71	26	680
E_8	3.5	78	18.5	675
E_9	3.2	81.8	15	680
E_{10}	15	81.2	3.8	730
E_{11}	18.9	77.5	3.6	750
E_{12}	22.4	63.2	14.4	680
E_{13}	26.6	54.7	18.7	690
E_{14}	28.8	52	19.2	700
E_{15}	20	32	48	645
E_{16}	4.5	30	65.5	615

Table 3. The temperature and compositions for ternary eutectic points in the BaO-Bi_2O_3-B_2O_3 system

3.2. DTA and X-ray characterisation of ternary stoichiometric glasses and glass ceramics from the BaO-Bi_2O_3-B_2O_3 system

The glasses corresponding to known sixth stoichiometric compounds in the BaO-Bi_2O_3-B_2O_3 system examined in the present study and following glass compositions (mol%) have been melted: $14.28BaO \bullet 7.14Bi_2O_3 \bullet 78.57B_2O_3$ ($BaBiB_{11}O_{19}$), $25BaO \bullet 25Bi_2O_3 \bullet 50B_2O_3$($BaBi_2B_4O_{10}$), $33.33BaO \bullet 33.33Bi_2O_3 \bullet 33.33B_2O_3$($BaBi_2B_2O_7$), $11.11BaO \bullet 55.55Bi_2O_3 \bullet 33.33B_2O_3$ ($BaBi_{10}B_6O_{25}$), $50BaO \bullet 25Bi_2O_3 \bullet 25B_2O_3$ ($BaBiBO_4$) and $60BaO \bullet 10Bi_2O_3 \bullet 30B_2O_3$ ($Ba_3BiB_3O_9$). These glasses DTA curves are shown in Fig. 19, giving the peaks due to the glass transition, crystallization, melting, and liquidus temperatures. The glass characteristics points T_g (glass transition), T_s(glass softening), T_c (peak of exothermal effects connected with crystalline phases crystallizations) and T_m (minimum of endothermic effects associated with these phases melting) observed on DTA curves (Fig. 19, curves 1-6) of all tested powder samples summarized on table 4.

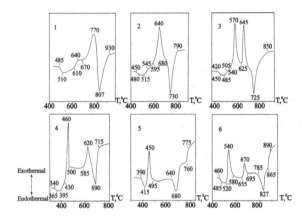

Figure 19. DTA curves (heating rate 10K/min) of glasses corresponding to ternary compounds in the BaO-Bi₂O₃-B₂O₃ system: 1-BaBiB₁₁O₁₉, 2-BaBi₂B₄O₁₀, 3-BaBi₂B₂O₇, 4-BaBi₁₀B₆O₂₅, 5-BaBiBO₄, 6-Ba₃BiB₃O₉.

##	Glass compositions, correspoding to stoichometric compounds	Dilatometric characteristics			DTA characteristics			
		TEC($\alpha_{20\text{-}300}$)•10⁷K⁻¹	T_g,°C	T_s,°C	T_g, °C	T_{cr}, °C	T_m, °C	T_L, °C
1	BaBiB₁₁O₁₉ (glass)	72	498	535	485	640; 770	807	807
	BaBiB₁₁O₁₉ (615°C 24h)	49.8						
2	BaBi₂B₄O₁₀ (glass)	96	445	475	450	545; 640	730	730
	BaBi₂B₄O₁₀ (640°C 24h)	77.9						
3	BaBi₂B₂O₇ (glass)	108	415	455	420	570; 645	725	725
	BaBi₂B₂O₇ (640°C 24h)	96						
4	BaBi₁₀B₆O₂₅ (glass)	99	350	380	340	460; 620	690	690
	BaBi₁₀B₆O₂₅(590°C 24h)	97						
5	BaBiBO₄ (glass)	120	400	450	390	450	680	760
	BaBiBO₄ (570°C 24h)	110.8						
6	Ba₃BiB₃O₉ (glass)	127	460	490	460	540; 670	827	865
	Ba₃BiB₃O₉ (690°C 24h)	109.8						

Table 4. Chemical compositions, DTA (glass transition -T_g, crystallization peak -T_{cr}, melting -T_m, liquidus- T_L) and dilatometric characteristics (glass transition temperature -T_g, softening point - T_s, thermal expansion coefficient -TEC) of BaO-Bi₂O₃-B₂O₃system glasses and crystallaised glasses.

Two exothermic effects were observed on DTA curve of 14.28BaO•7.14Bi₂O₃•78.57B₂O₃ (mol %) glass composition: first weak effect at 640°C and second strong effect at 770°C(Fig.19, curve1). The melting temperature (T_m) is equal to 807°C and corresponding to Egorisheva and Kargin's data [30]. X-ray patterns of this glass crystallization products show one Ba-$BiB_{11}O_{19}$ crystalline phase presence [30], which formed at powder samples crystallization in an temperature interval 640-770°C (Fig.20, curve1). It is possible to assume, that weak exothermic effect at 640°C apparently is connected with pre-crystallisation fluctuations taking place in glass matrix [50]. Diffuse character of second exothermic effect at 770°C testifies about dominating surface crystallisation of the given glass particles.

One sharp exothermic effect at 640°C and sharp endothermic effect at 730°C were observed on DTA curve of $BaBi_2B_4O_{10}$ glass composition (Fig. 19, curve2). The melting temperature (T_m) is equal to 730°C and corresponding to Egorisheva's data [30]. X-ray diffraction patterns of this glass crystallization products show one $BaBi_2B_4O_{10}$ crystalline phase crystallization[30], which formed at glass powder samples crystallization at temperature 640°C (Fig. 20, curve2) and its melting. T_m is equal to 730°C and corresponding to Egorisheva's data [30]. Hardly visible pre-crystallisation fluctuation exothermal effect is observed also at 545°C (Fig. 19, curve2).

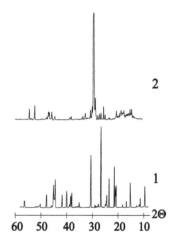

Figure 20. XRD-patterns of the crystallized glasses corresponding to ternary $BaBiB_{11}O_{19}$ (1- 760 °C 24h,cooling in the muffle) and $BaBi_2B_4O_{10}$ (2- 640°C 24h,cooling in the muffle) compounds

On the DTA curves of stoichiometric $BaBi_2B_2O_7$ and $BaBi_{10}B_6O_{25}$ glass compositions observed two exothermal effects at 570 and 645°C for $BaBi_2B_2O_7$ and at 460 and 620°C for $BaBi_{10}B_6O_{25}$ (Fig. 19, curves 3,4). But both compositions have one endothermic effect of melting at 725 and 690°C respectively for $BaBi_2B_2O_7$ and $BaBi_{10}B_6O_{25}$ testifying to one formed crystalline phase melting (Fig. 19, curves 3,4). X-ray data of these samples confirmed monophase crystallizations in each samples (Fig. 21, curves 1,2; Fig.22, curves 1,2).

According to [34] the X-ray powder diffraction patterns of formed $BaBi_2B_2O_7$ crystalline phase at stoichiometric glass composition (33.33BaO•33.33Bi_2O_3• 33.33B_2O_3 mol%) at second exothermic peaks temperature (640°C 24h) was indexed on an orthorhombic cell with following lattice parameters: a=11.818Å, b=8.753 Å, c=7.146Å, cell volume V=739.203 Å, Z=4 (Fig. 21, curve2). XRD-patterns of products of same glass crystallization at 570°C 24h keeps all diffraction lines of its analogue obtained at 640°C 24h (Fig. 21, curve1). Difference is observed only in sharp increasing of intensity (I/I$_o$) of [030] diffraction line from 4 to 43 at high temperature crystallization. That leads to reorientation of crystal structure, decreasing [030] diffraction line and accompanied with occurrence of the second exothermal effect on DTA curve (Fig. 19, curve 3).

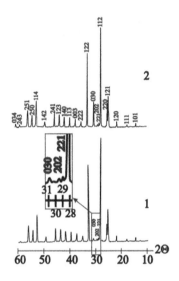

Figure 21. XRD-patterns of the crystallized glasses corresponding to ternary BaBi$_2$B$_2$O$_7$ composition:

1-570°C 24h, cooling in the muffle; 2-640°C 24h, casting in the cold water

X-ray powder diffraction patterns of BaBi$_{10}$B$_6$O$_{25}$ single crystals was indexed on an orthorhombic cell with following lattice parameters: a=6.434Å, b=11.763 Å, c=29.998Å, cell volume V=2270.34 Å, Z=8 [34]. XRD-patterns of products of same compositions (11.11BaO•55.55Bi$_2$O$_3$• 33.33B$_2$O$_3$ mol%) glass crystallization at both exothermal effects (420°C 24h and 620°C 24h) have all diffraction lines of the BaBi$_{10}$B$_6$O$_{25}$ single crystals (Fig. 22, curves1-3). Naturally, BaBi$_{10}$B$_6$O$_{25}$ single crystal has well generated planes and clear observed diffraction lines on XRD-patterns in contrast to crystalline phases formed at same composition glasses crystallization. However, the most intensive diffraction line (I/I$_o$=100) of single crystals is [133], whereas products of glass crystallizations have [203] strongest diffraction line and [133] diffraction line became 5-10 times less (Fig. 22, curves1-3). Now it is difficult to us only on the basis of XRD-patterns analysis of glass crystallizations products to assume the nature of the second exothermal effect at 620°C on DTA curve of BaBi$_{10}$B$_6$O$_{25}$ of glass composition (Fig.19, curve 4). Their XRD-patterns are identical each other and to single crystals, but contain slightly quantity of not indexed reflexes, which are absent in X-ray powder diffraction patterns of $BaBi_{10}B_6O_{25}$ single crystals (Fig. 22, curves1-3).

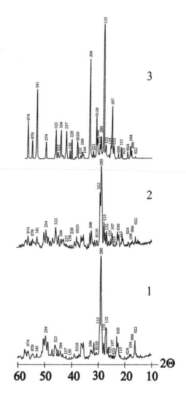

Figure 22. XRD-patterns of the crystallized glasses corresponding to ternary $BaBi_{10}B_6O_{25}$ composition:(1-460°C 24h,cooling in the muffle; 2-620°C 24h, casting in the cold water) and $BaBi_{10}B_6O_{25}$ single crystals (3).

The DTA curve of $50BaO \bullet 25Bi_2O_3 \bullet 25B_2O_3$ mol% ($BaBiBO_4$) glass composition contain exothermal effect of glass crystallization at 450°C and endothermic effect of this crystalline phase melting at 680°C (Fig. 19, curv.5). X-ray diffraction patterns of this glass crystallization products show one $BaBiBO_4$ crystalline phase formation at glass powder samples crystallization at temperature interval 450-640°C(Fig. 23, curve 1), which completely correspond to Barbier with co-authors data [29]. A second endothermic effect within the interval of 745-775°C with minimum at 760°C is associated with $BaBiBO_4$ incongruent melting (Fig. 19, curve 5).

We have revealed also, that the crystalline $BaBiBO_4$ compound is melted incongruently at 680°C with the melt and crystalline $BaBiO_3$ formation (Fig. 23, curve 2). The $BaBiO_3$ crystalline phase was observed on XRD-patterns of thermal treated at 720 °C and fast freeze in cold water products and identified according to X-ray database [43, fail # 01-074-7523]. The dissolution of this $BaBiO_3$ phase in a melt leads to the appearance on a DTA curve the second endothermic effect in an interval 745-775°C (Fig. 19, curve 5). Above 775°C we have glassforming $BaBiBO_4$ composition melt without presence of any crystalline phase.

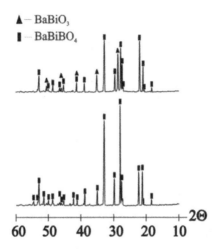

Figure 23. XRD-patterns of the crystallized glasses corresponding to ternary $BaBiBO_4$ composition:

1-450°C 24h, cooling in the muffle; 2-720°C 3h, casting in the cold water

Two exothermal effects of glass crystallization at 540 and 670°C and one endothermic effect of crystalline phase melting at 827°C are seen on the DTA curve of the $60BaO \bullet 10Bi_2O_3 \bullet 30B_2O_3$ mol% ($Ba_3BiB_3O_9$) glass composition (Fig. 19, curve 6). X-ray diffraction patterns of this glass crystallization products at 540 and 670°C show one $Ba_3BiB_3O_9$ crystalline phase formation (Fig. 24, curves 1,2) at glass powder samples crystallization at 540 and 670°C and fully correspond to Egorisheva with co-authors data, which synthesized for the first time and have describe $Ba_3BiB_3O_9$ compound [31]. However, we didn't indicate polymorphic transition of $Ba_3BiB_3O_9$ at 850°C as reported in [31]. Presence of second endothermic effect within the interval 840-890°C with minimum at 865°C is associated with $Ba_3BiB_3O_9$ incongruent melting (Fig. 19, curve 6). We have revealed that the crystalline $Ba_3BiB_3O_9$ compound is melted incongruently at 827°C with the glass forming melt and crystalline phase formation (Fig. 24, curve 3). The $Ba_2B_2O_5$ crystalline phase was observed in amorphous matrix on XRD-patterns of thermal treated at 830 °C and fast freeze in cold water products and identified according to X-ray database [43, fail # 024-0087]. For clear $Ba_2B_2O_5$ observation on XRD-patterns the preliminary crystallized at 670°C 24h sample have been exposed at 830°C 3 h (Fig. 23, curve3). Dissolution of this $Ba_2B_2O_5$ phase in a melt leads to the appearance on a DTA curve the second endothermic effect in an interval 840-890°C (Fig. 19, curve 6). Above 890°C we have glass-forming $Ba_3BiB_3O_9$ composition melt without presence of any crystalline phase at cooling rate 10^2 K/s.

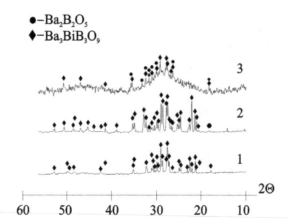

Figure 24. XRD-patterns of the crystallized glasses corresponding to ternary 6BaBi3BO$_4$ composition:1.540°C 24h, cooling in the muffle; 2-670°C 24h, casting in the cold water; 3-670°C 24h+830°C 3 h, casting in the cold water

3.3. TEC study of the stoichiometric compositions glasses in the BaO-Bi$_2$O$_3$-B$_2$O$_3$ system

The isolines diagram of BaO-Bi$_2$O$_3$-B$_2$O$_3$ system glasses TEC values is given on Fig. 25. It is clear observed common regularity, that the increase of barium and bismuth oxides amounts in glasses of binary BaO-Bi$_2$O$_3$ and Bi$_2$O$_3$-B$_2$O$_3$ systems leads to increase TEC of glasses. The same tendency is observed for glasses of ternary system: joint presence of BaO and Bi$_2$O$_3$ and increase their amounts leads to increase glasses TEC values from 70 to 127•10^{-7}K^{-1} (Fig. 25).

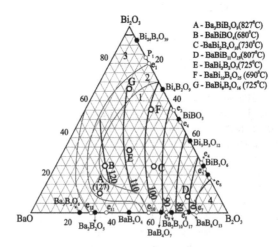

Figure 25. BaO-Bi$_2$O$_3$-B$_2$O$_3$ system's glasses TEC (α_{20-300}•10^{-7}K^{-1}) values isolines

The high boron content glass composition corresponds to $BaBiB_{11}O_{19}$ ($14.28BaO \bullet 7.14Bi_2O_3 \bullet 78.57B_2O_3$ mol %) have TEC=$72 \bullet 10^{-7}K^{-1}$ and T_g=498°C calculated from dilatometric curve (Table 4). Reduction the B_2O_3 amount together with increasing of BaO and Bi_2O_3 amounts in glass compositions leads to increase TEC and reduction Tg values: for glass composition $25BaO \bullet 25Bi_2O_3 \bullet 50B_2O_3$ mol % ($BaBi_2B_4O_{10}$) TEC=$96 \bullet 10^{-7}K^{-1}$ and T_g= 445°C; $33.3BaO \bullet 33.3 Bi_2O_3 \bullet 33.3B_2O_3$ mol %($BaBi_2B_2O_7$) TEC=$108 \bullet 10^{-7}K^{-1}$ and T_g=415°C; $11.1BaO \bullet 55.5Bi_2O_3 \bullet 33.3B_2O_3$ mol %($BaBi_{10}B_6O_{25}$) TEC=$97 \bullet 10^{-7}K^{-1}$ and T_g=350°C; $16.67BaO \bullet 66.67Bi_2O_3 \bullet 16.67B_2O_3$ mol%($BaBi_8B_2O_{16}$) TEC=$110 \bullet 10^{-7}K^{-1}$ and T_g=415°C. However, for $50BaO \bullet 25 Bi_2O_3 \bullet 25B_2O_3$ mol % ($BaBiBO_4$) and $60BaO \bullet 10Bi_2O_3 \bullet 30B_2O_3$ mol % ($Ba_3BiB_3O_9$) glass compositions simultaneous increase both TEC and T_g values were observed: TEC=$120 \bullet 10^{-7}K^{-1}$ and T_g=400°C; TEC=$127 \bullet 10^{-7} K^{-1}$ and T_g=460°C respectively for $BaBiBO_4$ and $Ba_3BiB_3O_9$ (Fig. 25, table4).

TEC values of crystallized glasses corresponding to the ternary barium bismuth borates given in Table 4. Crystallized barium bismuth borate glass samples have TEC values lower, than initial glasses and equals to: $49 \bullet 10^{-7}K^{-1}$ for $BaBiB_{11}O_{19}$ sample (750°C 24h), $78 \bullet 10^{-7}K^{-1}$ for $BaBi_2B_4O_{10}$ sample (630°C 24 h), $96 \bullet 10^{-7} K^{-1}$ for $BaBi_2B_2O_7$ sample (640°C 24h), $97 \bullet 10^{-7}K^{-1}$ for $BaBi_{10}B_6O_{25}$ sample (610°C 24h), $110 \bullet 10^{-7} K^{-1}$ for $BaBiBO_4$ sample (450°C24h) and $109 \bullet 10^{-7} K^{-1}$ for $Ba_3BiB_3O_9$ sample (690°C 24h). The same tendency, as well as for their glassy analogues, is observed for crystallized glass samples: increase of barium and bismuth oxides amounts in ternary compounds leads to their TEC values increase.

4. Ferroelectric properties of new ternary $BaBi_2B_2O_7$ and $BaBi_{10}B_6O_{25}$ stoichometric compositions glass ceramics.

The ferroelectric (polarization - electric field) hysteresis, is a defining property of ferroelectric materials. Thus, the most widely studied characteristics of ferroelectric hysteresis were those of interest for this particular application: the value of the switchable polarization (the difference between the positive and negative remanent polarization, $P_R - (-P_R)$), dependence of the coercive field Ec on sample thickness, decrease of remanent or switchable polarization with number of switching cycles, polarization imprint, endurance, retention [51].

Electric field induced polarization (P) and remanent polarization(P_r) were measured at room temperature for $BaBi_2B_2O_7$ and $BaBi_{10}B_6O_{25}$ glass tape samples crystallized using various regimes (Fig. 26).

a. $BaBi_2B_2O_7$ glass tape sample of 0.07 mm in thickness crystallized at 450°C 24h, $2P_r$ = 0.15 $\mu C/cm^2$;

b. $BaBi_{10}B_6O_{25}$ glass tape sample of 0.06 mm in thickness crystallized at 380 °C 12h, $2P_r$ = 0.32 $\mu C/cm^2$;

c. $BaBi_{10}B_6O_{25}$ glass tape sample of 0.06 mm in thickness crystallized at 410°C 12h, $2P_r$ = 0.62 $\mu C/cm^2$;

d. $BaBi_{10}B_6O_{25}$ glass tape sample of 0.05 mm in thickness crystallized at 410°C 24h, $2P_r = 0.9$
     $\mu C/cm^2$

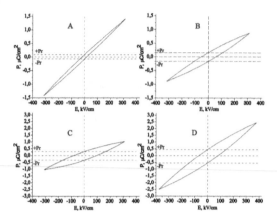

Figure 26. Dependence of polarization (P) on electric field (E) for crystallized stoichiometric glass compositions:

Linear P–E curves are observed up to fields of 40-120 kV/cm for all measured samples with
thickness 0.05-0.07mm. The polarization becomes nonlinear with increasing of applied electric field, and at 140-380 kV/cm the remanent polarization $2P_r$ values were found 0.15
$\mu C/cm^2$ for the $BaBi_2B_2O_7$ (Fig. 26, A). The remanent polarization $2P_r$ value for $BaBi_{10}B_6O_{25}$
crestallized glass tape samples encreasing with termal treatment temperature from 0,32 to 0,
64 $\mu C/cm^2$ (Fig. 26, B & C) and time (Fig. 26, D). The highest remanent polarization value
($2P_r$=0.9 $\mu C/cm^2$) has $BaBi_{10}B_6O_{25}$ glass tape sample crystallized at 410°C 24h (Fig.26, D). According to obtained results it is possible to conclude that samples are ferroelectrics.

5. Discussion

The pricipial diference of our methodology from traditional is a glass samples using as initial testing substance for phase diagram of very complex ternary $BaO-Bi_2O_3-B_2O_3$ system
construction. It is a very effective method, due possibility to indicate temperature intervals
of all processes taking place in glass samples: glass transition, crystallization, quantity of
formed crystalline phases and their melting. Whereas, samples prepared by traditional solid
phase synthesis are less informative and often lose a lot of information. Super cooling technique created by our group allowed us both to expand borders of glass formation and to
have enough quantity samples for DTA and X-ray investigations and $BaO-Bi_2O_3-B_2O_3$ system phase diagram construction (Fig.2).

The region of stable glasses includes the binary compounds BaB_4O_7, $Ba_2B_{10}O_{17}$, BaB_8O_{13},
$Bi_4B_2O_9$, $BiBO_3$, $Bi_3B_5O_{12}$, BiB_3O_6, and $Bi_2B_8O_{15}$ in the $BaO-B_2O_3$ and $Bi_2O_3-B_2O_3$ systems (Fig.

2). Binary BaB$_4$O$_7$, Ba$_2$B$_{10}$O$_{17}$, and BaB$_8$O$_{13}$ barium borates have melting temperatures (Tm) of 910, 905, and 890°C and can be found between the eutectics e$_7$, e$_8$, e$_9$, and e$_{10}$ with Tm = 878, 869, 895, and 899°C, respectively. The transition to a crystallization field of barium metaborate is accompanied by a sharp increase of liquidus temperature (T$_L$) and a decrease of the glass forming ability of the melts. The final compound, forming a stable glass, contains ~43 mol %BaO and has T$_L$ = 950°C. Compounds BaB$_2$O$_4$ and Ba$_2$B$_2$O$_5$, having higher Tm, which are 1095 and 1050°C, respectively [18, 21, 22], are found in the region of the compounds obtained in the form of glasses with a cooling rate of (10^3–10^4) K/s. Glass formation in the BaO–B$_2$O$_3$ binary system is limited by the eutectic e$_{12}$ with Tm = 1025°C (Fig. 2) because of the sharp increase of the liquidus temperature during the transition to a field of crystallization of the Ba$_3$B$_2$O$_6$ compound (Tm = 1383°C). BaB$_2$O$_4$ (Tm = 1095°C) is the dominating compound in the system and does not form stable glasses. Its considerable crystallization field narrows the region of stable glasses in the ternary system, which is only restricted by compounds with T$_L$ ~ 950°C (Fig. 2).

Binary bismuth borates Bi$_4$B$_2$O$_9$, BiBO$_3$, Bi$_3$B$_5$O$_{12}$, BiB$_3$O$_6$, and Bi$_2$B$_8$O$_{15}$ have Tm 675, 685, 722, 708, and 715°C can be found between the eutectics e$_1$, e$_2$, e$_3$, e$_4$, e$_5$, e$_6$ with a Tm of 622, 646, 665, 698, 695, and 709°C, respectively [2, 34]. The region of stable glasses in the Bi$_2$O$_3$–B$_2$O$_3$ system is limited by a compound containing ~70%mol Bi$_2$O$_3$ and having T$_L$ = 670°C. Compounds that are found in the range of 70–80mol% Bi$_2$O$_3$ (before the e$_1$ point) are obtained in the form of glasses during cooling at a rate of ~10^2 K/s. During the transition to the crystallization field of Bi$_{24}$B$_2$O$_3$ and Bi$_2$O$_3$, T$_L$ increases to 825°C (Tm of Bi$_2$O$_3$). Glasses in this part of the system are obtained during melt cooling at a rate of (10^3–10^4) K/s. These compounds have a low liquidus temperature; however, the structure factor essentially influences their glass forming ability, not allowing glass formation at low melt cooling rates.

Six ternary compounds are known in the BaO–Bi$_2$O$_3$–B$_2$O$_3$ system: Ba$_3$BiB$_3$O$_9$, BaBiBO$_4$, BaBi$_2$B$_4$O$_{10}$, and BaBiB$_{11}$O$_{19}$ synthesized by Barbie and Egorysheva in 2005–2006, and BaBi$_2$B$_2$O$_7$ and BaBi$_{10}$B$_6$O$_{25}$ revealed by our research group in 2008–2009. BaBi$_2$B$_4$O$_{10}$, BaBiB$_{11}$O$_{19}$, BaBi$_2$B$_2$O$_7$, and BaBi$_{10}$B$_6$O$_{25}$ melt congruently at 730, 807, 725, and 695°C respectively, and BaBiBO$_4$ melts incongruently at 680°C and has T$_L$ = 760 °C. All these five ternary compounds along with the eutectics formed between each other and with binary barium and bismuth borates form a "plateau" with low T$_L$, which is responsible for the formation of the region of stable glasses in the ternary system BaO–Bi$_2$O$_3$–B$_2$O$_3$.

Compounds joining low temperature eutectic e$_1$ (622°C) [2] in the Bi$_2$O$_3$–B$_2$O$_3$ binary system and the eutectics e$_{13}$ (~790°C) and e$_{14}$ (~750°C) in the BaO–Bi$_2$O$_3$ binary system 26 - 28] form glasses only at higher cooling rates of their melts (10^3–10^4) K/s. Glass formation in the BaO–Bi$_2$O$_3$ binary system stops at 45 mol% BaO content (T$_L$ ~930°C) [26 - 28]. Along with the factor of the liquidus temperature [52], a considerable contribution to the glass formation of the pointed compositions is made by the structural factor of the melt. The combination of the structural factors of the melt and the liquidus temperature is also considerable during the transition to the vitreous state of the compositions, which are found in the crystallization fields of BaB$_2$O$_4$ and Ba$_2$B$_2$O$_5$, where they show the tendency towards glass formation only at high rates of melt cooling.

There are very stable congruent melted binary BaB_2O_4 and ternary $BaBi_2B_4O_{10}$, $BaBiB_{11}O_{19}$, and $BaBi_{10}B_6O_{25}$ compounds in the studied ternary $BaO-Bi_2O_3-B_2O_3$ system. They have dominating positions in ternary diagram and occupied the biggest part of it (Fig.18). Mutual influence of these compounds and other binary and ternary compounds (BaB_4O_7, $Ba_2B_{10}O_{17}$, BaB_8O_{13}, $Bi_4B_2O_9$, $BiBO_3$, $Bi_3B_5O_{12}$, BiB_3O_6, $Bi_2B_8O_{15}$, $BaBi_2B_2O_7$, and $BaBi_8B_2O_{16}$) lead to formation of sixteen revealed at present time ternary eutectics (Fig.18& Table 3), which have essential influence on liquidus temperature decrease and to assist in glass formation. Ternary $BaBi_2B_4O_{10}$ compound forms eight eutectics with binary and ternary compounds, its neighbors: E_4(622°C), E_3(640°C), E_2(585°C), E_5(610°C), E_6(675°C), E_{12}(680°C), E_{13} (690°C), and E_{14} (700°C) (Fig.18& Table 3). $BaBiB_{11}O_{19}$ compound forms seven eutectics with its neighbors: E_6 (675°C), E_7 (680°C), E_8 (675°C), E_9 (680°C), E_{10} (730°C), E_{11} (750°C),and E_{12} (680°C) (Fig.18& Table 3). $BaBi_{10}B_6O_{25}$ compound forms five eutectics with its neighbors: E_1 (590°C), E_2 (585°C), E_3 (640°C), E_{15} (645°C), and E_{16} (615°C) (Fig.18& Table 3). Determined ternary eutectics together with binary eutectics e_1, e_2, e_3, e_4, e_5, and e_6 of $Bi_2O_3-B_2O_3$ system have allowed to outline the fields of binary $Bi_4B_2O_9$, $BiBO_3$, $Bi_3B_5O_{12}$, BiB_3O_6, and $Bi_2B_8O_{15}$ bismuth borates crystallisation, as well as together with binary eutectics e_7, e_8, e_9, and e_{10} of $BaO-B_2O_3$ system have allowed to outline the fields of binary BaB_4O_7, $Ba_2B_{10}O_{17}$, BaB_8O_{13} barium borates and partly BaB_2O_4 crystallization on the $BaO-Bi_2O_3-B_2O_3$ system phase diagram (Fig.18).

The clear correlation between glass forming and phase diagrams has been observed in studied system. The glass melting temperature and level of glass formation depending on the cooling rate of the studied melts are in good conformity with boundary curves and eutectic points (Fig.2& 18).

The phase diagram of the well known binary $Bi_2O_3-B_2O_3$ system has been corrected in the interval between the $Bi_4B_2O_9$ and $Bi_3B_5O_{12}$ compounds. The eutectic composition, $48.5Bi_2O_3 \cdot 51.5B_2O_3$ (mol%), between $BiBO_3$ and $Bi_3B_5O_{12}$, with m.p. 665±5°C, has been determined. It is shown that the compound $BiBO_3$ is congruently melting with a m.p. of 685±5°C.

The next unexpected results were obtained at phase diagram construction: two new ternary $BaBi_2B_2O_7$ and $BaBi_{10}B_6O_{25}$ compounds have been revealed at the same glass compositions crystallisation. X-ray characteristics of the new ternary compound $BaBi_2B_2O_7$, synthesized at the $33.33BaO \cdot 33.33Bi_2O_3 \cdot 33.33B_2O_3$ (mol%) glass composition crystallization at 640°C, 20 h. The x-ray powder diffraction patterns of $BaBi_2B_2O_7$ could be indexed on an orthorhombic cell with lattice parameters as follows: $BaBi_2B_2O_7$ a=11.818 Å, b=8.753 Å, c=7.146 Å, cell volume V=739.203 Å^3, Z=4.

Single crystals of $BaBi_{10}B_6O_{25}$ were obtained by cooling of a melt with the stoichiometric composition. Glass powder of composition $11.11BaO \cdot 55.55Bi_2O_3 \cdot 33.33B_2O_3$ (mol%) was heated in a quartz glass ampoule up to 750°C at a rate 10 K/min. After 2 h at high temperature, the melt was cooled at a rate 0.5 K/h. Single crystals with sizes up to 1.66×0.38×0.19 mm^3 were grown. The x-ray powder diffraction patterns of $BaBi_{10}B_6O_{25}$ could be indexed on an orthorhombic cell with lattice parameters as follows: a=6.434 Å, b=11.763 Å, c=29.998 Å, cell volume V=2270.34 Å^3, Z=8.

Common regularities of bulk glass samples TEC changes in studied BaO-Bi$_2$O$_3$-B$_2$O$_3$ system have been determined: the increase of barium and bismuth oxides amounts in glasses of binary BaO-Bi$_2$O$_3$ and Bi$_2$O$_3$-B$_2$O$_3$ systems leads to increase TEC of glasses. The same tendency is observed for glasses of ternary BaO-Bi$_2$O$_3$-B$_2$O$_3$system: joint presence of BaO and Bi$_2$O$_3$ and increase their amounts leads to increase glasses TEC values from 70 to 127•10^{-7}K^{-1} (Fig. 25).

Crystallized barium bismuth borate glass samples have TEC values lower, than initial glasses and equals to: 49•10^{-7}K^{-1} for BaBiB$_{11}$O$_{19}$ sample (750°C 24h), 78•10^{-7}K^{-1} for BaBi$_2$B$_4$O$_{10}$ sample (630°C 24 h), 96•10^{-7} K^{-1} for BaBi$_2$B$_2$O$_7$ sample (640°C 24h), 97•10^{-7}K^{-1} for BaBi$_{10}$B$_6$O$_{25}$ sample (610°C 24h), 110•10^{-7} K^{-1} for BaBiBO$_4$ sample (450°C24h) and 109•10^{-7} K^{-1} for Ba$_3$BiB$_3$O$_9$ sample (690°C 24h). The same tendency, as well as for their glassy analogues, is observed for crystallized glass samples: increase of barium and bismuth oxides amounts in ternary compounds leads to their TEC values increase.

Electric field induced polarization (P) and remanent polarization (P$_r$) were measured at room temperature for BaBi$_2$B$_2$O$_7$ and BaBi$_{10}$B$_6$O$_{25}$ glass tape samples crystallized at various regimes. All tested samples shown loop of hysteresis.

Linear *P–E* curves are observed up to fields of 40-120 kV/cm for all measured samples with thickness 0.05-0.07mm. The polarization becomes nonlinear with an increase of applied electric field, and at 140-400 kV/cm the remanent polarization 2P$_r$ values were found 0.15 μC/cm^2 for the BaBi$_2$B$_2$O$_7$ (Fig.26, A), and 0.32- 0.9 μC/cm^2 for the BaBi$_{10}$B$_6$O$_{25}$ (Fig.26, B-D), crystallized glass tape samples. According to obtained results it is possible to conclude that all tested samples are ferroelectrics.

5. Conclusion

Effective way of new system investigation and new compounds and characteristic points revealing via simultaneous glass forming and phase diagrams construction have been shown. Phase diagram of the ternary BaO-Bi$_2$O$_3$-B$_2$O$_3$ system have been constructed for the first time us result of fourteen pseudo-binary systems and sections phase diagrams investigations.

The phase diagram of the well known binary Bi$_2$O$_3$–B$_2$O$_3$ system has been corrected in the interval between the Bi$_4$B$_2$O$_9$ and Bi$_3$B$_5$O$_{12}$ compounds. The eutectic composition, 48.5Bi$_2$O$_3$•51.5B$_2$O$_3$ (mol%), between BiBO$_3$ and Bi$_3$B$_5$O$_{12}$, with m.p. 665±5°C, has been determined. It is shown that the compound BiBO$_3$ is congruently melting with a m.p. of 685±5°C.

Two new ternary BaBi$_2$B$_2$O$_7$ and BaBi$_{10}$B$_6$O$_{25}$ compounds have been revealed at the same glass compositions crystallisation. The new ternary compound BaBi$_2$B$_2$O$_7$ synthesized at the 33.33BaO•33.33Bi$_2$O$_3$•33.33B$_2$O$_3$ (mol%) glass composition crystallization at 640°C, 20 h. The x-ray powder diffraction patterns of BaBi$_2$B$_2$O$_7$ could be indexed on an orthorhombic cell with lattice parameters as follows: BaBi$_2$B$_2$O$_7$ a=11.818 Å, b=8.753 Å, c=7.146 Å, cell volume V=739.203 Å^3, Z=4.

Single crystals of BaBi$_{10}$B$_6$O$_{25}$ were obtained by cooling of a melt with the stoichiometric composition. Glass powder of composition 11.11BaO•55.55Bi$_2$O$_3$•33.33B$_2$O$_3$ (mol%) was heated in a quartz glass ampoule up to 750°C at a rate 10 K/min. After 2 h exposition at high temperature, the melt was cooled at a rate 0.5 K/h. Single crystals with sizes up to $1.66 \times 0.38 \times 0.19$ mm^3 were grown. The x-ray powder diffraction patterns of $BaBi_{10}B_6O_{25}$ could be indexed on an orthorhombic cell with lattice parameters as follows: a=6.434 Å, b=11.763 Å, c=29.998 Å, cell volume V=2270.34 Å^3, Z=8.

Ternary $BaBi_2B_4O_{10}$, $BaBiB_{11}O_{19}$, and $BaBi_{10}B_6O_{25}$ compounds have dominating positions in ternary diagram and occupied the biggest part of it. Mutual influence of these and other binary and ternary compounds (BaB_4O_7, $Ba_2B_{10}O_{17}$, BaB_8O_{13}, $Bi_4B_2O_9$, $BiBO_3$, $Bi_3B_5O_{12}$, BiB_3O_6, $Bi_2B_8O_{15}$, $BaBi_2B_2O_7$, and $BaBi_8B_2O_{16}$) lead to formation of sixteen ternary eutectics, which have essential influence on liquidus temperature decrease and to assist in glass formation.

The clear correlation between glass forming and phase diagrams has been observed: glass melting temperature and level of glass formation depending on the cooling rate of the studied melts are in good conformity with boundary curves and eutectic points.

Common regularities of bulk glass samples TEC changes in studied BaO-Bi_2O_3-B_2O_3 system have been determined: the increase of barium and bismuth oxides amounts in glasses of binary BaO-Bi_2O_3 and Bi_2O_3-B_2O_3 systems and their joint amounts increasing in ternary compositions leads to increase glasses TEC values from 70 to $127 \bullet 10^{-7}K^{-1}$.

Crystallized barium bismuth borate glass samples have TEC values lower, than initial glasses. Increase of barium and bismuth oxides amounts in ternary compounds leads to their TEC values increasing from 49 to $109 \bullet 10^{-7}K^{-1}$.

Electric field induced polarization (P) and remanent polarization (P$_r$) were measured at room temperature for $BaBi_2B_2O_7$ and $BaBi_{10}B_6O_{25}$ glass tape samples crystallized at various regimes. The remanent polarization 2P$_r$ values were found 0.15 µC/cm^2 for the $BaBi_2B_2O_7$, and 0.32- 0.9 µC/cm^2 for the $BaBi_{10}B_6O_{25}$ crystallized glass tape samples. According to obtained results it is possible to conclude that all tested samples are ferroelectrics.

Acknowledgements

This work was supported by the International Science and Technology Center (Projects # A-1591). Author is very gratefull to all project team and to Dr. Rafael Hovhannisyan , Dr. Nikolay Knyazyan and Prof. Heli Jantunen for effective cooperation and fruitful discussions.

Author details

Martun Hovhannisyan*

Address all correspondence to: eni_arm@yahoo.com

Address all correspondence to: raf.hovhannisyan@yahoo.com

Address all correspondence to: martun_h@yahoo.com

«ENI» Institute of Electronic Materials LTD, Armenia

References

[1] The American Ceramic Society & National Institute of Standards and Technology [ACerS & NIST]. (2004). Phase Equilibria Diagrams. *CD-ROM Database, Version 3.0*, 157498215.

[2] Levin, E. M., & Mc Daniel, C. L. (1962). The System Bi$_2$O$_3$-B$_2$O$_3$. *J. Am.Cer.Soc*, 45(8), 355-360.

[3] Masurin, O. V., et al. (1979). Properties of Glasses and Glass-Forming Melts. Handbook., In Russian, v.III, part 2, L.,, Nauka Press.

[4] Muehlberg, M., Burianek, M., Edongue, H., & Poetsch, C. (2002). Bi$_4$B$_2$O$_9$ crystal growth and some new attractive properties. *J. Cryst. Growth*, 237, 740-744.

[5] Becker, P., & Bohaty, L. (2003). Linear electro-optic properties of bismuth triborate, BiB$_3$O$_6$ (BIBO). *Phys. Chem. Glasses*, 44, 212-214.

[6] Ihara, R., Honma, T., Benino, Y., Fujiwara, T., & Komatsu, T. (2004). Second-order optical nonlinearities of metastable BiBO3 phases in crystallized glasses. *Optical Materials*, 27, 403-408.

[7] Oprea, I. I. (2005). Optical Properties of Borate Glass- Ceramic. *Dissertation zur Erlangung des Grades Doktor der Naturwissenschaften, Osnabruck University, Germany.*

[8] Pottier , M. J. (1974). Mise en evidence d'un compose BiBO3 et de son polymorphisme par spectroscopie vibrationnelle. *Bull. Soc. Chim. Belg.*, 83, 235-238.

[9] Becker, P., Liebertz, J., & Bohatý, L. (1999). Top-seeded growth of bismuth triborate, BiB$_3$O$_6$. *J.Cryst.Growth*, 203, 149-155.

[10] Becker, P., & Froehlich, R. (2004). Crystal growth and crystal structure of metastable bismuth ortoborate BiBO$_3$. *Z. Naturforsch. B*, 59, 256-258.

[11] Zargarova, M. I., & Kasumova , M. F. (2004). Liquidus surface of the system ZnO - Bi$_2$O$_3$-B$_2$O$_3$ projection. *Zh. Neorg. Mater*, In Russian, 26(8), 1678-1681.

[12] Kargin, Yu. F., Zhereb, V. P., & Egorysheva, A. V. (2002). Phase diagram of metastable stats of the Bi$_2$O$_3$-B$_2$O$_3$ system. *Zh. Neorg. Khim*, 47(8), 1362-1364, In Russian.

[13] Barbier, J., Penin, N., & Cranswick, L. M. (2005). Melilite-Type Borates Bi$_2$ZnB$_2$O$_7$ and CaBiGaB$_2$O$_7$. *Chem. Mater.*, 17, 3130-3136.

[14] Barbier, J., & Cranswick, L. M. (2006). The Non-Centrosymmetric Borate Oxides, MBi$_2$B$_2$O$_7$(M=Ca,Sr). *Solid State Chem*, 179, 3958-3964.

[15] Egorisheva, A. V., Skorikov, V. M., & Volodin, V. D. (2008). Calcium Bismuth Borates in the CaO-Bi$_2$O$_3$-B$_2$O$_3$ system. Zh.Nerg.Mater. In Russian)., 44(1), 76-81.

[16] Kargin, Yu. F., Ivicheva, S. N., Komova, M. G., & Krut'ko, V. A. (2008). Subsolidus Phase Relation in the SrO-Bi$_2$O$_3$-B$_2$O$_3$ system. *Zh.Nerg.Khim*, 53(3), 474-478, In Russian.

[17] Barbier, J., Davis, L. J. M., Goward, G. R., & Cranswick, L. M. D. (2009). Ab initio structure determination of SrBi$_2$OB$_4$O$_9$ by powder X-ray/neutron diffraction and NMR spectroscopy. *Powder Diffr*, 24(1), 35-40.

[18] Levin, E. M., & Mc Murdie, H. F. (1949). The System BaO-B2O3. *J. Res.Nat.Bur.Standards*, 42(2), 131-138.

[19] Levin, E. M., & Ugrinic, G. M. (1953). The System Barium Oxide-Boric Oxide-Silica. *J. Res. Nat.Bur. Stand*, 51, 37-56.

[20] Green, C. H., & Wahler, R. L. (1969). Crystallization of Barium Aluminum Borate Glasses-Rhodium as a Specific Catalyst. *In:Kinetics of reactions in ionic systems. Proc. Intern. Symp. 18-23 IV, New York*, 545-553.

[21] Hubner, K. H. (1969). Ueber die Borate 2BaO•5B$_2$O$_3$, tief-BaO•B$_2$O$_3$, 2BaO•B$_2$O$_3$ und 4BaO•B$_2$O$_3$. *Neues Jahrb. Mineral., Monatsch*, 335-343.

[22] Hovhannisyan, R. M. (2006). Binary alkaline-earth borates: phase diagrams correction and low thermal expansion of crystallized stoichiometric glass compositions. *Phys.Chem.Glasses: Eur. J.Glass Sci. Technol. B*, 47(4), 460-465.

[23] Crichton, S. N., & Tomozawa, M. (1997). Prediction of phase separation in binary borate glasses. *J. Non-Crystalline Solids*, 215, 244-251.

[24] Hageman, V. B. M., & Oonk, H. A. J. (1987). Liquid immiscibility in the B$_2$O$_3$-MgO, B$_2$O$_3$-CaO, B$_2$O$_3$-SrO and B$_2$O$_3$-BaO systems. *Physics and Chemistry of Glasses*, 28, 183-187.

[25] Klinkova, L. A., Nikolaychik, V. I., Barkovskiy, N. V., & Fedotov, V. K. (1999). Phase relations in the system Ba-Bi-O (20-80 mol% BiO$_{1.5}$) at p$_{O2}$=0.01, 0,21 and 1 atmosphere. *Zh.Neorg. Khim In Russian).*, 44(11), 1774-1782.

[26] Mueller, K., Meen, J. K., & Elthon, D. (1999). System BaO-Bi$_2$O$_3$ in O$_2$. *In ACerS & NIST, 2004*.

[27] Shevchuk, A. V., Skorikov, V. M., Kargin, Yu. F., & Konstantinov, V. V. (1985). The system BaO-Bi$_2$O$_3$. *Zh.Neorg. Khim*, 30(6), 1519-1522, In Russian.

[28] Witler, D., & Roth, R. S. (2004). System BaO-Bi2O3 in Air,. *in Phase Equilibria Diagrams, CD-ROM Database, A CerS-NITS*.

[29] Barbier, J., Penin, N., Denoyer, A., & Cranswic, L. M. D. (2005). BaBiBO$_4$, a novel non-centrosymmetric borate oxide. *Solid State Science*, 7, 1055-1061.

[30] Egorysheva, A.V, & Kargin Yu, F. (2006). Phase equilibrium in the sub-solidus parts of the Bi$_2$O$_3$-BaB$_2$O$_4$-B$_2$O$_3$ system. *Zh.Neorg. Khim ,In Russian*, 51(7), 1185-1189.

[31] Egorysheva, A.V, Skorikov, V.M, Volodin, V.D, Myslitskiy, O.E, & Kargin, Yu, F. (2006). Phase equilibrium in the BaO-B$_2$O$_3$-Bi$_2$O$_3$ system. *Zh.Neorg. Khim ,In Russian*, 51(12), 2078-2082.

[32] Bubnova, R. S., Krivovichev, S. V., Filatov, S. K., Egorysheva, A. V., & Kargin, Y. F. (2007). Preparation, crystal structure and thermal expansion if a new bismuth barium borate, BaBi$_2$B$_4$O$_{10}$. *Journal of Solid State Chemistry*, 180, 596-603.

[33] Elwell, D., Morris, A. W., & Neate, B. W. (1972). The flux system BaO-B$_2$O$_3$-Bi$_2$O$_3$. *J. of Crystal Growth*, 16, 67-70.

[34] Hovhannisyan, M. R., et al. (2009). A study of the phase and glass forming diagrams of the BaO-Bi$_2$O$_3$-B$_2$O$_3$ system. *Phys. Chem.Glasses: Eur. J.Glass Sci. Technol. B*, 50(6), 323-328.

[35] Hovhannisyan, M. R. (2009). Development of low melting sealing glasses on the bismuth borate and tellur molibdate systems basis. PhD Thesis. *State Engineering University of Armenia, Yerevan.*

[36] Hovhannisyan, M. R. (2008). Synthesis of new BaBi$_{10}$B$_6$O$_{25}$ compond in BaO-Bi$_2$O$_3$-B$_2$O$_3$ system. *Chemical J. of Armenia*, 62(2), 28-29.

[37] Hovhannisyan, R. M., et al. (2008). Mutual influence of barium borates, titanates and boron titanates on phase diagram and glass formation in the BaO-TiO2-B2O3 system. *Phys.Chem.Glasses: Eur. J.Glass Sci. Technol. B*, 49(2), 63-67.

[38] Hovhannisyan, R. M. (2010). Melts supercooling-effective way of inorganic systems glassforming and phase diagrams construction. *Proceedings of the II International conference of inorganic chemistry and chemical technology, 13-14th Sept., Yerevan*, 37-39, 978-9-99412-450-3.

[39] Hovhannisyan, R. M. (2011). New generation highly effective electro-and light-active nonlinear oriented tape glass ceramic on the basis of stoichiometric three-coordinated borates. *Final Technical Report of ISTC project # A-1486, Moscow.*

[40] Hovhannisyan, R. M., et al. (2011). Phase diagram, crystallization behavior and ferroelectric properties of stoichiometric glass ceramics in the BaO-TiO$_2$-B$_2$O$_3$ system. *in: Ferroelectrics- Physical Effects, InTech publishing, Croatia*, 49-76.

[41] Hovhannisyan, R. M. (2012). New Lead & Alkali Free Low Melting Sealing Glass Frits & Ceramic Fillers. *Final Technical Report of ISTC project # A-1591, Moscow.*

[42] Barseghyan, A. H., Hovhannisyan , R. M., Petrosyan , B. V., Aleksanyan , H. A., & Toroyan , V. P. (2012). Glass formation and crystalline phases in the ternary CaO-Bi$_2$O$_3$-Bi$_2$O$_3$ and SrO-Bi$_2$O$_3$-B$_2$O$_3$ systems. *Phys.Chem.Glasses: Eur. J.Glass Sci. Technol. B, in press.*

[43] International Center for Diffraction Data [ICDD]. (2008). Powder Diffraction Fails, PDF-2 release database,. *Pennsylvania, USA*, 1084-3116.

[44] Sawyer, C. B., & Tower, C. H. (1930). Rochelle Salt as a Dielectric. *Phys. Rev*, 35(3), 269-273.

[45] Imaoka, M., & Yamazaki, T. (1972). Glass-formation ranges of ternary systems. *III. Borates containing b-group elements, Rep. Inst. Ind. Sci. Univ. Tokyo*, 22(3), 173-212.

[46] Janakirama-Rao, Bh. V. (1965). Unusual properties and structure of glasses in the system $BaO-B_2O_3-SrO$, $BaO-B_2O_3-Bi_2O_3$, $BaO-B_2O_3-ZnO$, $BaO-B_2O_3-PbO$, Compt. *Rend. VII Congr. Intern. Du Verre, Bruxelles*, 1(104-104).

[47] Izumitani, T. (1958). Fundamental studies on new optical glasses,. *Rep. Governm. Ind. Res. Inst. Osaka* [311], 127.

[48] Kawanaka, Y, & Matusita, K. (2007). Various properties of bismuth oxide glasses for PbO free new solder glasses. *Proc.XXIth Intern.Congr.on Glass (in CD-ROM), Strasbourg* [I23].

[49] Milyukov, E. M., Vilchinskaya, N. N., & Makarova, T. M. (1982). Optical constants and some other characteristics of glasses in the systems $BaO-Bi_2O_3$ and $La_2O_3-Bi_2O_3-B_2O_3$. Fizika i Khimiya Stekla in Russian)., 8(3), 347-349.

[50] Sarkisov, P. D. (1997). The Directed Glass Crystallization- Basis for Development of Multipurpose Glass Ceramics,. *Mendeleev Russian Chemical Technological University Press, Moscow(In Russian*.

[51] Damjanovic, D. (2005). Hysteresis in Piezoelectric and Ferroelectric Materials,. In: The Science of Hysteresis, Mayergoyz, I & Bertotti, G Elsevier, 0-12480-874-3 , 3, 337-465.

[52] Rawson, H. (1967). Inorganic glass-forming systems. *Academic Press, London & New York*, 312.

Chapter 8

Ferroelectric Domain Imaging Multiferroic Films Using Piezoresponse Force Microscopy

Hongyang Zhao, Hideo Kimura, Qiwen Yao,
Lei Guo, Zhenxiang Cheng and Xiaolin Wang

Additional information is available at the end of the chapter

1. Introduction

Recently, multiferroic materials with the magnetoelectric coupling of ferroelectric (or anti-ferroelectric) properties and ferromagnetic (or antiferromagnetic) properties have attracted a lot of attention.[1-4] Among them, $BiFeO_3$(BFO) and $YMnO_3$has been intensively studied. For such ABO_3perovskite structured ferroelectric materials, they usually show antiferromagnetic order because the same B site magnetic element except $BiMnO_3$ is ferromagnet. While for the $A_2BB'O_6$ double perovskite oxides, the combination between B and B' give rise to a ferromagnetic coupling. They are also expected to be multiferroic materials. Several bismuth-based double perovskite oxides ($BiBB'O_6$) have aroused great interest like Bi_2NiMnO_6, La_2NiMnO_6, $BiFeO_3$-$BiCrO_3$. But as we know, few researches are focused on Bi_2FeMnO_6. We believe it is particular interesting to investigate Bi_2FeMnO_6 (BFM). The origin of the ferromagnetism in these compounds has been discussed in many reports. According to Goodenough-Kanamori's (GK)rules, many ferromagnets have been designed through the coupling of two B site ions with and without e_g electrons.In BFM, the 180 degree -Fe^{3+}-O-Mn^{3+}- bonds is quasistatic, partly because the strong Jahn-Teller uniaxial strain in an octahedral site.Because the complication of the double perovskite system, there are still some questions about the violation of GK rules in some cases and the origin of the ferromagnetism or antiferromagnetism.The other problem is the bad ferroelectric properties. In order to characterization of their ferroelectric/piezoelectric properties, scanning probe microscopy (SPM) techniques were used.

Multiferroic materials can be classified into two categories: one is single-phase materials; the other is multilayer or composite hetero-structures that contain more than one ferroic phases [5]. The most desirable multiferroic material is the intrinsic single-phase material, which is still rarely produced although significant advancements had been made recently. Therefore, it is essential to broaden the searching field for new candidates of multiferroics. This work focuses on both typesmultiferroic materials: Single-phase Bi_2FeMnO_6 and multilayered $YMnO_3/SnTiO_3$. Scanning probe microscopy (SPM) techniques were used for ferroelectric domain imaging of these multiferroic materials.

As we know, most piezoelectric/ferroelectric materials with good performances are based on the perovskite-type oxides of ABO_3, among them, the most extensively studied ones should be PbTiO3 and Pb(ZrTi)O$_3$ based materials due to their high dielectric constants and good piezoelectric/ferroelectric properties. However, these materials containing lead which leads to environmental problems. Thus, it is desirable to develop new lead-free piezoelectric materials to replace PZT based piezoelectrics for environmental protection. Through first principle calculations, $SnTiO_3$ containing Sn^{2+} ions was estimated to have excellent ferroelectric properties [6, 7]. The calculated results indicated that the spontaneous polarization and piezoelectric coefficients of $SnTiO_3$ is comparable with those of $PbTiO_3$. Moreover, the most stable structure is tetragonal perovskite with a=b=3.80Å and c=4.09Å [6, 7]. The metastable $SnTiO_3$ phase is very difficult to obtain, a good way to stabilize the $SnTiO_3$ phase is to mix it with other compounds. Several studies have been focused on the synthesis of Sn-doped ceramics in $BaTiO_3$ system, such as $(Ba_{0.6}Sr_{0.4})(Ti_{1-x}Sn_x)O_3$ [8], $(Ba_{1-x-y}Ca_xSn_y)TiO_3$ [9]. However, it is difficult to obtain Sn^{2+} by traditional bulk synthesis techniques. On the other hand, it is easier to fabricate $SnTiO_3$ using pulsed laser deposition method (PLD). Our interest was to find out whether the $SnTiO_3$ phase can be stabilized through the layer-by-layer deposition using PLD, and whether it is possible to obtain such metastable materials in non-equilibrium conditions.

The hexagonal manganite $YMnO_3$, which shows an antiferromagnetic transition at $T_N = 75$ K, and a ferroelectric transition $T_C = 913$ K, is one of the rare existing multiferroics [10-12]. It was chosen as the basic composition to sandwich $SnTiO_{3+x}$ phase and to form YST multilayer film. In this work, we reported the effect of substrate orientation and layer numbers of $YMnO_3$ and $SnTiO_{3+x}$ on the piezoelectric/ferroelectric and magnetic properties in the designed layered YST system. It was hoped that our method would serve as a model system to introduce the use of PLD techniques in the growth of multilayer multiferroic materials and metastable materials. Such techniques could be promising for device design and the searching in fabricating new materials.

2. SPM system

SPM has emerged as a powerful tool for high-resolution characterization of ferroelectrics for the first time providing an opportunity for non-destructive visualization of ferroelectric domain structures at the nanoscale. [13]These nanostructures can be used for studying the intrinsic size effects in ferroelectrics as well as for addressing such technologically important issues as processing damage, interfacial strain, grain size, aspect ratio effect, edge effect, domain pinning and imprint. [14] The system composes of a conducting tip in contact with the dielectric surface on a conductive substrate can be considered as a capacitor. SPM is a well-established field offering multiple opportunities for new discoveries and breakthroughs. [15-17]

Piezoelectric materials provide an additional response to the applied ac electric field due to the converse piezoelectric effect: $\Delta l = d_{33}V$, where Δl is the displacement, d_{33} is the effective longitudinal piezoelectric coefficient. [18, 19] If follows that both electrostatic and piezoelectric signals are linear with the applied voltage and thus contribute to the measured PFM response. These measurements are referred to as out-of-plane (OP) measurement. Measurements of local hysteresis loops are of great importance in inhomogeneous or polycrystalline ferroelectrics because they are able to quantify polarization switching on a scale significantly smaller than the grain size or inhomogeneity variation. Macroscopically, the switching occurs via the nucleation and growth of a large number of reverse domains in the situation where the applied electric field is uniform. Therefore, the d_{33} hysteresis reflects the switching averaged over the entire sample under the electrode. [20, 21] In the PFM experimental conditions, the electric field is strongly localized and inhomogeneous; therefore, the polarization switching starts with the nucleation of a single domain just under the tip.

3. Material designation and characterization

Multiferroic materials have been attracting considerable attention especially in the last ten years. [22-25] One of the most appealing aspects of multiferroics is their magnetoelectric coupling. Among them, the most studied is the provskite $BiFeO_3$(BFO) with room temperature multiferroic properties and $YMnO_3$. Based on BFO and $YMnO_3$, we designed new multiferroics and fabricated the films using pulsed laser deposition (PLD) method. The material system includes double perovskitemultiferroic Bi_2FeMnO_6 (BFM) and multilayered $YMnO_3/SnTiO_{3+x}$ (YST). Compared to BFO, the magnetic properties of BFM and YST were greatly improved. However, until now there is no report about their ferroelectric properties because the difficulty of obtaining well-shaped polarization hysteresis loops. It is important to study ferroelectric properties because the possible coupling between ferroelectric and antiferromagnetic domains and its lead-free nature. As is well known, the ferroelectric property is mainly determined by the domain structures and domain wall motions. The most exciting recent developments in the field of multiferroics and the most promise for future discoveries are in interfacial phenomena. [26] The interfaces include those that emerge spontaneously and those that are artificially engineered. Therefore, the domain structure and polarization switching were studied in these three new multiferroic films using piezoresponse force microscopy (PFM).

The emerging technique of PFM is proved to be a powerful tool to study piezoelectric and ferroelectric materials in such cases and extensive contributions have been published In PFM, the tip contacts with the sample surface and the deformation (expansion or contraction of the sample) is detected as a tip deflection. The local piezoresponse hysteresis loop and information on local ferroelectric behavior can be obtained because the strong coupling between polarization and electromechanical response in ferroelectric materials. In the present study, we attempts to use PFM to study the ferroelectric/piezoelectric properties in films of BFM and YST. PFM response was measured with a conducting tip (Rh-coated Si cantilever, k~1.6 N m^{-1}) by an SII Nanotechnology E-sweep AFM. PFM responses were measured as a function of applied DC bias (V_{dc}) with a small ac voltage applied to the bottom electrode (substrate) in the contact mode, and the resulting piezoelectric deformations transmitted to the cantilever were detected from the global deflection signal using a lock-in amplifier.

3.1. Characterization of BFM film

Magnetism and ferroelectricity exclude each other in single phase multiferroics. It is difficult for designing multiferroics with good magnetic and ferroelectric properties. Our interest is to design new candidate multiferroics based on $BiFeO_3$. $BiFeO_3$ is a well-known multiferroic material with antiferromagntic Neel temperature of 643K, which can be synthesized in a moderate condition. [22, 23, 25] In contrast, $BiMnO_3$ is ferromagnetic with Tc = 110K and it needs high-pressure synthesis. [27, 28] The possible magnetoelectric coupling has motivated a lot of studies on the ferroelectric and antiferromagnetic domains. The ferroelectric domain structures of BFO in the form of ceramics, single crystals and thin films have been intensively studied using PFM, TEM and other techniques.

Single phase BFM ceramics could be synthesized by conventional solid state method as the target. For BFM ceramics, the starting materials of Bi_2O_3, Fe_3O_4, $MnCO_3$ were weighed according to the molecular mole ratio with 10 mol% extra Bi_2O_3. They were mixed, pressed into pellets and sintered at 800 °C for 3 h. Then the ceramics were crushed, ground, pressed into pellets and sintered again at 880 °C for 1 h. BFM films were deposited on $SrTiO_3$(STO) substrate by pulsed laser deposition (PLD) method at 650°C with 500 ~ 600mTorr dynamic oxygen. [29, 30]

The structure of BFM was calculated [31] and it is connected with the magnetic configurations as shown in Figure 1. It has three possible space groups of Pm3m, R3 and C2 and the magnetic configurations were presumedfor each structure symmetry to be G-type antiferromagnetic (G-AFM) and ferromagnetic (FM) structures. The most stable structure of BFM is monoclinic with C2 space group. Mn tends to show 3+ valence which will induce a large distortion because it is Jahn-Teller ion. The valence of Mn and Fe has been studied in the former work. [30, 32]

Figure 2 shows the results of the PFM images of BFM film. Several features could be observed: firstly, the obvious contrast could be seen and the grains in PFM and topography are correspondence; secondly, the existence of contrasts on both OP and IP indicates multiple orientations of domains; thirdly, the IP contrast is not so clear as OP contrast, that is to say, the suppression of the in-plane response for heterostructures suggesting a constrained ferroelectric domain-orientation along the OP direction. In the former paper, the ferroelectric domain switching and typical butterfly loops were observed. [32]

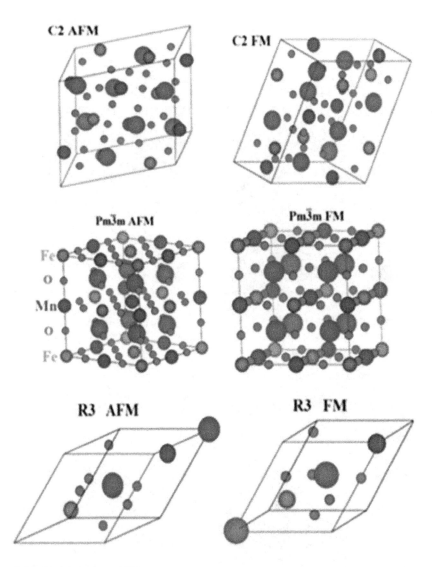

Figure 1. Calculated six structures of BFM.

The domains in BFO films have been analyzed in detail in Ref [19]: the bigger spontaneous ferroelectric domains were observed in BFO than in other ferroelectrics without multiferroic properties; the domains were irregular but the the model was predicted and consistent with the experimental results. According to the present results of BFM, further study is needed to obtain the domain morphology and do the calculation.

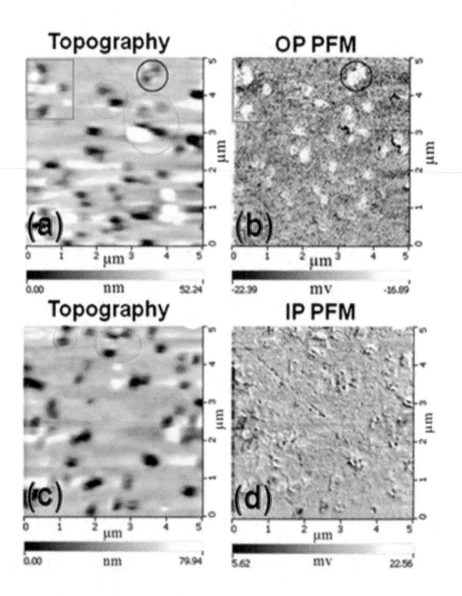

Figure 2. PFM images of BFM film

3.2. Characterization of YST film

Polycrystalline $YMnO_3$ and TiO_2 ceramics were synthesized by conventional solid state method as the targets. On addition the SnO and SnO_2 commercial targets were also used at the same time in our process. The deposition of these films includes the following major steps. YST films were deposited on (100) and (110) Nb: $SrTiO_3$ (STO) substrate using a pulsed laser deposition (PLD) system at 700°C with 10^{-1} ~10^{-5}Torr dynamic oxygen. The targets were alternately switched constantly and the films were obtained in a layer-by-layer growth mode. After deposition, the films were annealed in the same condition for 15 minutes at 700°C and then cooled to room temperature. [33]

In YST multilayered films, one layer is defined to be comprised of two sub-layers: (1) $YMnO_3$ and (2) $SnTiO_{3+x}$. The films deposited on (100), (111) and (110) STO are expressed as YST100, YST111 and YST110, respectively. The film YST110-4 denotes the films deposited on (110) STO with four layers, as shown clearly in Figure 3. The total deposition time for $YMnO_3$ and $SnTiO_{3+x}$ is the same of 40 and 20 minutes, respectively.

XRD patterns for the five films of YST100-2, YST100-4 were shown in Figure 4. The peaks were identified using the XRD results of $SnTiO_3$ (◎), $YMnO_3$ (Y) (shown in Figure 9, $FeTiO_3$ (▽) [34, 35]. $SnTiO_3$ is metastable and it showed two combined phases, one is $FeTiO_3$, and the other is the good ferroelectric phase which has tetragonal structure. XRD patterns of $SnTiO_3$ (◎) were shown in Figure 5 in supplementary materials, which is obtained using the calculated data[6, 7]. The symbol of "?" represents the phase which we cannot identify so far, it could be a peak from the (111) TiO_2 phase.

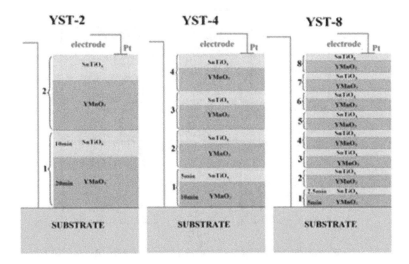

Figure 3. Schematic explanation of films with two, four and eight layers, each layer contains two sublayers of $YMnO_3$ and $SnTiO_3$.

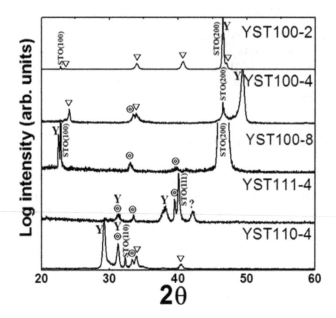

Figure 4. XRD patterns for the five films of YST100-2, YST100-4. [31]

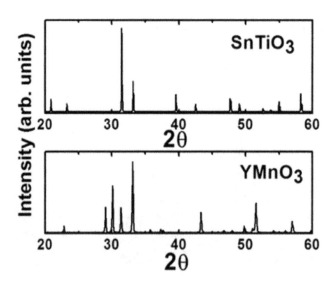

Figure 5. XRD patterns of calculated SnTiO₃ and YMnO₃ targets. [33]

Many reports about the YMnO$_3$ films can be found but most of them without ferroelectric characterization provided. This is mainly ascribed to the difficulty in obtaining a satisfactory ferroelectric measurement result. Figure 6 shows the electrical polarization hysteresis loops (P-E loops) of YST110-4 film. The ferroelectric type hysteresis loop was observed. It is obvious that the P-E loop is significantly different for different layer-number films, as well as for different substrate orientations. For the same STO orientation, the P-E loops were improved as the layer number increase, and the film fabricated on (110) STO shows improved properties compared to the film on (100) STO with the same layer number. As for the same deposition time for the four samples, it is found that more layers of SnTiO$_3$ in the YST system shows much better ferroelectric properties. It is suggested that the SnTiO$_3$ phase can exist only in ultra-thin thickness and can be stabilized by YMnO$_3$ sub-layers. That is to say, SnTiO$_3$ is indeed a ferroelectric materials but it is a great challenge [36] to obtain stable single phase SnTiO$_3$ films that is with good ferroelectric performance (it will be tried in our future works). Although the electric charge in the interface may affect its ferroelectric properties, the improved properties of YST110-4 proved that its effect was not significant. While the P-E loops were not improved in YST111-4, all the observations showed the anisotropy of the films.

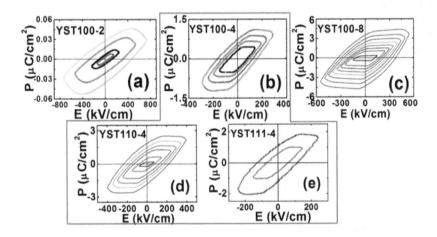

Figure 6. Ferroelectric polarization loop of YST films. [31]

Although well developed P-E loops were observed for YST100-8 and YST110-4, the loops still show the rounded features, indicating that contributions to the hysteresis loop from movable charges was significant. In addition, the P-E loops of the four films have revealed that the polarization is rather small (just a few µC/cm²) and they do not exhibit saturation. The fatigue properties were shown in Figure 7 for YST110-4. The polarization measurement was at 21V, the switching voltage was 14V and the frequency was 5k Hz. The remnant polarization decreased about 40% after 10⁷ read/write cycles for YST110-4. From all the comparisons, YST110-4 sample shows the comparable ferroelectric properties with YST100-8.

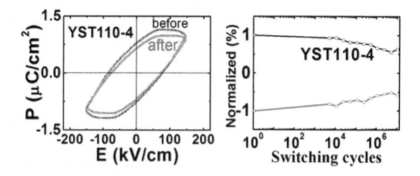

Figure 7. Fatigue measurement of YST110-4 film. [31]

Figure 8 shows TEM results of the YST110-4 film. It clearly shows four layers and one layer which is close to the substrate seems have some reaction with the substrate. The other three layers are homogeneous and we could see the domain structures through this image. The surface of the film was studied using AFM as shown in Figure 9. As expected, Figure 10 displays contrast over the polarized square after poled by positive and negative 10 V voltage, due to the different phases of the PFM response for the up and down domains. The obvious change of the contrast in YST110-4 film confirms that the polarization reversal is indeed possible and that the film is ferroelectric at room temperature.

Figure 8. TEM image of YST110-4 film.

Figure 9. AFM topography image of YST110-4 film.

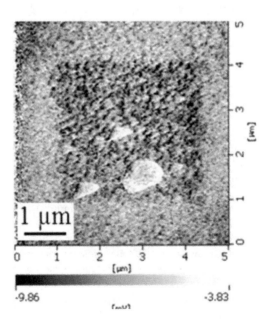

Figure 10. PFM image of YST110-4 film after poled using ± 10 V voltage

As seen in Figure 11, when a direct current voltage up to 10 V was applied on the samples, the sample exhibited "butterfly" loop. The loops were not symmetrical due to the asymmetry of the upper (tip) and bottom (substrate) electrodes. In addition, the substrate may also affect the d_{33}. According to the equation $d_{33} = \Delta l / V$, where Δl is the displacement, the effective d_{33} could be calculated. At the voltage of 10 V, both samples show the maximum effective d_{33} of about 6.21 pm/V for YST110-4 sample.

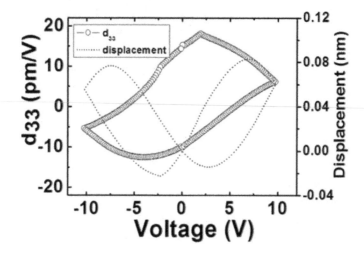

Figure 11. Butterfly and local piezoresponse hysteresis loops of YST110-4 film.

4. Conclusions

The novel multiferroic films of BFM and multilayered YST were successfully produced using PLD method. Both of them were characterized using PFM method. The special or excellent properties can often be found in the metastable materials. Through calculation, s4ix structures are presumed in BFM with two magnetic configurations of G-AFM and FM. In YST films, the metastable $SnTiO_3$ phase was obtained. The improved ferroelectric properties were observed through increasing the layer numbers and more $SnTiO_3$ phase was stabilized, moreover, the change in contrast after bias is applied indicates a change in polarization direction and hence ferroelectric switching. Although it is a great challenge in obtaining the $SnTiO_3$ thin films, we believe that through the optimization of fabrication process and conditions, the single phase $SnTiO_3$ or multilayer films, or composite materials containing $SnTiO_3$ could become a new generation lead-free piezoelectric/ferroelectric material. For a thorough understanding of the mechanisms of the films, knowledge of the domain structures is a prerequisite, which is of crucial importance to increase and tune the functionality of multiferroic films.

Acknowledgements

The authors gratefully acknowledge Dr. Minoru Osada, Dr. Kazuya Terabe in NIMS and Prof. H. R. Zeng in Shanghai Institute of Ceramics for their experimental help and discussions. Part of this work was supported by JST ALCA and MEXT GRENE.

Author details

Hongyang Zhao[1,2*], Hideo Kimura[1], Qiwen Yao[1], Lei Guo[1], Zhenxiang Cheng[3] and Xiaolin Wang[3]

*Address all correspondence to: zhao.hongyang@nims.go.jp

1 National Institute for Materials Science, Sengen 1-2-1, Tsukuba, Japan

2 Shanghai Institute of Ceramics, Chinese Academy of Sciences, Shanghai, China

3 Institute for Superconducting and Electronics Materials, University of Wollongong, Innovation Campus, Fairy Meadow, Australia

References

[1] Hill N. A., Why are there so few magnetic ferroelectrics? J. Phys. Chem. B 2000; 104: 6694.

[2] Fiebig M., T. Lottermoser, D. Frohlich, A. V. Goltsev, R. V. Pisarev, Observation of cou-pled magnetic and electric domains.Nature 2002; 419: 818.

[3] Eerenstein W., Mathur N. D. and Scott J.F., Multiferroic and magnetoelectric materials. Nature 2006; 442: 759-765.

[4] Singh M. K., Prellier W., Singh M. P., Katiyar R. S., Scott J. F., Spin-glass transition in single-crystal BiFeO3. Phys. Rev. B 2008; 77: 144403.

[5] J. Das, Y.Y. Song, M.Z. Wu, Electric-field control of ferromagnetic resonance in mon-olithic $BaFe_{12}O_{19}$–$Ba_{0.5}Sr_{0.5}TiO_3$heterostructures. J. Appl. Phys. 2010; 108: 043911.

[6] Konishi Y., Ohasawa M., Yonezawa Y., Tanimura Y., Chikyow T., Wakisaka T., Koi-numa H., Miyamoto A., Kubo M., Sasata K., Mater. Res. Soc. Symp. Proc. 2003; 748 U3.13.1.

[7] Uratani Y., Shishidou T., Oguchi T., First-Principles Study of Lead-Free Piezoelectric SnTiO3.Jpn. J. Appl. Phys. 2008; 47 (9), 7735-7739.

[8]   Ha J. Y., Lin L. W., Jeong D. Y., Yoon S. J., Choi J. W., Improved Figure of Merit of (Ba,Sr)TiO3-Based Ceramics by Sn Substitution.Jpn. J. Appl. Phys. 2009; 48 011402.

[9]   Suzuki S., Takeda T., Ando A., Takagi H., Ferroelectric phase transition in Sn2+ ions doped (Ba,Ca)TiO3 ceramics.Appl. Phys. Lett. 2010; 96: 132903.

[10] Marin L.W., Crane S.P., Chu Y-H., Holcomb M.B., Gajek M., Huijben M., Yang C-H., Balke N., Ramesh R., Multiferroics and magnetoelectrics: thin films andnanostructures.J. Phys.: Condens. Matter 2008; 20: 434220.

[11] Yakel H.L., Koehler W.C., Bertaut E.F., Forrat E.F., On the crystal structure of the manganese(III) trioxides of the heavy lanthanides and yttrium. ActaCrystallogr. 1963; 16: 957-962.

[12] Bertaut E. F., Pauthenet R., Mercier M., Proprietesmagnetiqueset structures du manganite d'yttrium. Phys. Lett. 1963; 7: 110-111.

[13] Gruverman A, Kholkin A, Nanoscale Ferroelectrics: Processing, Characterization and Future Trends. Rep. Prog. Phys. 2006; 69: 2443-2474.

[14] Catalan G., Noheda B, McAneney J, Sinnamon L J, Gregg J M Strain gradients in epitaxial ferroelectrics. Phys. Rev. B 2005; 72: 0201102.

[15] Muralt P. J. Micromech. Ferroelectric thin films for micro-sensors and actuatorsMicroeng. 2000;10: 136.

[16] Terabe K, Nakamura M, Takekawa S. Kitamura K, Higuchi S. Gotoh Y, Cho Y Microscale to nanoscale ferroelectric domain and surface engineering of a near-stoichiometric LiNbO$_3$ crystal. Appl. Phys. Lett. 2003; 82: 433-435.

[17] Fong D D, Stephenson G B, Streigger S K, Eastman J A, Auciello O, Fuoss P H, Thompson C Ferroelectricity in ultrathin perovskite films Science. 2004; 304: 1650.

[18] Kholkin A L, Bdikin I K, Shvartsman V V, Orlova A, Kiselev D, Bogomolov V, MRS Proc. E 2005; 838: O7.6.

[19] Catalan G., Bea H., Fusil S., Bibes M., Paruch P., Barthelemy A., Scott J. F., Fractal Dimension and Size Scaling of Domains in Thin Films of Multiferroic BiFeO$_3$. Phy. Rev. Lett. 2008; 100 027602.

[20] Kalinin S V, Gruverman A and Bonnell D A Quantitative analysis of nanoscale switching in SrBi2Ta2O9 thin films by piezoresponse force microscopy Appl. Phys. Lett. 2004; 85: 795-797

[21] Tybell T, Paruch P, Giamarchi T, Triscone J-M Domain wall creep in epitaxial ferroelectric Pb(Zr$_{0.2}$Ti$_{0.8}$)O$_3$ thin films 2002 Phys. Rev. Lett. 89 097601

[22] Catalan G., Scott J. F., Adv. Mater. Physics and Applications of Bismuth Ferrite.2009; 21, 2643.

[23] Wang J., Neaton B. J., Zheng H., Nagarajan V., Ogale S. B., Liu B., Viehland D., Vaithyanathan V., Schlom D. G.,. Waghmare U. V, Spaldi N. A.n, Rabe K. M., Wutting

M., Ramesh R., Epitaxial BiFeO3 Multiferroic Thin Film Heterostructures. Science 2003; 299: 1719-1722.

[24] Kimura T., Goto T., Shintani H., Ishizaka K., Arima T., Tokura Y., Magnetic Control of Ferroelectric Polarization. Nature 2003; 55: 426.

[25] Cheng Z. X., Wang X. L., Dou S. X., Ozawa K., Kimura H., Improved ferroelectric properties in multiferroic $BiFeO_3$ thin films through La and Nbcodoping. Phys. Rev. B 2008; 77: 092101.

[26] Spaldin N. A., Cheong S. W., Ramesh R.,Multiferroics: Past,present, and future. Physics Today 2010; 63: 38-43.

[27] T. Atou, H. Chiba, K. Ohoyama, Y. Yamaguchi, Y. Syono, Structure determination of ferromagnetic perovskite $BiMnO_3$. J. Solid State Chem. 1999: 145(2) 639-642.

[28] T. Kimura, S. Kawamoto, I. Yamada, M. Azuma, M. Takano, Y. Tokura, Magnetoca-pacitance effect in multiferroic $BiMnO_3$, Phys. Rev. B 2003: 67 (R), 180401-180404.

[29] Zhao, H.Y.; Kimura, H.; Cheng, Z.X.; Wang, X.L. &Nishida, T. Room temperature multiferroic properties of Nd:BiFeO3/Bi2FeMnO6 bilayered films.Appl. Phys. Lett. 2009; 95(23)232904

[30] Zhao, H.Y.; Kimura, H.; Cheng, Z.X.; Wang, X.L.; Ozawa, K. &Nishida, T. Magnetic characterization of Bi_2FeMnO_6 film grown on (100) $SrTiO_3$ substrate Phys. Status Sol-idi RRL 2010; 4(11) 314.

[31] Bi, L.; Taussig, A.R.; Kim, H-S.; Wang, L.; Dionne, G.F.; Bono, D.; Persson, K.; Ceder, G.&Ross, C.A. Structural, magnetic, and optical properties of $BiFeO_3$ and Bi_2FeMnO_6 epitaxial thin films: An experimental and first-principles study. Phys. Rev. B 2008; 78 (10) 104106.

[32] ZhaoH Y, Kimura H.; Cheng Z.X.; Wang X.L., New multiferroic materials: Bi_2FeM-nO_6. Ferroelectrics-Material Aspects. InTech; 2011.

[33] Zhao H Y, Kimura H.; Cheng Z.X.; Wang X.L., Yao Q W, Osada M, Li B W, Nishida T A new multiferroicheterostructure of YMnO3/SnTiO3+x. Scrip. Mater. 2011; 65: 618-621

[34] Harrison R.J., Redfern S.A.T., Smith R.I., Thermodynamics of the R(3)over-bar to R(3)over-barc phase transition in the ilmenite-hematite solid solution.Am. Mineral. 2000; 85: 1694-1705.

[35] Leinenweber K., Utsmi W., Tsuchida Y., Yagi T., Kuita K., Unquenchable High-Pres-sure Perovskite Polymorphs Of $MnSnO_3$ And $FeTiO_3$. Phys. Chem. Miner. 1991; 18: 244.

[36] Venkatesan S., Daumont C., Kooi B.J., Noheda B., J Hossoneff Th.M. De, Nanoscale domain evolution in thin films of multiferroic $TbMnO_3$. Phys. Rev. B 2009; 80: 214111.

Chapter 9

Optical Properties of Ferroelectrics and Measurement Procedures

A.K. Bain and Prem Chand

Additional information is available at the end of the chapter

1. Introduction

It is well known that the optical properties of ferroelectric materials find wide ranging applications in laser devices. Particularly in the recent years, there has been tremendous interest in the investigation of the nonlinear optical properties of ferroelectric thin films [1-5] for planar waveguide and integrated –optic devices. A new class of thin film waveguides has been developed using $BaTiO_3$ thin films deposited on MgO substrates [6]. Barium strontium titanate $Ba_{1-x}Sr_xTiO_3$ (BST) is one of the most interesting thin film ferroelectric materials due to its high dielectric constant, composition dependent Curie temperature and high optical nonlinearity. The composition dependent T_c enables a maximum infrared response to be obtained at room temperature. The BST thin films in the paraelectric phase, have characteristics such as good chemical and thermal stability and good insulating properties, due to this nature they are often considered the most suitable capacitor dielectrics for successful fabrication of high density Giga bit (Gbit) scale dynamic random access memories (DRAMs). Compositionally graded ferroelectric films have exhibited properties not previously observed in conventional ferroelectric materials. The most notable property of the graded ferroelectric devices or graded Functionally Devices (GFDs) is the large DC polarization offset they develop when driven by an alternating electric field. Such GFDs can find applications as tunable multilayer capacitors, waveguide phase shifters and filters [7]. Recently, BST thin films were used in the formation of graded ferroelectric devices by depositing successive layers of BST with different Ba/Sr ratios [8].

In our work, the Barium strontium titanate ($Ba_{0.05}Sr_{0.95}TiO_3$) ferroelectric thin films were prepared on single crystal [001] MgO substrates using the pulsed laser deposition method. The refractive index of BST ($Ba_{0.05}Sr_{0.95}TiO_3$) thin films is determined in the wavelength range between 1450-1580 nm at the room temperature. The dispersion curve is found to decrease gradually with increasing wavelength. The average value of the refractive index is found to be 1.985 in the wavelength range between 1450-1580 nm which is important for optoelectronic device applications [9].

1: Present address: Director, PM College of Engineering, Kami Road, Sonepat-131001.

2: Present Address: 99 Cyril Road, Small Heath, Birmingham B10 0ST.

Lithium heptagermanate $Li_2Ge_7O_{15}$ (LGO) is regarded as a weak ferroelectric and its curie point T_c is 283.5K [10,11]. Due to its intermediate behaviour between order-disorder and displacive types in a conventional grouping of ferroelectric materials LGO remains a subject of interest from both the theoretical and the application point of view. The paraelectric phase above T_c is orthorhombic D^{14}_{2h} ~ pbcn and below T_c the ferroelectric phase is C^5_{2v} ~ pbc2$_1$ with four formula units in a unit cell in both the phases. Below T_c LGO shows dielectric hysteresis loop and the permittivity shows a sharp peak at T_c [10-12]. Below T_c the spontaneous polarization appears along the c-axis. Many interesting physical properties of LGO such as birefringence [13], elastic behaviour [14], thermal expansion [11], dielectric susceptibility [12,15, 16] and photoluminescence [17] exhibit strong anomalies around T_c. The optical properties, however vary only to such a small degree that the transition could not be detected with the aid of a standard polarization microscope [13]. Employing a high resolution polarization device, Kaminsky and HaussÜhl [13] studied the birefringence in LGO near T_c and observed anomalies at the phase transition.

The study of piezo-optic dispersion of LGO (un-irradiated and x-irradiated) in the visible region of the spectrum of light at room temperature (RT=298 K) shows an optical zone/window in between 5400Å and 6200Å with an enhanced piezo-optical behavior [18]. The temperature dependence of the photoelastic coefficients of the ferroelectric crystals $Li_2Ge_7O_{15}$ (both un-irradiated and x-irradiated) in a cooling and a heating cycle between room temperature and 273K shows an interesting observation including the lowering of the T_c under uniaxial stress contrary to the increase of T_c under hydrostatic pressure and observation of thermal photoelastic hysteresis similar to dielectric behavior [19]. The study of fluorescence spectra of the crystals $Li_2Ge_7O_{15}:Cr^{3+}$ in the temperature interval 77-320 K shows the sharply decrease of intensities of the R_1 and R_2 lines (corresponding to the Cr^{3+} ions of types I and II) during cooling process near the temperature $T_c = 283.5$ K[16].

The present chapter includes optical properties of the ferroelectric BST thin films and the Lithium heptagermanate ($Li_2Ge_7O_{15}$) single crystals, fabrication methods, measurement procedures of the refractive index of BST thin films on MgO substrates, the fluorescence spectra and the photoelastic coefficients of LGO single crystals (un-irradiated and x-irradiated) at different wave lengths and temperatures around the phase transition temperature T_c. The potential of these materials for practical applications in the opto-electronic devices will also be discussed.

1.1. Optical property of Barium strontium titanate thin films

The Barium strontium titanate ($Ba_{0.05}Sr_{0.95}TiO_3$) ferroelectric thin films were prepared on single crystal MgO substrates using the pulsed laser deposition (PLD) method at a substrate temperature of 780 °C and then annealed at 650 °C for 55 min. The x-ray diffraction (XRD) analysis revealed that the films are oriented with [001] parallel to the substrate [001] axis and thus normal to the plane of the films [9]. The films were grown to a thickness of 430 nm.

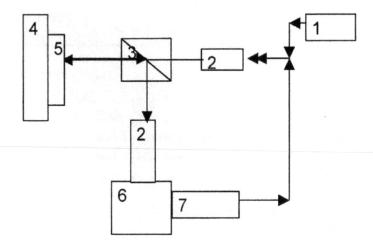

Figure 1. Schematic diagram of the experimental setup for the measurement of refractive index of the $Ba_{0.05}Sr_{0.95}TiO_3$ thin films at room temperature. He-Ne Laser for alignment (1), Lenses (2), Polarizer (3), sample holder (4), BST sample (5), Agilent light wave measurement system (6) and Tunable Laser (7).

Figure 1 shows the schematic diagram of the experimental setup for the measurement of the refractive index of $Ba_{0.05}Sr_{0.95}TiO_3$ thin films on MgO substrates through a reflection method. The He-Ne laser beam is used as a source of light to setup the alignment of the reflected beam of light from the samples to the detector. The incident beam is allowed to pass through a polarizer onto the sample. The reflected light is then passed through the same polarizing beam splitter oriented at 45° relative to the incident light and finally allowed to fall on the detector that is at 90° to the reflected/incident beam of light. The reflectivity measurement of the black metal, mirror, MgO substrate and $Ba_{0.05}Sr_{0.95}TiO_3$ thin films were carried using the Agilent 8164A Light Wave Measurement system in the wavelength region of 1450-1580 nm at room temperature.

The refractive index of substrate MgO is taken to be 1.7 [9]. The reflectivity of $Ba_{0.05}Sr_{0.95}TiO_3$ film is then normalized with respect to the mirror. The value of refractive index is derived from model described in ref. [20]. The fitting is done with the calculated data of the reflectivity of $Ba_{0.05}Sr_{0.95}TiO_3$ in the wavelength range between 1450-1580 nm. Figure 2 shows the refractive index of $Ba_{0.05}Sr_{0.95}TiO_3$ thin films as a function of wavelength at room temperature. The refractive index of $Ba_{0.05}Sr_{0.95}TiO_3$ with $Ba_xSr_{1-x}TiO^3$ (BST) and other materials of ferroelectric thin films at different wavelengths are presented in table 1.

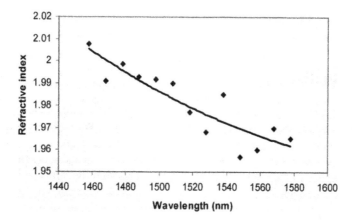

Figure 2. The variation of refractive index of $Ba_{0.05}Sr_{0.95}TiO_3$ thin films as a function of wavelength.

Sample Name	Refractive index	Wavelength	Remarks
$Ba_xSr_{1-x}TiO3(x=0\%)$	2.37	600 nm (RT)	Ref. [21]
$Ba_xSr_{1-x}TiO3(x=30\%)$	2.42	600 nm (RT)	Ref. [21]
$Ba_xSr_{1-x}TiO3(x=50\%)$	2.37	600 nm (RT)	Ref. [21]
$Ba_xSr_{1-x}TiO3(x=70\%)$	2.34	600 nm (RT)	Ref. [21]
$Ba_xSr_{1-x}TiO3(x=50\%)$	2.45	470 nm (RT)	Ref. [22]
$Ba_xSr_{1-x}TiO3(x=80\%)$	2.27	430 nm (RT)	Ref. [23]
$PbZr_{1-x}Ti_xO3$ (PZT)	2.87	400 nm (300 °C)	Ref. [24]
$PbZr_{1-x}Ti_xO3$ (PZT)	2.82	400 nm (50 °C)	Ref. [24]
$PbZr_{1-x}Ti_xO3$ (PZT)	2.66	500 nm (300 °C)	Ref. [24]
$PbZr_{1-x}Ti_xO3$ (PZT)	2.67	500 nm (50 °C)	Ref. [24]
$PbTiO_3$	3.10	470 nm (RT)	Ref. [25]
$PbTiO_3$	2.80	490 nm (RT)	Ref. [25]
$PbTiO_3$	2.75	550 nm (RT)	Ref. [25]
$Sr_xBa_{1-x}Nb_2O_6$ (x=61%)	2.35	400 nm (RT)	Ref. [26]

Table 1. The refractive index of BST and other ferroelectric thin films.

As shown in Figure 2, the dispersion curve decreases gradually with increasing wavelength. The average value of the refractive index is found to be ~ 1.985 in the wavelength range of 1450-1580 nm which is important for optoelectronic device (optical waveguide) applications. The variation of refractive index is attributed predominantly to the changes of electronic structure associated with the larger lattice parameter and variations in atomic co-ordination [27] that is local relaxations.

1.2. Growth and structure of Li₂Ge₇O₁₅ Crystals

Single crystals of $Li_2Ge_7O_{15}$ are grown in an ambient atmosphere by Czochralski method from stoichiometric melt, employing a resistance heated furnace. Stoichiometric mixture of powdered Li_2CO_3 and GeO_2 in the ratio of 1.03 and 7.0 respectively was heated at 1100 K for 24 hours to complete the solid state reaction for the raw material for the crystal growth. The crystals were grown by rotating the seed at the rate of 50 rpm with a pulling rate of 1.2 mm/hour. The cooling rate of temperature in the process of growth was 0.8-1.2 K/hour. The crystals grown were colourless, fully transparent and of optical quality. The crystal axes were determined by x-ray and optical methods.

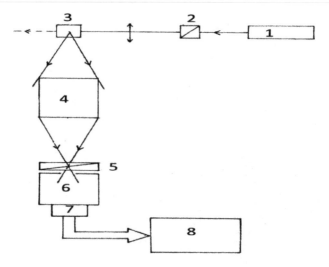

Figure 3. Schematic diagram of the experimental setup for the measurement of fluorescence spectra of the crystals $Li_2Ge_7O_{15}:Cr^{3+}$. He-Ne Laser for radiation (1), Glen prism (2), crystal sample (3), condenser (4), polarizer (5), spectrograph (6) Multichannel analyzer(7) and computer (8).

The desired impurities such as Cr^{+3}, Mn^{+2}, Bi^{+2}, Cu^{2+} and Eu^{+2} etc are also introduced in desired concentration by mixing the appropriate amount of the desired anion salt in the growth mixture. The crystal structure of LGO above T_c is orthorhombic (psedohexagonal) with the space group D^{14}_{2h} (Pbcn). The cell parameters are a: 7.406 Å, b: 16.696 Å, c: 9.610 Å, Z = 4 and b~√3c. Below T_c a small value of spontaneous polarization occurs along c-axis and the ferroelectric phase belongs to C^5_{2v} (Pbc2₁) space group. The crystal structure contains strongly packed layers of GeO_4 tetrahedra linked by GeO_6-octahedra to form a three dimensionally bridged frame work in which Li atoms occupy the positions in the vacant channels extending three dimensionally [14, 28, 29]. The size of the unit cell (Z = 4) does not change at the phase transition and ferroelectric phase transition is associated with a relaxational mode as well as the soft phonon [30]. Activation of the pure crystals with impurity ions will demand charge compensating mechanism through additional defects in the pure lattice.

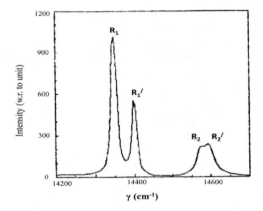

Figure 4. Fluorescence Spectra of the crystals $Li_2Ge_7O_{15}:Cr^{3+}$ at T=77 K.

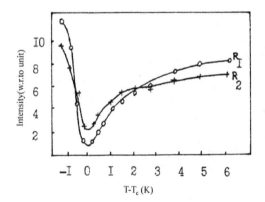

Figure 5. Temperature dependence of intensity of R_1 and R_2 Lines of fluorescence in $Li_2Ge_7O_{15}:Cr^{3+}$ crystals near the phase transition temperature T_c.

1.3. Study of fluorescence spectra of $Li_2Ge_7O_{15}:Cr^{3+}$ crystals

The fluorescence spectra of the crystals $Li_2Ge_7O_{15}:Cr^{3+}$ were studied in the temperature interval 77-320 K including the phase transition temperature $T_c = 283.5$ K. The experimental set to record the fluorescence spectra of the crystals $Li_2Ge_7O_{15}:Cr^{3+}$ is shown in fig.3. A laser with the pair of mode ($\lambda_1=510.6$ nm, $\lambda_2=578.2$ nm) was used as a source of excitation of the crystal sample. The recording of fluorescence spectra were carried out by the optical multichannel analyzer in combination with the polychromator. The radiation beam was initially polarized with glen prism. The plane Polaroid was used as an analyzer that was placed before the input aperture of the polychromator.

The fluorescence spectra consist of narrow intensity lines referred to R_1 and R_2 with frequencies $\gamma_1 \sim 14348$ cm^{-1} and $\gamma_2 \sim 14572$ cm^{-1}. These lines split further into two components each $R_1{}'$ and $R_2{}'$ respectively at lowering the temperature towards 77 K. Besides this, a wide long wavelength region/zone is observed in the spectra. It may be related with the effect of electron-photon interaction. It is known that Cr^{3+} doping ions in the structure of $Li_2Ge_7O_{15}:Cr^{3+}$ crystals substitute the Ge^{4+} host ions within oxygen octahedral (GeO_6) complexes [31-36]. The optical spectra of Cr^{3+} ions shows the existence of two types of Cr^{3+} centre (type I and II with different values of effective g-factor) as observed in EPR (Electron Paramagnetic Resonance) spectra of Cr^{3+} ions in ferroelectric phase of the crystals $Li_2Ge_7O_{15}:Cr^{3+}$ [32, 33]. Two pair of R lines $4A_2$ - ^2E (at T=77 K, its positions are R_1=14348 cm^{-1}, $R_1{}'$=14402 cm^{-1}, R_2=14572 cm^{-1} and $R_2{}'$=14593 cm^{-1}) are observed at low temperature region (T<190 K) in the optical spectra of the crystals $Li_2Ge_7O_{15}:Cr^{3+}$ as shown in Fig. 4. Actually the two different types of Cr^{3+} centers (R and R') with pretty different positions below $\bar{E}$ and above $2\bar{A}$ levels of the excited E^2 level are duplicate [37, 38] and conform to the EPR observations.

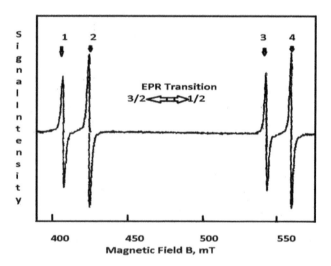

Figure 6. Part of EPR spectrum of Cr^{3+} doped LGO crystal in an arbitrary orientation at RT. The four EPR signals are attributed to four distinct Cr^{3+} sites per unit cell of LGO.

The intensity of fluorescence of the R_1 and R_2 lines of the crystals $Li_2Ge_7O_{15}:Cr^{3+}$ were studied near the phase transition temperature T_c at the direction $E \perp [001]$. It is observed that the intensity of R_1 and R_2 lines are decreased sharply near the phase transition temperature T_c but at high temperature region (T>T_c) the intensity again increases as shown in figure 5. Such nature of suppression of R_1 and R_2 lines was not observed previously and it may be related with the mechanism of interaction of excitation spectra of light in the crystals $Li_2Ge_7O_{15}{:}Cr^{3+}$ near the phase transition temperature T_c[17].

The Crystals doped with chromium impurity give EPR (Electron Paramagnetic Resonance) signals characteristics of the trivalent chromium ions [Fig.6]. It is known that impurity Cr^{3+} ions substitute the Ge^{4+} host ions within oxygen octahedral in the basic structure of (LGO) crystal [31-36]. Incorporation of tri-positive chromium ions into GeO_6-octahedra changes the local symmetry of the lattice site from monoclinic C_2 group to triclinic C_1 group. The local symmetry lowering is attributed to the effect of the additional Li^+ defect required for compensating the charge misfit of Cr^{3+} ion at the Ge^{4+} site. Taking into account a weak coupling of lithium ions with the germanium – oxygen lattice framework, the interstitial Li^+ is considered to be the most probable charge compensating defect, located within the structural cavity near the octahedral CrO_6 complex (Fig.7). Subsequent measurements of optical spectra have confirmed the model of Cr^{3+}– Li^+ pair centers in the LGO crystal structure [32, 33].

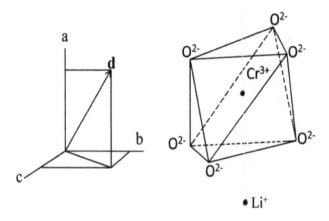

Figure 7. Physical model of Cr^{3+} centers in $Li_2Ge_7O_{15}{:}Cr^{3+}$ crystals and its dipole moment d.

The available data make it possible to assume that electric dipole moments of Cr^{3+}– Li^+ pairs are directed along the crystal axis "a" of the crystal. Interstitial Li^+ ions locally break the symmetry axis C_2 of the sites within the oxygen octahedral complexes [34]. As a result, there are two equivalent configurations of the pair centers which are conjugated by broken C_2 axis and have dipole moments with opposite orientations. It may be assumed that pair centers can reorient due to thermal activation. Reorientation of the pair centers should be accompanied by: i) shortening of the configuration life time and ii) switching of defect dipole moments [35]. This is reflected in the typical temperature dependence of the imaginary part of dielectric permittivity of chromium doped LGO single crystals [35] along the a-axis of the crystal shown in fig.8.

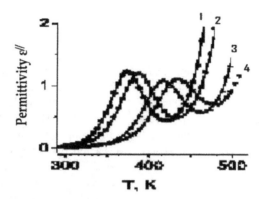

Figure 8. Temperature dependence of Imaginary part ε'' of permittivity $\varepsilon = \varepsilon' - j\varepsilon''$ of LGO:Cr^{3+}, measured along crystal a - axis at the following frequencies : 0.5 kHz (1); 1 kHz (2); 5 kHz (3); 10 kHz (4). Distinct peak is observed in ε'' in the temperature range 350-450 K and the peak is observed to shift to higher temperature for higher frequencies. In contrast no such peak is observed for the real part ε' of the permittivity. The typical behavior is observed only along crystal a –axis which incidentally coincides with the proposed Cr^{3+}- Li^+ centers.

1.4. Principle of Photoelasticity

If a rectangular parallelepiped with edges parallel to x[100], y[010] and z[001] axes is stressed along z-axis and observation is made along y-axis, as shown in Fig.9, then the path retardation δ_{zy} introduced per unit length due the stress introduced birefringence is given by

$$\delta_{zy} = (\Delta n_z - \Delta n_x) = C_{zy} P_{zz} \tag{1}$$

where Δn_z and Δn_x are the changes in the corresponding refractive indices, $(\Delta n_z - \Delta n_x)$ is the corresponding stress induced birefringence, P_{zz} is the stress along z-axis and C_{zy} is a constant called the Brewster constant or the relative photoelastic coefficient. In general the Brewster constant is related to the stress optical and strain optical tensors of forth rank [39] and is a measure of the stress induced (piezo-optic) birefringence. It is conveniently expressed in the unit of 10^{-13} cm^2/dyne per cm thickness along the direction of observation is called a Brewster [39].

1.5. Measurement procedure of photoelastic constants

To study the piezo-optical birefringence the experimental set up consists of a source of light (S), a lens (L) to render the rays parallel, a polarizer (P), an analyzer Polaroid (A), a Babinet compensator (B) and a detector (D), as shown in Fig.10. The P and A combination are adjusted for optimal rejection of light. The sample with stressing arrangement and a Babinet compensator are placed between P and A. A monochromator and a gas flow temperature controlling device are used to obtain the piezo-optic coefficients (C_λ) at different wavelengths and temperature. The subscript λ in the symbol C_λ denotes that the piezo-optic coefficient depends on the wavelength of light used to measure it. The experiments are carried out for different wavelengths using white light and a monochromator and the monochromatic sodium yellow light. An appropriate stress along a desired direction of the sample is applied with the help of a stressing apparatus comprising a mechanical lever and load.

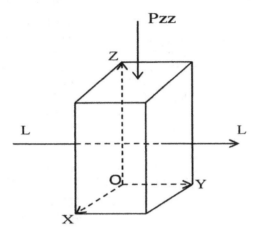

Figure 9. A solid under a linear stress of stress-optical measurements (P_{zz} is the applied stress and LL is the direction of light propagation and observation).

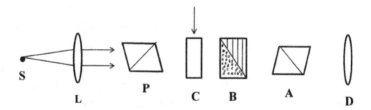

Figure 10. A schematic diagram of the experimental setup for the measurement of photoelastic constants of the crystals at room temperature. Source of light (S), Lense (L), Polarizer (P), Crystals (C) under stress, Babinet Compensator (B), Analyzer (A) and Detector (D).

To start with, the Babinet compensator is calibrated and the fringe width is determined for different wavelengths of light in the visible region. The crystal specimen is placed on the stressing system so that the stress could be applied along vertical axis and observation made along horizontal axis. A load on the crystal shifts the fringe in the Babinet compensator and this shift is a measure of the piezo-optic behavior. The piezo-optic coefficients (C_λ) are now calculated using the calibration of the Babinet compensator. The experiment is repeated for other orientations of the crystals and the results are obtained.

1.6. Piezo-optic Dispersion of $Li_2Ge_7O_{15}$ Crystals

The experimental procedure for the piezo-optic measurements is described in section 1.5. The polished optical quality samples worked out to dimensions i) 5.9 mm, 9.4 mm and 5.0 mm; ii) 3.17 mm, 5.88 mm and 6.7 mm, along the crystallographic a, b and c axes respectively. The stress was applied with an effective load of ~23 kg in each case [40].

The values of C_λ thus obtained at different wavelengths are given in Table 2 and the results are plotted in Fig. 11. Here C_{pq} is the piezo-optic coefficient with the stress direction being p and observation direction being q. The results show an interesting piezo-optic behavior. A survey of literature indicates that the piezo-optic behavior of materials studied till now shows a reduction of C_λ with increasing wavelength in the visible region [39]. In the present case, C_λ decreases with wavelength up to a certain wavelength as in other normal materials and then suddenly shows a peak and later on the usual behavior of reduction in the values of piezo-optic coefficients is observed.

		Wavelengths				
Obs.	C_{pq}	4358Å	4880Å	5390Å	5890Å	6140Å
1	C_{xy}	4.024	3.819	3.722	4.328	3.677
2	C_{xz}	5.243	4.895	4.770	5.552	4.451
3	C_{yx}	4.084	3.525	3.092	3.562	2.913
4	C_{yz}	4.353	4.118	3.946	4.261	3.866
5	C_{zy}	4.179	2.814	3.177	3.713	3.172
6	C_{zx}	3.312	2.991	2.650	4.190	2.618

Table 2. Stress optical coefficients c_{pq} (in Brewster) for $Li_2Ge_7O_{15}$ at different wave lengths.

To the best knowledge of the authors this behavior is unique to the LGO crystals. For the sake of convenience we denote C_λ measured at λ = 5890 Å as C_{5890} and so on. The results show that sometimes the value of C_{5890} is even higher than that at C_{4400}, the value of piezo-optic coefficient obtained at the lowest wavelength studied here. This is the case with C_{xy}, C_{zx} and C_{xz}. For other orientations the value is lower than that at 4400 Å. Further, C_λ is found to have increased to more than 50% in the case of stress along [001] and observation along [100]. Also, it is interesting to note that the value of C_{6140} is less than that of C_{5390}, in tune with usual observation of piezo-optic dispersion. Thus one can see an "optical window" in between 5400 Å and 6200 Å. The height of this optical window is different for various orientations, though the width seems approximately the same. The maximum height of about 1.5 Brewster was found for C_{zx} followed by C_{xz} with about 0.9 Brewster. It should be noted here that z-axis is the ferroelectric axis for LGO. It is also interesting to note that the change in height is more in the former while the actual value of C_λ is less compared to that of the latter. The percentage dispersion also is different for various orientations. It is very high, as high as 25% for C_{zy}, while it is just 10% for C_{xy}.

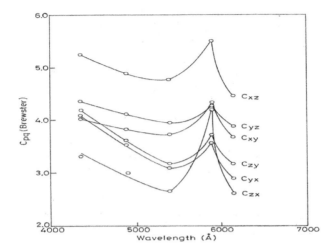

Figure 11. Stress optical dispersion of Li$_2$Ge$_7$O$_{15}$ crystals with wavelength at room temperature (298 K).

Figure 12 shows the variation of $C_{zx}(\lambda)$ at the temperatures ranging from 298K to 283K on cooling process of the sample LGO. It is clear from the figure that the distinct peak of $C_{zx}(\lambda)$ appears only at the sodium yellow wavelength of 5890 Å for the whole range of temperatures (298 K–283 K) investigated. It is also interesting to note that a temperature anomaly is also observed around 283 K. LGO undergoes a second order phase transition at 283.5 K from the high temperature paraelectric phase to the low temperature ferroelectric phase. So this anomaly is related to this phase transition of the LGO crystal.

The observed peculiarity of piezo-optic behavior could be due to many factors, viz., i) anomalous behavior of refractive index or birefringence ii) anomalous ferroelastic transformation at some stage of loading iii) shift of absorption edge due to loading. The following have been done to identify the reasons for this peculiar behaviour.

Birefringence dispersion has been investigated in the visible region and no anomalies in its behavior has been observed. This rules out the first of the reasons mentioned. The reason due to ferroelastic behavior also is ruled out since the effect would be uniform over all the wavelengths investigated. It was not possible to investigate the effect of load on the absorption edge. Hence an indirect experiment has been performed. If there is a shift in the absorption edge due to loading the sample, the peak observed now at sodium yellow light would shift with load. No clear shift of the peak could be observed within the experimental limits. Another interesting experiment was done to identify the source of the anomaly. It is well known that T_c of LGO changes under uniaxial stress. The measurements were made near T_c under different stress (loads). Although T_c was found to shift a little with load the dispersion peak did not show any discernible shift. No particular reason could be established as to why a dispersion peak appears around sodium yellow region. Another interesting work in this direction is on $Gd_2(MoO_4)_3$ — where an anomalous peak was recorded in spontaneous bire-fringence at 334.7 nm [41], an observation made for the first time.

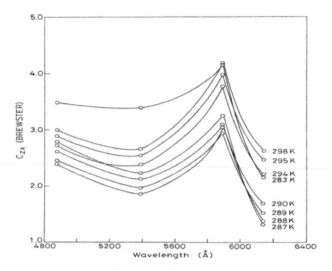

Figure 12. The variation of $C_{zx}(\lambda)$ at the temperatures ranging from 298 K to 283 K on cooling process of the sample $Li_2Ge_7O_{15}$.

It is well known that the photoelasticity in crystals arises due to change in number of oscilla-tors, effective electric field due to strain and the polarisability of the ions. In the present case, as the wavelength approaches around 5400 Å, the ionic polarisability seems to be changing enormously. There is no optical dispersion data available on LGO. We have conducted an experiment on transmission spectra of LGO along x, y and z-axes, which shows a strong ab-sorption around 5400 Å. The observed anomaly in the piezo-optic dispersion may be attrib-uted to the absorption edge falling in this region. This explanation needs further investigation in this direction. It is also known that the strain optical dispersion arises due to the shift in absorption frequencies and a change in the oscillator strength caused by the physical strain in the crystal.

1.7. Irradiation Effect on Piezo-optic Dispersion of $Li_2Ge_7O_{15}$ Crystals

The ferroelectric single crystals $Li_2Ge_7O_{15}$ was irradiated by x-ray for one hour and the ex-perimental processes described in section 1.5 were repeated for the crystal (irradiated) LGO in order to understand the radiation effect on piezo-optical birefringence dispersion [18]. The values of C_λ of the crystal (irradiated) LGO thus obtained at different wavelengths are given in Table 3 and the results are plotted in Fig. 13.

		Wavelengths				
Obs.	C'_{pq}	4358Å	4880Å	5390Å	5890Å	6140Å
1	C'_{xy}	4.08	3.87	3.72	4.33	3.73
2	C'_{xz}	5.35	5.00	4.88	5.59	4.55
3	C'_{yx}	4.02	3.47	3.01	3.50	2.83
4	C'_{yz}	4.39	4.19	4.01	4.26	3.90
5	C'_{zx}	4.63	4.46	4.41	4.66	4.29
6	C'_{zy}	3.71	3.26	2.97	3.43	2.72

Table 3. Stress Optical Coefficients C_{pq} (in Brewsters) for $Li_2Ge_7O_{15}$ (irradiated) at different wavelengths.

Some interesting results are obtained in the case of irradiated crystal LGO. The peak value of C'_{zx} has decreased about 18% and that of C'_{zy} has increased about 25% at the wave length $\lambda = 5890$ Å. Also, it is interesting to note that the value of C_{6140} is less than that of C_{5390} for the un-irradiated and irradiated sample of LGO crystal, in tune with usual observation of piezo-optic dispersion.

Irradiation of crystals can change physical properties of the crystals. Irradiation brings about many effects in the crystal such as creating defects, internal stress and electric fields etc. These irradiation effects in turn are supposed to affect the physical properties of the irradiated crystal as compared to un-irradiated crystal. While there was no appreciable change in the lattice parameters, a significant drop in the value of dielectric constant and tan δ was observed upon x-irradiation of ferroelectric glycine phosphate. An appreciable shift in the phase transition temperature towards the lower temperature was observed. These changes are attributed to the defects produced in it by irradiation [42]. The studies of triglycine sulphate (TGS) showed that very small doses of x-irradiation can give large changes of the ferroelectric properties. The direct evidence of domain clamping by defects was obtained from optical studies. With increasing dosage the dielectric constant peak and polarization curve broaden and move to lower temperatures. In our present studies, the x-irradiation is believed to produce internal stress and electric fields inside the crystals LGO due to defects that can change the values of piezo-optic constants [43].

1.8. Piezo-optic Birefringence in $Li_2Ge_7O_{15}$ Crystals

The temperature dependence of the photoelastic coefficients of the ferroelectric crystals $Li_2Ge_7O_{15}$ in a cooling and heating cycle between 298 K and 273 K was carried out with the experimental procedure described in section 1.5 [19]. A special arrangement was made to vary the temperature of the sample. The temperature was recorded with a digital temperature indicator and a thermocouple sensor in contact with the sample.

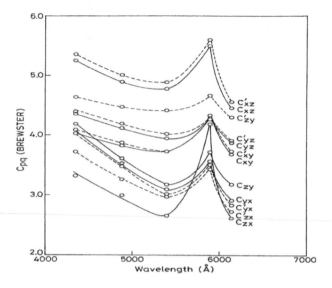

Figure 13. Stress optical dispersion of $Li_2Ge_7O_{15}$ crystals (un-irradiated and irradiated) with wavelength at room temperature (298 K).

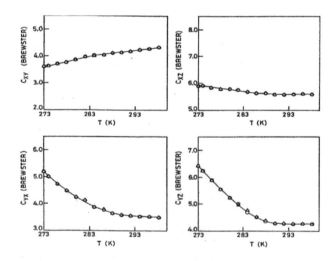

Figure 14. Temperature dependence of the piezo-optic coefficients C_{xy}, C_{xz}, C_{yz} and C_{yx} of the crystals LGO in a cooling (0) and heating (Δ) cycle.

The temperature dependence of piezo-optic coefficients C_{pq} of the crystals $Li_2Ge_7O_{15}$ between 298 K and 273 K were determined and are shown in Fig. 14 and Fig. 15. The values of C_{pq} at

291 K and 278 K were reported in paper [44] and it was observed that there were large changes in the values of C_{zy} and C_{yz} at 278 K and 291 K as compared to other components and C_{zy} did not show a peak in its temperature dependence between 291K and 278 K.

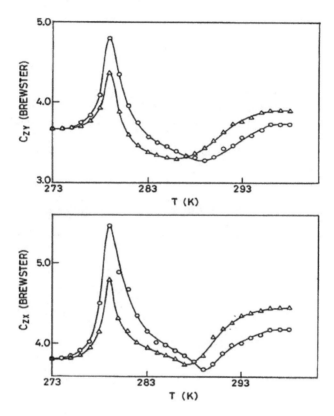

Figure 15. Anomalous temperature dependence of the piezo-optic coefficients C_{zx} and C_{zy} of the crystals LGO in a cooling (0) and heating (Δ) cycle.

Here in contrast we observed a peak in the temperature dependence of both C_{zy} and C_{zx} at 279 K. The temperature dependence of C_{pq} are quite interesting, for example the piezo-optic coefficients C_{yz}, C_{yx} and C_{xz} have negative temperature derivatives but C_{xy} has a positive temperature derivative. In complete contrast both C_{zy} and C_{zx} have both positive and negative temperature derivatives at different temperature intervals between 298 K and 273 K (Table: 4). Besides a clear thermal hysteresis is observed in C_{zy} and C_{zx} in a complete cooling and heating cycle (Fig. 15) whereas no discernible hysteresis is observed in rest of the piezo-optic coefficients (Fig. 14). The two distinct anomalies in the temperature dependence of C_{zy} and C_{zx} are characterized by a valley at T_m (~289 K) and a peak at T_c (~279 K). Anomalous temperature dependence of C_{zx} at different wave lengths is also shown in Fig. 16. The temperature dependence of the dielectric permittivity along the c-axis of LGO shows a sharp peak at T_c (283.5 K) and the Curie-Weiss law holds only for a narrow range of temperature ($T_c \pm 4$ K) [11,15, 16]. The peak for piezo-optic coefficient is attributed to the paraelectric to ferroelectric phase transition of LGO at T_c. To check the curie-Weiss law like dependence near T_c the following relation is used.

$$C^T_{pq} - C^0_{pq} = K_{pq}/(T - T_c) \tag{2}$$

Where C^T_{pq} and C^0_{pq} denote the value of the corresponding piezo-optic coefficients at temperature T and 273 K respectively and K_{pq} is a constant. Fig. 17 shows the $(C^T_{pq} - C^0_{pq})^{-1}$ vs ($T-T_c$) curve for C_{zx} and C_{zy}. It is clear from these curves that like dielectric constant the relation fits well only within a narrow range of temperature near T_c($T_c \pm 4$ K). The solid lines denote the theoretical curves with the following values $K_{zx} = 1.05$; $K_{zy} = 0.92$ for $T > T_c$, $K_{zx} = -0.40$; $K_{zy} = -0.34$ for $T < T_c$ and $T_c = 279$ K.

C_{pq}	Value of Derivative (Brewster/K)	Temperature Range Ratio
C_{zx}	0.013	296K-289K -1.69
	-0.022	289K-283K
	-0.090	282K-279K -2.0
	0.181	279K-276K
	~0	276K-273K
C_{xz}	-0.003	293K-273K
C_{zy}	0.020	296K-289K -0.75
	-0.015	289K-283K
	-0.095	282K-279K -1.9
	0.179	279K-276K
	~0	276K-273K
C_{yz}	-0.026	293K-273K
C_{xy}	0.007	293K-273K
C_{yz}	-0.023	293K-273K

Table 4. The temperature derivative $[dl_n C_{pq}/dT]$ of the piezo-optic coefficients of $Li_2Ge_7O_{15}$.

Furthermore the magnitudes of the ratio of the temperature derivatives below and above T_m and T_c are given in Table 4 and we can see that the ratio near T_c comes out to be about 2. Therefore it satisfies the law of two for the ratio of such derivatives of quantities which are coupled with the spontaneous polarization in second order ferroelectric phase transition such as in the case of triglycine sulphate [45] and LGO. Therefore the peak around T_c is [13, 15, 16] attributed to the paraelectric to ferroelectric phase transition of LGO. The smallness of K_{pq} and the applicability of relation (2) above only in a narrow range of temperature suggest that LGO may be an improper ferroelectric. The law of two does not hold for the ratio at T_m (Table 4). Therefore this anomaly is not related to the spontaneous polarization.

From the behaviour that only C_{zx} and C_{zy} show anomalous it is obvious that birefringence $(\Delta n_z - \Delta n_y)$ and $(\Delta n_z - \Delta n_x)$ show steep increase around T_c and below T_c show a $(T - T_c)^{1/2}$ behaviour correlated to the spontaneous polarization which is parallel to the z-axis (crystallographic c-axis). From the behaviour of C_{xy} and C_{yx} which do not show any temperature anomaly we may say that only n_z is responsible for the anomaly in accordance with the behaviour of the dielectric properties where only ε_{33} is strongly affected by the phase transition. These observations are in accordance with the results of Faraday effect and birefringence in LGO [13].

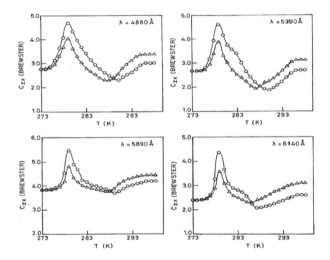

Figure 16. Anomalous temperature dependence of piezo-optic coefficient C_{zx} of the crystals LGO at different wave lengths in a cooling (0) and heating (Δ) cycle.

As mentioned by Lines and Glass [43], under an external pressure T_c of a ferroelectric phase transition may be shifted. This shift may be to the higher or the lower side of normal T_c. Wada et al. [46] studied the pressure effect on the ferroelectric phase transition in LGO through the dielectric and Raman scattering measurements and found a positive pressure coefficient $dT_c/dp = 14.6$ K/kbar. Preu and HaussÜhl [12] studied the dependences of dielectric constants on hydrostatic and uniaxial pressure as well as temperature. They observed a shift of T_c at a rate of 14.02 K/kbar for the hydrostatic pressure and ~7 K/kbar for the uniaxial pressure. In the present case the position of the peak of C_{zy} is found to depend on the stress applied. If the peak position is believed to represent the T_c, it appears to shift to the lower side under the uniaxial stress. To see whether T_c shifts linearly with uniaxial stress similar to the earlier observations [12, 46], we used different stresses within the elastic limits of LGO for

C_{zx} and found a linear relationship (Fig. 18). However, a negative stress coefficient $dT_c/dp \sim$ -22 K/kbar is obtained in this case which agrees only in magnitude with the hydrostatic pressure coefficient. The linear curve (Fig. 18) extrapolates to a $T_c = 281.5$ K in the unstressed state instead of 283.5K as determined by dielectric measurements [11, 15, 16]. This may be due to a non linear dependence of shift of T_c under stress near 283.5 K.

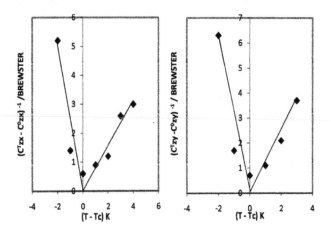

Figure 17. Plots of $(C^T_{pq} - C^0_{pq})^{-1}$ vs $(T-T_c)$ curve for C_{zx} and C_{zy}.

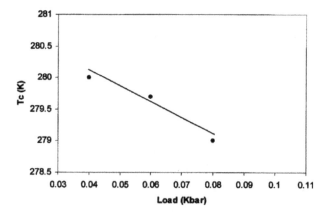

Figure 18. The stress dependence of the shift of T_c for C_{zx}.

Now we turn to the anomaly around T_m. Morioka et al. [47] proposed that there is an interaction between the soft phonon mode and a relaxational mode in the paraelectric phase in the temperature interval 300 K to T_c. The critical slowing down of the relaxational mode near T_c is expected to cause the increase of the fluctuation of the spatially homogeneous polarization and thereby the increase of the fluctuation of the hyperpolarizability with $k_c = 0$. Wada et al. [48] measured the soft phonon mode with the help of their newly designed FR-IR spectrometer and proposed that as T_c is approached from above soft phonon mode becomes over damped and transforms to a relaxational mode.

On the other hand there may exist a relaxational mode with an independent degree of freedom as well as the soft phonon mode and the character of the softening transfers from the phonon to the relaxational mode. This is an important problem in determining the dynamics of the peculiar ferroelectric phase transition of LGO, where both the dielectric critical slowing down characteristic of the order-disorder phase transition and the soft phonon mode characteristic of the displacive phase transition are observed [11, 14]. In the light of the above discussion we may say that the change up to T_m is caused by the softening of mode and the softening character transforms to the relaxations mode near T_m causing a change in the trend below T_m and near T_c the relaxational mode becomes dominant. The valley around T_m is perhaps caused by the interplay between the competitive relaxational mode and the soft phonon mode. It has been observed that softening of the velocity and rise of the damping of acoustic phonon occur in the paraelectric phase of LGO even quite far from T_c, i.e.$(T-T_c) \sim 30$ K and the effect is attributed to the fluctuation induced contributions [49].

Obs.	Cpq	Paraelectric (PE) phase (RT)	At T_c = 279 K
1	C_{xy}	4.38	3.85
2	C_{xz}	5.55	5.85
3	C_{yx}	3.60	4.46
4	C_{yz}	4.26	5.50
5	C_{zy}	3.71	4.83
6	C_{zx}	4.19	5.45

Table 5. Stress optical coefficients c_{pq} (in Brewster's) of $Li_2Ge_7O_{15}$ at RT=298 K and at T_c = 279 K.

Another interesting aspect is the observation of a significant thermal photoelastic hysteresis (Fig. 15). Although the peak position does not shift in the heating cycle the values of the photoelastic constants get reduced significantly in the heating cycle as compared to the corresponding values in a cooling cycle. A similar kind of hysteresis was observed in the dielectric behaviour of LGO and the appearance of the dielectric hysteresis is attributed to the internal space charge (electrets state) effects which produce an internal electric field in LGO on heating from the ferroelectric phase [15-17]. It was possible to compensate the internal electric field effects in dielectric measurements by an external electric field [15-17]. It is suspected that the photoelastic hysteresis also occurs due to similar effects. Although it was not possible to try to compensate the electric field effects in the present investigation, it is possible to attempt experiment under the simultaneous application of a suitable electric field and stress along z-direction.

Obs	C_{pq}	Rochelle Salt (RS)	KDP	ADP	Remarks
1	C_{xz}	3.74	0.28	1.25	Ref. [50] for RS
2	C_{yz}	4.29	0.28	1.25	a- polar axis
3	C_{yx}	3.56	1.04	4.30	Ref. [51] for KDP
4	C_{zx}	0.85	1.54	3.50	Ref. [52] for ADP
5	C_{zy}	2.61	1.54	3.50	
6	C_{xy}	3.04	1.04	4.30	

Table 6. Piezo-optic coefficients c_{pq} (in Brewsters) for some ferroelectric crystals in their paraelectric (PE) phases.

The Stress optical coefficients C_{pq} of the crystals $Li_2Ge_7O_{15}$ at paraelectric phase (RT = 298 K) and at T_c = 279 K are presented in Table 5. It is important to compare the values of C_{pq} for $Li_2Ge_7O_{15}$ with other ferroelectric crystals given in Table 6 particularly with Rochelle-salt (RS) which belongs to the orthorhombic class like LGO [44]. The values of C_{pq} are significantly higher for LGO as compared to these ferroelectric systems. So, the large photoelastic coefficients and the other properties like good mechanical strength, a transition temperature close to room temperature and stability in ambient environment favour LGO as a potential candidate for photoelastic applications.

The EPR (Electron Paramagnetic Resonance) spectroscopy of the transition metal ion doped crystals of LGO (Mn^{2+}, Cr^{3+}) has also been studied both in Paraelectric (PE) and ferroelectric (FE) phases in the temperature interval from 298 K to 279 K during cooling and heating cycles [17, 36, 53]. It is observed that on approaching T_c in a cooling cycle, the EPR lines are slightly shifted to the high field direction and undergo substantial broadening. At the temperature T_c (≈ 283.4 K), the EPR lines are splitted into two components which are shifted to the higher and lower field directions progressively as a result of cooling the sample below T_c as shown in Fig.19.

During heating cycle (i.e. approaching T_c from below), the phenomena occurred were just opposite to the above processes observed in the cooling cycle. However, the EPR line width (peak to peak ΔH_{pp}) for H ∥ c, H⊥a was found to decrease to about one third of its value at T_c in a heating cycle as compared to its value in the cooling cycle. The shape of the EPR resonance lines far from T_c has a dominant Lorentzian character (a Lorentzian line shape) but very near to T_c, the line shape has been described mainly by Gaussian form of distribution (a Gaussian line shape). All the peculiarities observed are attributed to the PE ↔ FE phase transition of the LGO crystals. The line width reduction near T_c is attributed to the internal space charge (electret state) effects which produce an internal electric field inside the crystals on heating process from the ferroelectric phase. This observation is similar to the photoelastic hysteresis behavior of the crystals LGO near T_c.

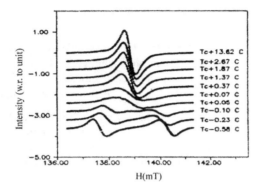

Figure 19. Temperature dependence of EPR lines of $Li_2Ge_7O_{15}$:Cr^{+3} crystals for $|M| = \frac{1}{2} \leftrightarrow 3/2$, $H \parallel a$, $H \perp c$ near T_c during cooling process.

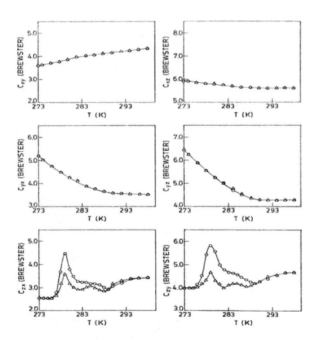

Figure 20. Temperature dependence of photoelastic coefficients C_{xy}, C_{xz}, C_{yz}, C_{yx}, C_{zx} and C_{zy} of the crystal (x-irradiated) LGO in a cooling (0) and heating (Δ) cycle.

1.9. Irradiation Effect on piezo-optic Birefringence in $Li_2Ge_7O_{15}$ Crystals

The photoelastic coefficients C_{pq} of the ferroelectric crystals $Li_2Ge_7O_{15}$ (x-irradiated) in a cooling and heating cycle between 298 K and 273 K was carried out with the experimental procedure described in section 1.5 and are shown in Fig. 20 [54]. The results show an interesting photoelastic behaviour.

Peaks are observed in the temperature dependence of the photoelastic coefficients C_{zy} and C_{zx} at temperature ~ 279 K in a complete cooling and heating cycle whereas no discernible hysteresis is observed in rest of the photoelastic coefficients. Anomalous temperature dependence of C_{zx} of the crystal (x-irradiated) LGO at different wave lengths are shown in Fig.21.

It is observed that the peak value of C_{zy} has increased about 25% and that of C_{zx} has decreased about 18% at the wave length λ=5890 Å during cooling process of the crystal (Fig.15 and Fig.20). The peak value of C_{zx} of the crystal (un-irradiated and x-irradiated) LGO thus obtained at different wave lengths (Fig.16 and Fig.21) are given in Table 7 and the results are plotted in Fig.22.

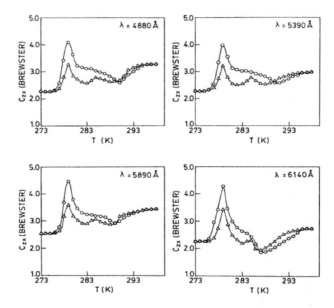

Figure 21. Temperature dependence of photoelastic coefficient C_{zx} of the crystal (x-irradiated) LGO at different wave lengths in a cooling (0) and heating (Δ) cycle.

It has been observed that the changes in the value of photoelastic coefficients C_{zy} and C_{zx} of the crystal (x-irradiated) LGO in a cooling and heating cycle occur only if the crystal is stressed along the polar axis (z-axis). It is known that the irradiation of crystals can change physical properties of the crystals.

Wave lengths	C_{zx} (un-irradiated)		C_{zx} (x-irradiated)	
(Å)	Cooling	Heating	Cooling	Heating
4880	4.8	4.0	4.05	3.3
5390	4.7	3.9	3.95	3.2
5890	5.6	4.8	4.6	3.7
6140	4.5	3.6	4.3	3.4

Table 7. The peak value of C_{zx} (in Brewster) for the Crystal (un-irradiated and x-irradiated) LGO at different wave lengths in the cooling and heating cycles.

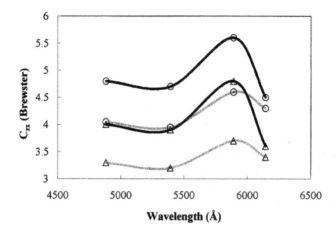

Figure 22. The peak value of C_{zx} for the un-irradiated (black colour) and x-irradiated (ash colour) crystal LGO at different wave lengths in a cooling (0) and heating (Δ) cycle.

Irradiation brings about many effects in the crystal such as creating defects, internal stress and electric fields etc [43]. In our present studies, the x-irradiation is believed to produce internal stress and electric fields inside the crystals $Li_2Ge_7O_{15}$ due to defects that can change the values of photoelastic coefficients.

2. Summary

It is known that the Barium strontium titanate $Ba_{1-x}Sr_xTiO_3$(BST) is one of the most interesting thin film ferroelectric materials due to its high dielectric constant, composition dependent curie temperature and high optical nonlinearity. The wavelength dependence of refractive index of BST ($Ba_{0.05}Sr_{0.95}TiO_3$) thin films has shown a nonlinear dependence in the 1450-1580 nm wavelength range at room temperature as described in section 1.1. The dispersion curve decreases gradually with increasing wavelength. The average value of the refractive index is found to be 1.985 in the 1450-1580 nm wavelength range which is considered to be important for optoelectronic device applications.

The study of fluorescence spectra of the crystals $Li_2Ge_7O_{15}:Cr^{3+}$ in the temperature interval 77-320 K shows the sharply decrease of intensities of the R_1 and R_2 lines (corresponding to the Cr^{3+} ions of types I and II) during cooling process near the temperature $T_c = 283.5$ K as described in section 1.3. Such nature of suppression of R_1 and R_2 lines was not observed previously and it may be related with the mechanism of interaction of excitation spectra of light in the crystals $Li_2Ge_7O_{15}:Cr^{3+}$ at the temperature T_c. The doping of chromium in LGO is believed to create Cr^{3+}- Li^+ defect pairs in the host LGO lattice at Ge^{4+} sites creating dipoles in two conjugate directions. The EPR, optical, dielectric and fluorescence studies conform each other and pose more scope for further studies.

The high optical quality, good mechanical strength and stability in ambient environment and large photoelastic coefficients in comparison with other ferroelectric crystals like Ro-chelle-salt, KDP and ADP favour the crystals LGO as a potential candidate for photoelastic applications. The piezo-optic dispersion of the crystals (un-irradiated and x-irradiated) LGO in the visible region of the spectrum of light at room temperature (298 K) have been described in sections 1.6 and 1.7. It shows an "optical zone or optical window" in between the wavelengths 5400 Å and 6200 Å with an enhanced piezo-optical behavior. This peculiar optical window can have a technical importance for example this window region can act as an optical switch for acousto-optical devices. From the studies undertaken it may be concluded that LGO is an attractive acousto-optic material which deserves further probe. It may be possible to understand the observed behavior if extensive piezo-optic and refractive index data become available over an extended range of wavelengths.

The temperature dependence of the photoelastic coefficients of the crystals (un-irradiated and x-irradiated) LGO in a cooling and heating cycle between room temperature (298 K) and 273 K have shown an interesting observations: lowering of the T_c under uniaxial stress contrary to the increase of T_c under hydrostatic pressure and observation of thermal photoelastic hysteresis similar to dielectric hysteresis behavior as described in sections 1.8 and 1.9. In our studies, the x-irradiation is believed to produce internal stress and electric fields inside the crystals LGO due to defects that can change the values of photoelastic coefficients, as described in sections 1.7 and 1.9.

Author details

A.K. Bain* and Prem Chand

Department of Physics Indian Institute of Technology Kanpur, Kanpur-208016, INDIA

References

[1] Lepingard, F. A., Kingston, J. J., & Fork, D. K. (1996). Second harmonic generation in LiTaO$_3$ thin films by modal dispersion and quasi phase matching. *Appl. Phys. Lett.*, 68, 3695-3697.

[2] Hewing, G. H., & Jain, K. (1983). Frequency doubling in a LiNbO$_3$ thin film deposited on sapphire. *J. Appl. Phys.*, 54, 57-61.

[3] Lee, S. H., Kim, D. W., Noh, T. W., Lee, J. H., Lim, M. J., & Lee, S. D. (1996). Nonlinear optical properties of epitaxial LiNbO$_3$ film prepared by pulsed laser deposition. *J. Korean Phys. Soc.*, 29, S628-S631.

[4] Moon, S. E., Back, S. B., Kwun, S. I., Lee, Y. S., Noh, T. W., Song, T. K., & Yoon, J. G. (2000). Orientational Dependence of Electro-optic Properties of SrBi$_2$Ta$_2$O$_9$ Ferroelectric Thin Films. Jpn. J. Appl. Phys., 39, 5916-5917.

[5] Reitze, D. H., Haton, E., Ramesh, R., Etemad, S., Leaird, D. E., & Sands, T. (1993). Electrical and Electro-optic Properties of Single Crystalline, Epitaxial Thin Films Grown on Silicon Substrates. *Appl. Phys. Lett.*, 63(5), 596-598.

[6] ORNL's. *Thin film Waveguide and the Information Highway by Carolyn Krause.*

[7] Varadam, V. K., Varadam, V. V., Selmi, F., Ounaise, Z., & Jose, K. A. (1994). Multilayer Tunable Ferroelectrci Materiasl and Thin Films. SPIE-Int. *Soc. Opt. Eng.*, 2189, 433-447.

[8] Mohammed, M. S., Naik, R., Mantese, J. V., Schubring, N. W., Micheli, A. L., & Catalan, A. B. (1996). Microstructure and ferroelectric properties of fine-grained Ba$_x$Sr$_{1-x}$TiO$_3$ thin films prepared by metalorganic decomposition. *J. Mater. Res.*, 11, 2588-2593.

[9] Bain, A. K., Jackson, T. J., Koutsonas, Y., Cryan, M., Yu, S., Hill, M., Varrazza, R., Rorison, J., & Lancaster, M. J. (2007). Optical Properties of Barium Strontium Titanate (BST) Ferroelectric Thin Films. *Ferroelectric Letters*, 34, 149-154.

[10] Wada, M., Swada, A., & Ishibashi, Y. (1981). Ferroelectricity and Soft Mode in Li$_2$Ge$_7$O$_{15}$. *J. Phys. Soc. Jpn.*, 50(6), 1811-1812.

[11] Wada, M., & Ishibashi, Y. (1983). Ferroelectric Phase Transition in Li$_2$Ge$_7$O$_{15}$. *J.Phys. Soc. Jpn .*, 52(1), 193-199.

[12] Preu, P., & Hauss Ühl , S. (1982). Dielectric Properties and Phase Transition in Li$_2$Ge$_7$O$_{15}$. *Solid State Commun*, 41(8), 627-630.

[13] Kaminsky, W., & Hauss, Ühl. S. (1990). Faraday effect aid birefringence in orthorhombic Li$_2$Ge$_7$O$_{15}$ near the ferroelectric phase transition. *Ferroelectrics Lett.*, 11(3), 63-67.

[14] Hauss, Ühl. S., Wallrafen, F., Recker, K., & Eckstein, J. (1980). Growth, Elastic Properties and Phase Transition of Orthorombic $Li_2Ge_7O_{15}$. *Z.krist.*, 153, 329-337.

[15] Kudzin, A., Yu, Volnyanskii M. D., & Bain, A. K. (1994). Temperature Hysteresis of the Permittivity of $Li_2Ge_7O_{15}$. *Phys. Solid State*, 36(2), 228-230.

[16] Kudzin, A., Yu, Volnyanskii M. D., & Bain, A. K. (1995). Influence of Space Charges on Ferroelectric Property of Weak Ferroelectric $Li_2Ge_7O_{15}$. *Ferroelectrics*, 164(1), 319-322.

[17] Bain, A.K. (1994). Study of Pecularities of Ferroelectric Phase Transition of the Crystals $Li_2Ge_7O_{15}$ (Ph.D. Thesis). *Dniepropetrovsk State University, Ukraine*, 128.

[18] Bain, A. K., Chand, P., Rao, K. V., Yamaguchi, T., & Wada, M. (2008). Irradiation Effect on Piezo-optic Dispersion of $Li_2Ge_7O_{15}$ Crystals. *Ferroelectrics*, 377(1), 86-91.

[19] Bain, A. K., Chand, P., Rao, K. V., Yamaguchi, T., & Wada, M. (2009). Anomalous Temperature Dependence of Piezo-optic Birefrigence in $Li_2Ge_7O_{15}$ Crystals. *Ferroelectrics*, 386(1), 152-160.

[20] O'Mahony, M.J. (1988). Semiconductor laser optical amplifers for use in future fiber systems. *J. Light wave Technology*, 6(4), 531-544.

[21] Miranda, F. A., Vankeuls, F. W., Romanofsky, R. R., Mueller, C. H., Alterovitz, S., & Subramanyam, G. (2002). Ferroelectric thin-film based technology for frequency and phase agile microwave communication applications. *Integrated Ferroelectrics*, 42, 131-149.

[22] Tian, H. Y., Luo, W. G., Pu, X. H., Qiu, P. S., He, X. Y., & Ding, A. L. (2001). Synthesis and characteristics of strontium-barium titanate graded thin films at low temperature using a sol-gel technique. *Solid State Communications*, 117, 315-319.

[23] Pontes, F. M., Leite, E. R., Pontes, D. S. L., Longo, E., Santos, E. M. S., Mergulhao, S., Pizani, P. S., Lanciotti Jr, F., Boschi, T. M., & Varela, J. A. (2002). Ferroelectric and optical properties of $Ba_{0.8}Sr_{0.2}TiO_3$ thin film. *J. Appl. Phys.*, 91(9), 5972-5978.

[24] Deineka, A., Jastrabik, L., Suchaneck, G., & Gerlach, G. (2002). Optical Properties of Self-Polarized PZT Ferroelectric Films. *Ferroelectrics*, 273, 155-160.

[25] Gu, Y. Z., Gu, D. H., & Gan, F. X. (2001). Optical Nonlinearity in $PbTiO_3$ Thin Film Deposited on $Al2O_3$ with RF-Sputtering System. *Nonlinear Optics*, 28(4), 283-289.

[26] Fernandez, F. E., Gonzalez, Y., Liu, H., Martinez, A., Rodriguez, V., & Jia, W. (2002). Structure, morphology, and properties of strontium barium niobate thin films grown by pulsed laser deposition. *Integrated Ferroelectrics*, 42, 219-233.

[27] Wohlecke, M., Marrello, V., & Onton, A. (1977). Refractive index of $BaTiO_3$ and $SrTiO_3$ films. *J. Appl. Phys.*, 48, 1748-1750.

[28] Wada, M., Swada, A., & Ishibashi, Y. (1984). The Oscillator Strength of the Soft Mode in $Li_2Ge_7O_{15}$. *J. Phys. Soc. Jpn.*, 53(10), 3319-3320.

[29] Iwata, Y., Shibuya, I., Wada, M., Sawada, A., & Ishibashi, Y. (1987). Neutron Diffraction study of structural Phase Transition in Ferroelectric $Li_2Ge_7O_{15}$. *J. Phys. Soc. Jpn.*, 56(7), 2420-2427.

[30] Wada, M. (1988). Soft Mode Spectroscopy Study of Ferroelectric Phase Transition in $Li_2Ge_7O_{15}$. *Ind. J. Pure and Appl. Phys.*, 26, 68-71.

[31] Galeev, A. A., Hasanova, N. M., Bykov, A. B., Vinokurov, B. M., Nizamutdinov, N. M., & Bulka, G. R. (1990). EPR of Cr^{3+} and Fe^{3+} in $Li_2Ge_7O_{15}$ single crystal. In: Spectroscopy, a crystal chemistry and a realstructure of minerals and their analogues. *Kazan' state university*, 77-87, in Russian).

[32] Basun, S. A., Kaplyanski, A. A., & Feofilov, S. P. (1994). Dipolar centres in $Li_2Ge_7O_{15}$ crystal activated with ($3d^3$) ions: a microstructure and spectroscopic effects of an internal and external electric field. *Solid State Phys.*, 36(11), 3429-3449.

[33] Kaplyanski, A. A., Basun, S. A., & Feofilov, S. P. (1995). Ferroelectric transition induced dipole moments in probe ions in $Li_2Ge_7O_{15}$ crystals doped with Mn^{4+} and Cr^{3+} . *Ferroelectrics*, 169, 245-248.

[34] Iwata, Y., Koizumi, H., Koyano, N., Shibuya, I., & Niizeki, N. (1973). Crystal structure determination of ferroelectric phase of $5PbO_3GeO_2$. *J. Phys. Soc. Jap.*, 35(1), 314-315.

[35] Volnianskii, M. D., Trubitsyn, M. P., & Obaidat, Yahia A. H. (2007). EPR and dielectric spectroscopy of reorienting Cr^{3+}- Li^+ pair centres in $Li_2Ge_7O_{15}$ Crystal. *Condensed Matter Physics*, 10(49), 75-78.

[36] Trubitsyn, M. P., Volnianskii, M. D., & Busoul, I. A. (1998). EPR study of the ferroelectric phase transition in $Li_2Ge_7O_{15}{:}Cr^{3+}$ crystal. *Solid State Phys.*, 40, 6, 1102-1105.

[37] Powell, R.C. (1968). Temperature Dependence of the Widths and Positions of the R Lines in $Li_2Ge_7O_{15}{:}Cr^{3+}$. *J. Appl. Phys.*, 39, 4517-4521.

[38] Powell, R.C. (1968). Energy Transfer between Chromium Ions in Nonequivalent Sites in $Li_2Ge_7O_{15.}$ *Phys. Rev.*, 173, 358-366.

[39] Narsimhamurty, T.S. (1981). Photoelastic and Electro-optic properties of Crystals. *New York: Plenum Press.*, 514.

[40] Bain, A. K., Chand, P., Rao, K. V., Yamaguchi, T., & Wada, M. (2008). Piezo-optic Dispersion of $Li_2Ge_7O_{15}$ Crystals. *Ferroelectrics*, 366(1), 16-21.

[41] Saito, K., Ashahi, T., Takahashi, N., Hignao, M., Kamiya, I., Sato, Y., Okubo, K., & Kobayashi, J. (1994). Optical activity of $Gd_2(MoO_4)_3$. *Ferroelectrics*, 152(1), 231-236.

[42] Vanishri, S., & Bhat, H. L. (2005). Irradiation Effects on Ferroelectric Glycine Phosphite Single Crystal. *Ferroelectrics*, 323(1), 151-156.

[43] Lines, M. E., & Glass, A. M. (2004). Principles and Applications of Ferroelectrics and Related Materials. *Oxford: Clarendon press*, 680.

[44] Bain, A. K., Chand, P., Rao, K. V., Yamaguchi, T., & Wada, M. (1998). Determination of the Photoelastic Coefficients in Lithium heptagermanate Crystals. *Ferroelectrics*, 209(1), 553-559.

[45] Hauss, Ühl. S., & Albers, J. (1977). Elastic and thermoelastic constants of triglycine sulphate (TGS) in the paraelectric phase. *Ferroelectrics*, 15(1), 73-75.

[46] Wada, M., Orihara, H., Midorikawa, M., Swada, A., & Ishibashi, Y. (1981). Pressure Effect on the Ferroelectric Phase Transition in $Li_2Ge_7O_{15}$. *J. Phys. Soc. Jpn.*, 50(9), 2785-2786.

[47] Morioka, Y., Wada, M., & Swada, A. (1988). Hyper-Raman Study of Ferroelectric Phase Transition of $Li_2Ge_7O_{15}$. *J. Phys. Soc. Jpn.*, 57(9), 3198-3203.

[48] Wada, M., Shirawachi, K., & Nishizawa, S. (1991). A Fourier Transform Infrared Spectrometer with a Composite Interferometer for Soft Mode Studies. *Jpn. J. Appl. Phys.*, 30(5), 1122-1126.

[49] Sinii, I. G., Fedoseev, A. I., & Volnyanskii, M. D. (1990). Relaxation- and fluctuation-induced damping of hypersound in the presence of dispersion. *Sov. Phys. Solid State*, 32(10), 1817-1818.

[50] Narasimamurty, T. S. (1969). Piezooptical constants of Rochelle salt crystals. *Phys. Rev.*, 186, 945-948.

[51] Veerabhadra, Rao K., & Narasimamurty, T. S. (1975). Photoelastic behavior of KDP. *J. Mat. Sci.*, 10(6), 1019-1021.

[52] Narasimamurty, T. S., Veerabhadra, Rao. K., & Petterson, H. E. (1973). Photoelastic constants of ADP. *J. Mat. Sci.*, 8(4), 577-580.

[53] Trubitsyn, M. P., Kudzin, A., Yu, Volnyanski. M. D., & Bain, A. K. (1992). Critical Broadening of EPR lines near the Ferroelectric Phase Transition in $Li_2Ge_7O_{15}$:Mn^{2+} . Sov. Phys. Solid State , 34(6), 929-932.

[54] Bain, A. K., & Chand, P. (2011). Irradiation Effect on Photoelastic Coefficients in Ferroelectric $Li_2Ge_7O_{15}$ Crystals. *Integrated Ferroelectrics*, 124, 10-18.

Chapter 10

Electronic Ferroelectricity in II-VI Semiconductor ZnO

Akira Onodera and Masaki Takesada

Additional information is available at the end of the chapter

1. Introduction

II-VI semiconductor Zinc oxide (ZnO) is a well-known electronic material [1-3]. Because of large piezoelectric constant and electromechanical-coupling constant, ZnO has been applied to ultrasonic transducer, SAW filter, gas sensor *etc* [3-5]. In addition, ZnO has been used widely as pigment and UV cut cosmetics extensively, so it is a safe material for our living environment, compared with heavy metals used in semiconducting process and materials science. Recently, ZnO has been studied as a material for the solar cell, transparent conductors formed on liquid crystal displays, and the blue laser [6]. Then, *p*-type ZnO was found by Joseph *et al* [7].

Ferroelectricity is recognized to appear mainly by the delicate balance between a long-range dipole-dipole interaction and a short-range interaction. When electrons should be well localized in dielectrics, the electronic distribution in the unit cell is determined by atomic positions of constituent ions. Slight distortion of electronic distribution due to structural changes in dielectrics gives a rise of dipole moments. From this point of view, the ferroelectric phase transition should be understood as a structural phase transition from the paraelectric phase with high symmetry to the ferroelectric phase with low symmetry. The atomic displacements are generally 0.01~0.1 Å, which are small compared with Bohr radius (0.53 Å). Therefore ferroelectrics are considered as a group of materials which are sensitive to structural changes. In this sense, the ferroelectric phase transition is classified into one of the structural phase transitions.

The basic Hamiltonian is generally given as

H= H(phonon) + H(electron) + H(electron-phonon),

where H(phonon) and H(electron) are due to motions of ion cores and valence electrons, and H(electron-phonon) presents interactions between ions and valence electrons. In dielectric materials, the contribution of electron systems is usually omitted since the band gap is generally wide. However, it should be necessary to consider the electron-phonon interaction in the case of ferroelectric semiconductors, since the correlation energy of dipole-dipole interaction (about 0.2 eV) is comparable with band-gap energy in some narrow-gap and wide-gap semiconductors. For this type of compounds, the electronic contribution due to bond charges and conduction electrons should play a key role and must be taken into account for understanding the nature of ferroelectricity. For long time, the importance of electronic contribution has been pointed out in the field of ferroelectrics, since the simple superposition of electronic polarizability does not hold in many ferroelectric substances. Recently, the novel ferroelectricity was discovered in narrow-gap and wide-gap semiconductors such as $Pb_{1-x}Ge_xTe$ [8], $Cd_{1-x}Zn_xTe$ [9] and $Zn_{1-x}Li_xO$ [10].The appearance of ferroelectricity is primarily due to electronic origin in $Zn_{1-x}Li_xO$. Although the ferroelectric phase transition accompanies with structural distortions in usual ferroelectrics, only small structural changes of the order of 10^{-3} Å are observed in $Zn_{1-x}Li_xO$. The change in d-p hybridization caused by Li-substitution is responsible for the novel ferroelectricity and dielectric properties. In this chapter, we summarized recent works mainly onZnO.

2. Zinc oxide

The crystal Structure of ZnO is wurtzite-type ($P6_3mc$) as shown in Fig. 1, which belongs to hexagonal system. It does not have the center of symmetry, and is polar along the c-axis. Although it has been pointed out that the structure of pure ZnO has the possibility to exhibit ferroelectricity, the polarization switching does not observed until its melting point (1975°C) because of large activation energy accompanied by dipole switching process. The lattice constants are a = 3.249858 Å, c= 5.206619 Å at room temperature (298 K) [11]. Zn and O ions are bonded tetrahedrally and form ZnO_4 groups. This tetrahedron is not perfect: the apical bond length of Zn-O is 1.992 Å (parallel to the c-axis) and the basal one is 1.973 Å. Bond length along the c-axis is longer than other three bonds by 0.96%: the ZnO_4tetrahedra distort along the c-axis, which result in dipole moment. According to recent first-principles studies, it was reported that the hybridization between the Zn $3d$-electron and the O $2p$-electron plays an important role for dielectric properties of ZnO [12, 13].The energy gap E_g is 3.44 eV [14]. Although stoichiometric ZnO is an insulator (intrinsic semiconductor), it exhibits n-type conductivity because of excess Zn atoms. The resistivity is about 300 Ωcm, which, however, changes drastically by doping various impurities. By doping of trivalent ions, such as Al^{3+}, In^{3+}, Ga^{3+}, it reduces to the order of 10^{-4}Ωcm and shows the conductivity near that of metals [15].

Although p-type conductivity is expected by doping of monovalent Li^+ ions, ZnO becomes an insulator and the resistivity increases to as much as 10^{10}Ωcm [16-18]. It is remarkable that the change in resistivity reaches over the order of 10^{14}, which covers from metal-like to insulating region. It suggests that the physical nature in this material changes drastically by a little amount of dopant.

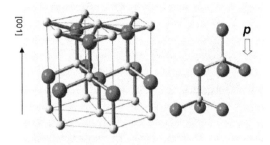

Figure 1. Crystal structure of ZnO (Wurtzite structure).

3. Electronic structure

Electronic structure of ZnO has been studied by LMTO (linear muffin-tin orbital) method [19] and by LAPW (linearized augmented-plane-wave) method using LDA (local density approximation). Band structure and the density of states is shown in Figs. 2 and 3 [20]. Around -17 eV, there are two bands originating from the O 2s-states. The narrow bands between -6 and -4 eV consist mainly of the Zn 3d-orbitals, and moderately dispersive bands from -4 to 0 eV consist mainly of the O 2p-orbitals. Figure 4 shows significant d-p hybridization. The 3d derived bands split into two groups, leading to double-peak structure in DOS (density of states). The lower peak is characterized by a strong d-p hybridization. The sharp upper peak between -4.8 and -4.2 eV has strong Zn 3d character and the hybridization with the O 2p-state is very small. Band gap is 0.77 eV. Discrepancy with the experimental value is larger than that of LMTO. It is due to the LDA gap error [21].

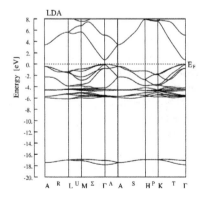

Figure 2. Band structure of ZnO using LAPW.

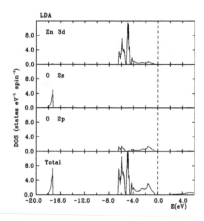

Figure 3. Density of states of ZnO using LAPW.

4. Ferroelectricity in binary semiconductors

Recently, the novel ferroelectricity was found in Li-doped ZnO [10, 22, 23], although pure ZnO has not shown any evidence of ferroelectricity. Generally representative ferroelectrics have complicate crystal structures such as Rochelle salt ($KNaC_4H_4O_6$ $4H_2O$), TGS (tri-glycine sulfate, $(NH_3CH_2COOH)_3H_2SO_4$) and $BaTiO_3$. Since ZnO is a simple binary compound, it is very convenient to study the microscopic mechanism and its electronic contribution for the appearance of ferroelectricity. Besides ZnO, IV-VI narrow-gap semiconductor $Pb_{1-x}Ge_xTe$ and II-VI wide-gap semiconductor $Cd_{1-x}Zn_xTe$, are known as materials of binary crystals accompanying ferroelectricity.

4.1. IV-VI Narrow-gap semiconductor

Among IV-VI semiconductor, $Pb_{1-x}Ge_xTe$ has been investigated extensively about its ferroelectricity [8]. $Pb_{1-x}Ge_xTe$ has a NaCl (rock-salt) type structure at room temperature. Although any ferroelectric activities have not been observed in pure PbTe and GeTe crystals, the ferroelectric phase transition is induced only in solid solution $Pb_{1-x}Ge_xTe$. In the low temperature phase of $Pb_{1-x}Ge_xTe$, the crystal becomes rhombohedral and exhibits ferroelectric activity. In the case of $x = 0.003$, it shows a large dielectric anomaly of the order of 10^3 at T = 100 K as shown in Fig. 5 and the softening of TO mode was observed.

The energy gap of $Pb_{1-x}Ge_xTe$ is no more than 0.3 eV, which is 3000 K in temperature. This value is comparable to the energy of the Lorentz field of dielectrics, $(4\pi/3)P$. Therefore, conduction electrons can couple strongly with phonons in this solid solution. The electron-phonon interaction decreases the frequency of TO phonon mode which results in the ferroelectric phase transition. In $Pb_{1-x}Ge_xTe$, doped Ge ions displace from the center position of Pb ion by 0.8 Å and behave as *off-center ions* because of the ionic size-mismatch between Pb ion (the ionic radius: 1.2 Å) and Ge ion (the ionic radius: 0.73 Å) [24]. The direction of displacement is the eight equivalent [111] directions, which is spatially vacant and polar in the rock-salt structure. Above T_c, the ions shift toward any one of these directions at random, but its displacement would be ordered along the trigonal axis below T_c. The ordering of the *off-center ion* triggers the softening of TO phonon mode and induces rhombohedral distortion in the way such as $\cdots Pb^{2+}-Te^{2-}\cdots Pb^{2+}-Te^{2-}\cdots Pb^{2+}-Te^{2-}\cdots$ or $\cdots Te^{2-}-Pb^{2+}\cdots Te^{2-}- Pb^{2+}\cdots Te^{2-}-Pb^{2+}\cdots$. This structural ordering generates spontaneous polarization. However, in $Pb_{1-x}Ge_xTe$, the *D-E* hysteresis loop has not been observed, while it is a direct evidence of the ferroelectricity. Whereas traditional dielectric measurements are performed on a parallel-plate capacitor, it is difficult to measure dielectric constant of such lossy $Pb_{1-x}Ge_xTe$ because of current leaks. Therefore dielectric constant is determined by the *C-V* measurement or the optical reflectiveity, which are commonly used in semiconductors. The large dielectric anomaly and soft mode of $Pb_{1-x}Ge_xTe$ suggest the ferroelectric activity.

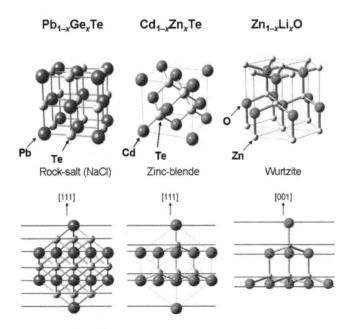

Figure 4. Crystal structures of $Pb_{1-x}Ge_xTe$, $Cd_{1-x}Zn_xTe$ and $Zn_{1-x}Li_xO$. The lower figures are plots along the polar [111] directions for $Pb_{1-x}Ge_xTe$ and $Cd_{1-x}Zn_xT$, and polar [0 0 1] direction for $Zn_{1-x}Li_xO$.

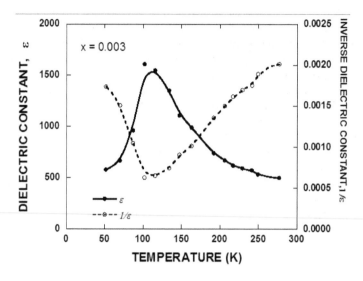

Figure 5. Temperature dependence of dielectric constant in $Pb_{1-x}Ge_xTe$ (x=0.003).

4.2. II-VI Wide-gap semiconductor $Cd_{1-x}Zn_xTe$

Among many wide-gap semiconductors, the ferroelectricity of $Cd_{1-x}Zn_xTe$ was discovered by R. Weil *et al* in 1988 [9, 25]. The E_g of $Cd_{1-x}Zn_xTe$ is 1.53 eV. It is a wide-gap semiconductor unlike the narrow-gap IV-VI $Pb_{1-x}Ge_xTe$. The crystal structure is cubic zinc-blende-type. Cd(Zn) ion is surrounded by four Te ions tetrahedrally. The center of symmetry in this compound does not exist. When x=0.1, the crystal exhibits a dielectric anomaly at 393 K as shoen in Fig. 6. The peak value of dielectric constant (ε_{peak}) is only 50, which is smaller by about two orders than that of $BaTiO_3$ (~14000) and that of $Pb_{1-x}Ge_xTe$ (~1000). Because of the large E_g and the resistivity (~kΩ), a ferroelectric *D-E* hysteresis loop was successfully observed in the low-temperature phase. The direction of spontaneous polarization is along the apex of a tetrahedron, [111], which is reported as P_s= 0.0035 μC/cm^2 in their first paper, and 5 μC/cm^2 in the second paper. Because the ionic radius of Zn ion (0.83 Å) is smaller than that of Cd ion (1.03 Å), Zn ion also locates at the *off-center* position which deviates from the center of tetrahedra toward the apex. According to the result of EXAFS (X-ray absorption fine structure), it shifts by 0.04 Å [26]. It is considered that the ordering of *off-center ions* causes rhombohedral distortion of the cubic lattice by about 0.01 Å, and induces ferroelectricity.

In $Cd_{1-x}Zn_xTe$, the soft mode has not been observed and the dielectric anomaly is small. These dielectric properties are different from those found in $Pb_{1-x}Ge_xTe$, which are summarized in Table 1. These facts are consistent each other when one consider the LST (Lyddane-Sachs-Teller) relation ($\omega_{LO}^2/\omega_{TO}^2 = \varepsilon/\varepsilon_\infty$). In $Pb_{1-x}Ge_xTe$, the existence of soft mode is responsible for large dielectric anomaly. Although the behavior of *off-center ions* plays an important role in this ferroelectricity like $Pb_{1-x}Ge_xTe$, the occurrence of phase transition seems to be driven in different way from that of $Pb_{1-x}Ge_xTe$, because of the different dielectric properties described above. Furthermore, the band gap is much larger than that of the narrow-gap ferroelectric semiconductors, which suggests that the electron-phonon coupling is not so strong in $Cd_{1-x}Zn_xTe$ solid solutions.

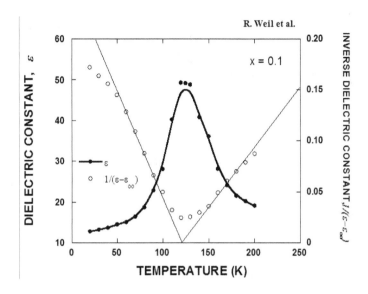

Figure 6. Temperature dependence of dielectric constant in $Cd_{1-x}Zn_xTe$ ($x=0.1$).

4.3. $Zn_{1-x}Li_xO$

Dielectric constants of $Zn_{1-x}Li_xO$ ceramics with $x=0.09$ show an anomaly at 470 K (T_c) (Fig. 7), although a high purity of ZnO does not show any anomaly from 20 K to 700 K [27]. The peak value of dielectric anomaly is 21 ($x=0.1$), which is the same order with $Cd_{1-x}Zn_xTe$ ($\varepsilon \sim 50$), but much smaller than ordinal ferroelectrics by 2~4 orders. This means that the dipole-dipole correlation accompanied with this ferroelectric phase transition is not so large in $Zn_{1-x}Li_xO$. In the measurement of D-E hysteresis loop, a small and clear hysteresis curve was observed. The spontaneous polarization varies by samples from 0.05 $\mu C/cm^2$ to 0.59 $\mu C/cm^2$. This is due to that the preferred orientation of the c-axis, the direction of P_s of the samples varies depending on samples. Naturally, powder of ZnO has egg-shaped grains which are elongated along the c-axis due to the anisotropy of elastic constants. Therefore, the orientation of the ceramic sample depends on the condition of the pressing process in the sample preparation. After the correction of the preferred orientation by using X-ray diffraction, the value of spontaneous polarizations of each sample converges to 0.9 $\mu C/cm^2$ as shown in Fig. 8 [28, 29]. The transition temperature, T_c, depends on the Li concentration. The phase diagram between T_c and x is shown in Fig. 9. As temperature increases, T_c becomes higher, which reminds us the phase diagram of quantum ferroelectrics such as $KTa_{1-x}Nb_xO_3$ and $Sr_{1-x}Ca_xTiO_3$.

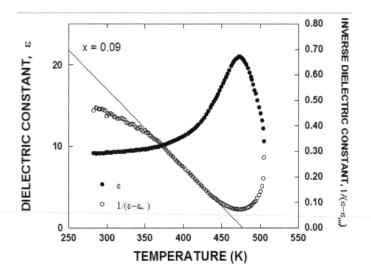

Figure 7. Temperature dependence of dielectric constant of $Zn_{1-x}Li_xO$ ceramics (x= 0.09).

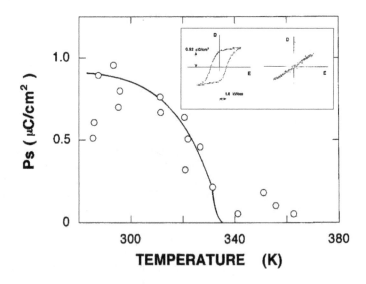

Figure 8. Temperature dependence of spontaneous polarization of $Zn_{1-x}Li_xO$ ceramics(x = 0.06). The inset is D-Ehyteresis loop observed below (left) and above T_c (right).

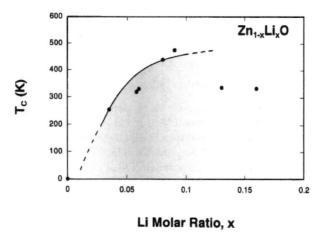

Figure 9. The T_c-x phase diagram of $Zn_{1-x}Li_xO$.

Specific heat anomaly is small at the ferroelectric-paraelectric phase transition temperature. The transition entropy ΔS is almost 0. If Li ions which substitute Zn ions occupy the off-center position and behave an order-disorder motion, ΔS must be about $R\ln 2$. If the nature of the transition is displacive type, $\Delta S \simeq 0$. The observed thermal behavior suggests displacive nature, but, however, no soft mode was observed in Raman scattering measurement [30]. The small dielectric anomaly and the absence of soft mode are consistent each other, when we consider the LST-relation. Relaxation mode corresponding to an order-disorder motion has not been found also by Raman scattering [31]. These peculiar dielectric properties of $Zn_{1-x}Li_xO$ is resemble to those of $Cd_{1-x}Zn_xTe$. These evidences suggests a new type of ferroelectric phase transition.

	$Pb_{1-x}Ge_xTe$	$Cd_{1-x}Zn_xTe$	$Zn_{1-x}Li_xO$
Crystal Structure of Paraelectric Phase	Rock-salt (NaCl) (($Fm3m$))	Zinc-blende (($F\bar{4}3m$))	Wurtzite (($P6_3mc$))
Ferroelectric Phase	Rhombohedral (($R3m$))	Rhombohedral (($R3m$))	Wurtzite (($P6_3mc$))
T_c (K) (x=0.1)	200	390	470
E_g (eV)	0.3	1.53	3.2
ρ (Ωm)	10	10^3	10^{10}
P_s (μC/cm^2)	-	5	0.9
Soft mode	o	x	x
ε_{peak}	10^3	50	21

Table 1. Dielectric properties of binary ferroelectric semiconductors.

5. Electronic ferroelectricity in ZnO

It is considered that the replacement of host Zn atoms by substitutional Li atoms plays an primary role for the appearance of ferroelectricity in ZnO. The problem is the effect of Li-doping. Here we consider the following two models [32].

i.   Structural size-mismatch model

Pure ZnO is polar along the c-axis and has dipole moments in crystals. Because of the size-mismatch between Zn ion (the ionic radius: 0.74 Å) and Li ion (the ionic radius: 0.60 Å), substituted Li ions displace from the Zn positions. These displacements may force to induce extra dipole moments locally. It is considered that these local dipoles couple with dipoles of the mother ZnO crystal and the structural ordering of local dipoles triggers to induce a ferroelectric phase transition.

ii.   Electronic model

According to the first-principle study by Corso *et al* [10], the hybridization between Zn $3d$ electron and O $2p$ electron plays an important role in dielectric properties of ZnO. As the Li atom has no d-electrons, it is considered that the partial replacement of Zn ions by Li ions changes the nature of d the d-p hybridization. Therefore this doping is considered to induce local extra dipole moments. These extra local dipoles couple with parent dipoles and induce a new type of ferroelectricity.

To clarify which model is effective in the appearance of ferroelectricity in ZnO, the following structural and dielectric measurements were performed.

5.1. Rietveld analysis of structural changes in $Zn_{1-x}Li_xO$ ceramics

Structural changes associated with Li-substitution were studied by X-ray powder diffraction [33]. Ceramic samples of $Zn_{1-x}Li_xO$ prepared by using a SPS (spark plasma sintering) method were used in this experiment. The nominal value of x is 0.1 for the first sample. The T_c of the sample was 470 K which is determined by dielectric constant measurements. The systematic check of the absence of possible reflections (h-k=$3n$ and l= odd for (hkl), n = integer) suggests that the space group of the ferroelectric phase still remains $P6_3mc$, which is same as pure ZnO. The obtained pattern was analyzed by the Rietveld method including the Li concentration, x, as a refinable parameter. All parameters refined are shown in Table 2. The lattice constants are a = 3.2487(1) and c = 5.2050(1) Å at 293 K. The actual Li concentration x is determined to be 0.09. The final discrepancy factors were R_{wp}=7.4%, R_p=5.4% and R_I=5.9%. The Zn-O bond lengths are 1.988_7 Å along the c-axis and 1.973_5Å in the basal plane, respectively. This means that the bond length along the c-axis shows a decrease by an amount of 0.003 Å by Li-substitution in Li-doped ZnO. This lattice distortion is the order of 10^{-3} Å only, while the displacement of Ti ions is the order of 0.1 Å in well-known ferroelectric $BaTiO_3$. In this phase transition, the crystal symmetry does not change associated with Li-substitution and structural change is considerably small compared with the structural phase transition in typical ferroelectrics.

Atom	x	y	z	B	Occupation
Zn	1/3	2/3	0.3821(3)	0.52	0.91
Li	1/3	2/3	03821(3)	0.31	0.09
O	1/3	2/3	0	0.40	1.0

Table 2. Positional (x,y,z) and isotropic thermal (B) parameters in $Zn_{1-x}Li_xO$ at 293K.

5.2. Effect of Be and Mg dopants on T_c

The effect of dopants on T_c was studied based on the two models mentioned above. If the ionic size-mismatch between the substituted and the host ions is important for ferroelectricity, the introduction of Be ions should be more effective than Li and Mg doping. The ionic radii of Li^+, Be^{2+} and Mg^{2+} ions are 0.60 Å, 0.3 Å and 0.65 Å, respectively. If the changes in electronic configuration are important, the Mg^{2+} ion $(1s^22s^22p^6)$ should play a different role from the isoelectronic Li^+ and Be^{2+} ions $(1s^2)$. Dielectric constant measurements were done on Mg- and Be-doped ZnO ceramics [32]. A dielectric anomaly of $Zn_{0.9}Be_{0.1}O$ was found at 496 K, which is very similar to those observed in $Zn_{1-x}Li_xO$. However, $Zn_{1-x}Mg_xO$ samples show a clear decrease of T_c. The dielectric anomaly in 30% Mg-doped ZnO found at 345 K is 150 K lower than the Li-doped one. The concentration dependence of T_c of Be- and Mg-doped ZnO is summarized in Fig. 10. The series of dielectric measurements show that the introduction of Mg^{2+} suppresses T_c, while isoelectronic Be^{2+} shows almost the same T_c.

It is considered that the appearance of ferroelectricity in ZnO is primarily due to electronic origin. The change in d-p hybridization caused by Li-substitution is responsible for this novel ferroelectricity and dielectric properties.

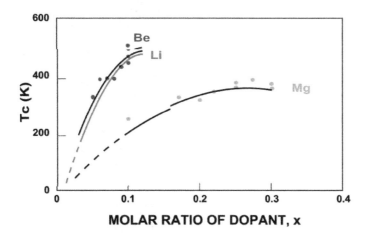

Figure 10. T_c vs. molar concentrations(x) of dopants, Li, Be and Mg in ZnO.

In the next section, we shall study the change in electronic distribution, especially in the nature of d-p hybridization by Li-doping directly by X-ray diffraction, and discuss new electronic ferroelectricity in ZnO.

6. X-ray study of electronic density distribution

X-ray diffraction measurements were performed for single crystals of pure ZnO and $Zn_{1-x}Li_xO$ at 297 K and 19 K in order to investigate the changes in the crystal structure and the electronic density distribution by Li-substitution in detail. Single crystal of pure ZnO was prepared by the hydrothermal method [34]. The content of excess Zn ions of obtained single crystal was 1.7 ppm and color was light yellowish (Fig. 11). In this hydrothermal method, it is rather difficult to add a large amount of Li ions into single crystal. In order to dope Li ions, several c-plate samples (0.16 mm thick) were annealed at an atmosphere of Li ions at 920 K for 24 hours. The light yellowish color of as-grown single crystal became transparent after this doping as shown in Fig. 12. Li concentration x was measured by chemical analysis and it is confirmed that $x = 0.082$~0.086.

Figure 11. Single crystal of as grown ZnO.

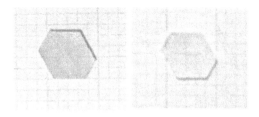

Figure 12. Single crystal of pure (left) and Li-doped ZnO (right).

Table 3 shows an experimental data of X-ray diffraction performed at room temperature (293 K) and low temperature (19 K) by using a He-gas closed-cycle cryostat (Cryogenic RC-110) mounted on a Huber off-center four-circle diffractometer to reduce the effect of thermal vibration. The space group is confirmed to be P6$_3$mc for both pure and Li-doped crystals. The Lorentz and polarization, thermal diffuse scattering (TDS) (the elastic constants used in TDS collection are C$_{11}$=20.96, C$_{33}$=21.09, C$_{12}$=12.10, C$_{13}$=10.51, C$_{44}$=4.243 [10^{11}dyn/cm^2], absorption (spherical) and extinction (anisotropic, type I) corrections were made. Crystal structures were refined using a full-matrix least squares program (RADIEL) [35].

	ZnO		Zn$_{1-x}$Li$_x$O	
Temperature	293 K	19 K	293 K	19 K
X-ray Radiation	AgKα	AgKα	MoKα	AgKα
(sinθ/λ)$_{max}$	1.36	1.36	1.00	1.27
Number of Reflections ($|F_o|$ "/> 3σ $|F_o|$)	3157	2533	1296	2377
μR	1.91	2.84	7.67	3.96
R(F) (%)	2.61	3.84	3.05	3.62
R$_w$(F) (%)	3.38	4.86	3.90	4.54
S	1.00	1.06	1.09	1.01

Table 3. Experimental data.

Accurate electron densities of single crystals of paraelectricZnO and ferroelectric Li-doped ZnO at 19 K were analyzed by the maximum-entropy method (MEM) [36]. The MEM analyses were performed using the MEED (maximum-entropy electron density) program [37].

6.1. Crystal structures of single crystals of ZnO and Li-doped ZnO at 293 K

The final positional parameters and thermal factors at 293 K are given with their estimated standard deviation in Table 4 [38]. The final discrepancy factors are R(F)=2.61 %, R$_w$(F)=3.38 % for ZnO, and R(F)=3.05 %, R$_w$(F)=3.90 % for Zn$_{1-x}$Li$_x$O. The lattice constants a and c of Zn$_{1-x}$Li$_x$O become smaller than those of pure ZnO by 0.002 ~ 0.003 Å. Lattice distortion in Zn$_{1-x}$Li$_x$O is only the order of 10^{-3} Å along the polar axis, which is consistent with that of ceramic sample measured by thermal expansion [39]. These changes are regarded as the changes associated with ferroelectric phase transition because Li-doped ZnO is ferroelectric phase and pure ZnO is paraelectric phase. In BaTiO$_3$, the lattice constants change by 0.02 Å. In ZnO, changes in the lattice constants are smaller than that of BaTiO$_3$ by one order. The thermal factors U$_{33}$ of Zn and O atoms in Zn$_{1-x}$Li$_x$O are smaller than U$_{11}$ by 25-35 %, although those values in pure ZnO are almost same. This implies that the thermal vibration in Zn$_{1-x}$Li$_x$O is suppressed along the c-axis. But in ZnO, the effect from bonding electron is also included in the thermal factors since strong covalence exists. The effects of thermal vibration and bonding electron cannot be distinguished at 297 K. The Li concentration x, the positional parameter u and isotropic thermal factor U of Li are refined after all the other parameters are determined. Two cases are assumed for the Li position: the first is the interstitial position and the second is host Zn position. The discrepancy factor became larger when Li locates at interstitial Li position than Zn position. Therefore, present results support that Li substitutes Zn. The precise refinement shows that Li locates at the off-center position from the Zn position by 0.02 Å in $Zn_{1-x}Li_xO$ along the c-axis.

	x	y	z	U_{11}	U_{33}	U_{eq}	Occupation
ZnO (a=3.2489(1) Å, c=5.2049(3) Å) at 293 K							
Zn	1/3	2/3	0.3815(1)	0.0086(1)	0.0088(1)	0.0087(1)	1.0
O	1/3	2/3	0	0.0085(1)	0.0088(1)	0.0086(1)	1.0
$Zn_{1-x}Li_xO$ (a=3.2467(3) Å, c=5.2032(1) Å) at 293 K							
	x	y	z	U_{11}	U_{33}	U_{eq}	Occupation
Zn	1/3	2/3	0.3804(3)	0.0075(1)	0.0055(1)	0.0068(1)	0.914
Li	1/3	2/3	0.376(36)			0.0036(48)	0.086
O	1/3	2/3	0	0.0088(3)	0.0057(41)	0.0078(3)	0.957

Table 4. Crystal structure of ZnO and $Zn_{1-x}Li_xO$ at 293 K.(The form of thermal factors are exp$[-2\pi^2(U_{11}a^{*2}h^2 + U_{22}b^{*2}k^2 + U_{33}c^{*2}l^2 + 2U_{12}a^*b^*hk + 2U_{23}b^*c^*kl + 2U_{31}c^*a^*lh)]$ for anisotropic (Zn and O) and exp$(-8\pi^2U_{eq}\sin^2\theta/\lambda)$ for isotropic (Li) atoms.)

The bond lengths and angles of ZnO_4 group are shown in Table 5. The Zn-O bond length does not change in the basal plane, but becomes short along the c-axis by an amount of 0.007 Å by Li-substitution.

	Bond length		Bond Angle	
	Apical	Basal	Apical	Basal
ZnO	1.986(1) Å	1.975(1) Å	108.20(3)°	110.71(3)°
$Zn_{1-x}Li_xO$	1.979(1) Å	1.975(1) Å	108.37(4)°	110.55(4)°

Table 5. Bond lengths and angles in ZnO and $Zn_{1-x}Li_xO$ at 293 K.

The distributions of electronic density around Zn and O atoms at 293 K are obtained by using Fourier analysis (Fig. 13). The difference Fourier maps between observed distribution ϱ_{obs} and calculated distribution ϱ_{cal} in the (110) plane are shown in Fig. 14. Since the bonding electron is not distributed spherically in ZnO, the section of the bonding electron appears in the difference map. It is seen that there are large negative distributions (blue region) around Zn atom along the c-axis in $Zn_{1-x}Li_xO$. This suggests that the core electrons disappear from the Zn atom. The broadening of map due to anharmonic thermal vibrations is also appreciated. The positive electronic density is observed near the O atom. It corresponds to the antibonding orbital of O $2p$-electron.

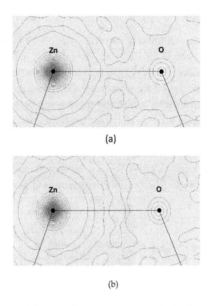

(a)

(b)

Figure 13. The charge density maps of (a) ZnO and (b) Zn$_{1-x}$Li$_x$O at 293 K in the (110) plane obtained by Fourier analysis.

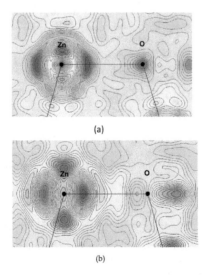

(a)

(b)

Figure 14. The difference fourier maps of charge densities of (a) ZnO and (b) Zn$_{1-x}$Li$_x$O at 293 K in the (110) plane with a contour increment of 0.2 e/ Å^3. Bluish cold color means negative charge density and reddish warm region is positive charge density.

6.2. Crystal structures at 19 K

Crystal structures at low temperature are shown in Table 6 [38]. The final discrepancy factors are $R(F)$=3.84 %, $R_w(F)$=4.86 % for ZnO, and $R(F)$=3.62 %, $R_w(F)$=4.54 % for $Zn_{1-x}Li_xO$. Li ion shifts from the Zn position by 0.08 Å at 19 K, which is four times larger than that at r. t.. The Zn-O bond lengths decrease by an amount of 0.002 Å along both basal and apical axes in $Zn_{1-x}Li_xO$, while their bond angles are almost the same in both crystals (Table 7 and Fig. 15). The Fourier and difference maps of electronic distribution are shown in Figs. 16 and 17. Comparing with the result of 293 K, bonding electrons are clearly observed around the center of Zn-O bond in both crystals. In $Zn_{1-x}Li_xO$, negative distribution is observed around Zn atom, whose shape corresponds to Zn-$3d_z{}^2$-orbital. It suggests the disappearance of $3d$-electrons from Zn atom in $Zn_{1-x}Li_xO$. The positive density near O atoms corresponds to the $2p$-orbital of O atom.

	x	y	z	U_{11}	U_{33}	Occupation	
			ZnO (a=3.2465(8) Å, c=5.2030(19) Å) at 19 K				
Zn	1/3	2/3	0.3812(1)	0.0033(1)	0.0032(1)	1.0	
O	1/3	2/3	0	0.0041(1)	0.0048(2)	1.0	

	x	y	z	U_{11}	U_{33}	U_{eq}	Occupation
			$Zn_{1-x}Li_xO$ (a=3.2436(5) Å, c=5.1983(30) Å) at 19 K				
Zn	1/3	2/3	0.3811(2)	0.0032(1)	0.0026(1)	0.0030(1)	0.914
Li	1/3	2/3	0.366(12)	-	-	0.0036(48)	0.086
O	1/3	2/3	0	0.0053(3)	0.0043(3)	0.0050(2)	0.957

Table 6. Crystal structure of ZnO and $Zn_{1-x}Li_xO$ at 19 K.(The form of thermal factors are $\exp[-2\pi^2(U_{11}a^{*2}h^2 + U_{22}b^{*2}k^2 + U_{33}c^{*2}l^2 + 2U_{12}a^*b^*hk + 2U_{23}b^*c^*kl + 2U_{31}c^*a^*lh)]$ for anisotropic (Zn and O) and $\exp(-8\pi^2U_{eq}\sin^2\theta/\lambda)$ for isotropic (Li) atoms.)

	0.957		0.957	
	Apical	Basal	Apical	Basal
ZnO	1.983(2) Å	1.974(1) Å	108.25(4)°	110.66(4)°
$Zn_{1-x}Li_xO$	1.981(2) Å	1.972(1) Å	108.27(3)°	110.65(3)°

Table 7. Bond lengths and angles in ZnO and $Zn_{1-x}Li_xO$ at 19 K.

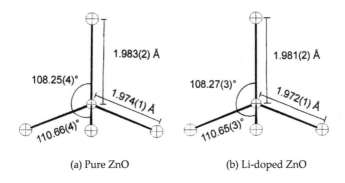

(a) Pure ZnO   (b) Li-doped ZnO

Figure 15. Structural Changes of ZnO_4 group at 19 K. The shift of Li ion is 0.08 Å from the Zn position.

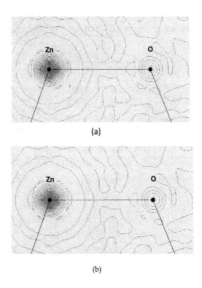

Figure 16. The charge density maps of (a) ZnO and (b) $Zn_{1-x}Li_xO$ at 19 K in the (110) plane obtained by Fourier analysis.

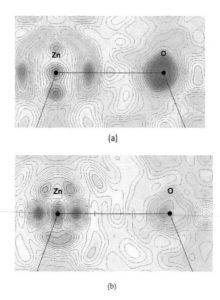

(a)

(b)

Figure 17. The difference maps of charge densities of (a) ZnO and (b) $Zn_{1-x}Li_xO$ at 19 K in the (110) plane with a contour increment of 0.2 e/ Å^3. Bluish cold color means negative charge density and reddish warm color region is positive charge density.

6.3. Bond electron densities at 19 K

The discrepancy factors of MEM analysis for pure ZnO at 19 K are $R(F)$=1.20 % and $R_w(F)$=1.39 % [40]. The electron density map of ZnO in the (110) plane is shown in Fig. 18(b). The covalent character of Zn-O bonds is clearly observed. Charge densities at the center of Zn-O bonds are 0.58 e/ Å^3 for the apical bond, and 0.56 e/Å^3 for the basal one. The density of bonding electron of the apical bond is larger than that of the basal by 3.6 %. The electron density of O atom distorts toward the Zn atom along the c-axis. The electron density of Zn atom elongates to the third nearest O atom.

The final discrepancy factors for the ferroelectric $Zn_{1-x}Li_xO$ at 19 K are $R(F)$=0.87 % and $R_w(F)$=0.85%. The electron density map of $Zn_{1-x}Li_xO$ in the (110) plane is shown in Fig. 18(c). The covalent character of Zn-O bonds is also seen in $Zn_{1-x}Li_xO$. Charge densities of the Zn-O at the saddlepoint are 0.46 e/ Å^3 for the apical bond, and 0.49 e/ Å^3 for the basal one. Each value is smaller than that of pure ZnO. It is considered that this may be due to the decrease of total charge by Li-substitution. The density of bonding electron of the basal bond is larger than that of the apical by 6.5 %, on the contrary to the case of ZnO. The electron density of O atom is distorted anisotropically, similar to the pure ZnO, but the direction is opposite. The extension of Zn atom toward the third nearest O atom, which is observed in ZnO, was not detected in $Zn_{1-x}Li_xO$.

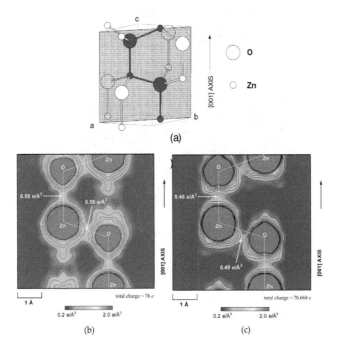

Figure 18. The electronic Distribution by MEM method. (a) The (110) plane of ZnO. Shaded atoms are included in the plane.(b)The MEM charge density map of ZnO at 19 K in the (110) plane. (c) The MEM charge density map of $Zn_{0.914}Li_{0.086}O$ at 19 K in the (110) plane. Contours are drawn from 0.4 e/ Å³ at 0.2 e/ Å³ intervals.

6.4. Difference between Li-doped ZnO and pure ZnO

The difference of the charge densities between two crystals is calculated by subtracting the MEM charge densities of pure ZnO from those of $Zn_{1-x}Li_xO$ as in Fig. 19 [40]. The values of MEM charge densities of pure ZnO are normalized by multiplying 0.930, because the total charges of two crystals are different. The positive and negative peaks were observed around the O atom. This is due to the shift of O atom from the host position opposite to the Zn atom along the c-axis in $Zn_{1-x}Li_xO$. Four negative peaks are observed around the Zn atom. These peaks correspond to the Zn $3d$-orbitals and suggest that the $3d$-electrons disappear from the Zn site, compared with wave functions of Zn $3d$- and O $2p$-orbitals of ZnO obtained by DV-Xα calculation in Fig. 20. This result is the same as that calculated by the Fourier synthesis previously. The bonding region around the apical and the basal Zn-O bonds is covered by positive electron distribution. This implies that the Zn $3d$-electrons transfer to the bonding electrons in $Zn_{1-x}Li_xO$.

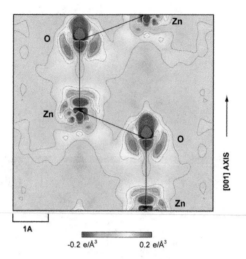

Figure 19. The difference map between MEM charge densities of $Zn_{1-x}Li_xO$ and ZnO at 19 K in the (110) plane. Contours are drawn from -5.0 e/ $Å^3$ to 5.0 e/ $Å^3$ at 0.5 e/ $Å^3$ intervals. Shaded area eith reddish color indicates positive charge.

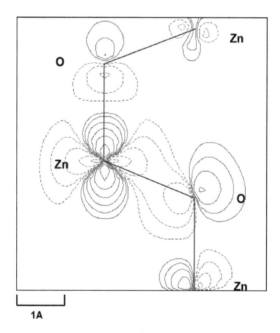

Figure 20. Wave functions of Zn 3*d*- and O 2*p*-orbitals of ZnO obtained by DV-Xα calculation. Red lines indicate the positive and Blue lines indicate the negative area.

In Li-doped ZnO, negative peaks around Zn atom corresponding to $3d_z^2$ orbital were observed in difference Fourier map, and also in the MEM difference map between the ferroelectric phase and the paraelectric phase. These results may be related to the existence of localized d-holes in Li-doped ZnO. It is considered that $3d_z^2$ electrons transfer to bonding region and play a role for contribution for the covalence observed from the result of the MEM analysis. Furthermore, the antibonding orbital of $2p$ electron is observed in difference Fourier map of Li-doped ZnO. It is suggested that the interaction between d-holes and p-electrons should be closely related to the appearance of the ferroelectricity in Li-doped ZnO.

7. Discussion

The ferroelectric phase transition accompanies with structural distortions in usual ferroelectrics. Portengen, Ostreich and Sham reported the theory of electronic ferroelectricity [41], which examined possibilities of electronic ferroelectricity, based on the spinless Falikov-Kimball (FK) model [42] with a k-dependence hybridization in Hartree-Fock approximation. The FK model introduces two types of electrons, itinerant d-electrons and localized f-electrons. The valence transition is driven by on-site Coulomb repulsion between the d-electrons and f-electrons. They found that the Coulomb interaction between itinerant d-electrons and the localized f-electrons give rise to an excitonic$<d^+f>$ expectation value, which breaks the center of symmetry of the crystal and leads to electronic ferroelectricity in mixed-valent compounds. In this electronic model, the transition involves a change in the electronic structure rather than the structural one. The estimated spontaneous polarization is of the order of 10 $\mu C/cm^2$, which is comparable to those in displacive type ferroelectrics such as $BaTiO_3$.

In the case of Li-doped ZnO, the d-p hybridization should be changed by Li-doping. The d-holes and itinerant electrons should be closely related to the appearance of Li-doped ZnO. It should be further detailed studies would be necessary whether this proposed theory is applicable for the electronic ferroelectricity found in ZnO or not.

Recently Glinchuk et al propose the mechanism of impurities induced ferroelectricity in nonperovskite semiconductor matrices due to indirect dipole interaction via free carriers [43]. They showed that the ferroelectricity in Li-doped ZnO might appear due to indirect interaction of dipoles, formed by off-center impurities, via free charge carriers, namely, the Ruderman–Kittel–Kasuya–Yosida (RKKY)-like indirect interaction of impurity dipoles via free charge carriers. They estimated that the typical semiconducting concentration of the carriers like 10^{17} cm^{-3} is sufficient for the realization of the ferroelectricity. The finite conductivity does not mean complete destruction of possible ferroelectric order and shows rather many interesting effects. In this theory, they have succeeded to obtain the spontaneous polarization and the phase diagram of T_c vs. impurity molar ratio (x) as shown in Fig. 21.

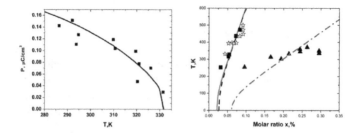

Figure 21. The calculated spontaneous polarization and the phase diagram of T_c vs. impurity molar ratio (x) after Glinchuk et al [36].

Experimental results suggest the existence of localized d-holes and p-electrons in Li-doped ZnO. We consider that the interaction between d-holes and p-electrons may be related to the appearance of the ferroelectricity in Li-doped ZnO. However, there are many proposals for this peculir novel ferroelectricity in ZnO. Scott and Zubko have pointed out the possibility of a classic electret mechanism for Li-doped ZnO[44]. Tagantsev discussed a Landau theory, where a crystal on cooling from a state with polar symmetry exhibits a maximum of dielectric permittivity and D-E hysteresis loops [45]. He proposed that these ferroelectric like phenomena corresponds to the case for Li-doped ZnO. Furthermore, the multiferroic behavior has been reported in impurity-doped ZnO [46]. At present, further detailed experiments should be expected to clarify the nature of Li-doped ZnO.

8. Conclusion

Dielectric properties and crystal structures and electron density distributions studied on pure ZnO and $Zn_{1-x}Li_xO$ by the precise X-ray diffraction are reviewed in comparison with ferroelectric semiconductors $Pb_{1-x}Ge_xTe$ and $Cd_{1-x}Zn_xTe$. It is considered that the appearance of ferroelectricity in ZnO is primarily due to electronic origin. The change in d-p hybridization caused by Li-substitution is considered to be responsible for this novel ferroelectricity and dielectric properties. Although the ferroelectric phase transition accompanies with structural distortions in usual ferroelectrics, the structural changes observed in $Zn_{1-x}LiO$ are the order of 10^3 Å. The clear change of Zn $3d$-electron is observed. It suggests Zn $3d$-electron of $Zn_{1-x}LiO$ transfers from Zn atom to the bonding. The positive electronic density is observed near the O atom. It corresponds to the antibonding orbital of O $2p$-electron. These results suggest the existence of localized d-holes and p-electrons in Li-doped ZnO. It may be probable that the interaction between d-holes and p-electrons may be related to the appearance of the ferroelectricity in Li-doped ZnO.

For long time, the importance of electronic contribution has been pointed out in the field of ferroelectrics, since the simple superposition of electronic polarizability does not hold in many ferroelectric substances. Many efforts have been done in vain because of complexity of crystal structures of ferroelectrics. As the crystal structure of ZnO is very simple, the electronic contribution could be observed rather easily. It is considered that this result is the first example to discuss the electronic contribution for ferroelectricity.

Author details

Akira Onodera and Masaki Takesada

Department of Physics, Faculty of Science, Hokkaido University, Sapporo, Japan

References

[1] Klingshirn C F, Meyer B K, Waag A, Hoffmann A and Geurts J. Zinc Oxide From Fundamental Properties Towards Novel Applications. Springer, 2010.

[2] Yao T (ed.). ZnO Its Most Up-to-date Technology and Application, Perspectives (in japanese). CMC Books, 2007.

[3] Heiland G. Mollwo E and Stockmann F. Solid State Phys 1959; 81: 193.

[4] Campbell C. Surface Acoustic Wave Devices and Their Signal Processing Application. San Diego: Academic Press; 1989.

[5] Hirshwald W. Current Topics Mater Sci 1981; 7, 143.

[6] Tsukazaki A, Ohtomo A, Onuma T, Ohtani M, Makino T, Sumiya M, Ohtani K, Chichibu S F, Fuke S, Segawa Y, Ohno H, Koinuma H and Kawasaki M. Nature Materials 2005; 4: 42.

[7] Joseph M, Tabata H and Kawai T. Jpn J Appl Phys 1999; 38: L1205.

[8] Bilz H, Bussmann-Holder A, Jantsch W and Vogel P. Dynamical Properties of IV-VI

[9] Compounds. Berlin: Springer-Verlag; 1983.

[10] Weil R, Nkum R, Muranevich E and Benguigui L. Phys Rev Lett 1989; 62: 2744.

[11] Onodera A, Tamaki N, Kawamura Y, Sawada T and Yamashita H. Jpn. J Appl Phys 1996; 35: 5160.

[12] Abrahams C S and Bernstein L J. Acta Cryst 1969; B25: 1233.

[13] Corso D A, Posternak M, Resta R and Baldereshi A. Phys Rev 1994; B50: 10715.

[14] Zakharov O, Rudio A, Blase X, Cohen L M and Louie G S. Phys Rev1994; B50: 10780.

[15] Massida S, Resta R, Posternak M and Baldereschi A. Phys Rev 1995; B50.

[16] Osikiri M and Aryasetiawan F. J Phys Soc Jpn 2000; 69: 2123.

[17]   Minami T, Nanto H and Tanaka S. Jpn J Appl Phys 1984; 23: L280.

[18]   Laudise A R, Kolb D E and Caporaso J A. J Am Ceram Soc 1964; 47: 9.

[19]   Kolb D E and Laudise A R. J Am Ceram Soc 1966; 49: 302.

[20]   Yang CK and Dy S K. Solid State Commun 1993; 88: 491.

[21]   Usuda M, Hamada N, Kotani T and Schilfgaarde M. Phys Rev 2001; B66: 075205.

[22]   Zhang B S, Wei S -H and Zunger A. Phys Rev 2001; B63: 075205.

[23]   Tamaki N, Onodera A, Sawada T and Yamashita H. J Kor Phys 1966; 29: 668.

[24]   Onodera A, Tamaki N, Jin K and Yamashita H. Jpn J Appl Phys 1997; 36: 6008.

[25]   Islam T Q and Bunker A B. Phys Rev Lett 1987; 59: 2701.

[26]   Benguigui L, Weil R, Muranevich E, Chack A and Fredj E. J Appl Phys 1993; 74:513.

[27]   Terauchi H, Yoneda Y, Kasatani H, Sakaue K, Koshiba T, Murakami S, Kuroiwa Y,
       Noda Y, Sugai S, Nakashima S and Maeda H. Jpn J Appl Phys 1993; 32: 728.

[28]   Onodera A, Yoshio K, Satoh H, Yamashita H and Sakagami N. Jpn. J. Phys. Phys.
       1998; 37; 5315.

[29]   Onodera A, Tamaki N, Satoh H, Yamashita H and Sakai A. Ceramic Transactions
       1999; 100; 77.

[30]   Onodera A and Satoh H. Frontiers in Science and Technology – Ferroelectrics Vol. 1,
       Stefan University Press, 2002; 93.

[31]   Islam E, Sakai A and Onodera A. J Phys Soc Jpn 2001; 70: 576.

[32]   Kagami D, Takesada M, Onodera A and Satoh H. J Kor Phys 2011; 59: 2532.

[33]   Hagino S, Yoshio K, Yamazaki T, Satoh H, Matsuki K and Onodera A. Ferroelectrics
       2001; 264: 235.

[34]   Onodera A, Yoshio K, Satoh H, Tamaki T, Takama T, Fujita M and Yamashita H. Fer-
       roelectrics 1999; 230: 465.

[35]   Sakagami N. J Cryst Growth 1990; 99: 905.

[36]   Coppens P, Guru Row N T, Leung P, Stevens D E, Becker J P and Yang W Y. Acta
       Crystallogr 1979; A35: 63.

[37]   Sakata M and Sato M. Acta Crystallogr 1990; A46: 263.

[38]   Kumazawa S, Kubota Y, Tanaka M, Sakata M and Ishibashi Y. J Appl Cryst 1993; 26:
       453.

[39]   Yoshio K, Onodera A, Satoh H, Sakagami N and Yamashita H. Ferroelectrics 2001;
       264; 133.

[40]   Onodera A, Tamaki N, Satoh H and Yamashita H. Ferroelectrics 1998; 217: 9.

[41] Yoshio K, Onodera A, Satoh H, Sakagami N and Yamashita H. Ferroelectrics 2002; 270; 357.

[42] Portengen T, Ostreich Th and Sham J L. Phys Rev 1996; B54: 17452.

[43] Falicov M L and Kimball C J. Phys Rev Lett 1969; 22: 997.

[44] Glinchuk D M, Kirichenko V E, Stephanovich A V and Zaulychny Y B. Appl Phys 2009; 105: 104101.

[45] Scott F J and Zubko P. IEEE Xplore 2005; ISE-12 (12th International Symposium on Electrets): 113-116.

[46] Tagantsev K A. Appl Phys Let 2008; 93: 202905.

[47] Yang C Y, Zhong F C, Wang H X, He B, Wei Q S, Zeng F and Pan F. Appl Phys 2008; 104: 064102.

Chapter 11

Gelcasting of Ferroelectric Ceramics: Doping Effect and Further Development

Dong Guo and Kai Cai

Additional information is available at the end of the chapter

1. Introduction

Ferroelectric ceramics may be seen as the most important type of ferroelectric materials, which have been used in a wide spectrum of electrical and microelectronic devices, including underwater transducers, micro-pumps and valves, ultrasonic motors, thermal sensors, probes for medical imaging and non-destructive testing, and accelerometers, etc (Cross LE, 1996; Setter et al., 2000).

Ferroelectric ceramics used in the devices have various shapes. A certain shape formation technique is required to make ceramics with the desired shapes. Dry pressing is the most commonly used ceramic forming technique. In this technique, dry powders containing organic binder are filled into a solid mold, then dry ceramic green bodies with the shape of the mold cavity are formed under mechanical or hydraulic compacting presses selected for the necessary force and powder fill depth. The pressure is around several tens of MPa or higher. Therefore, large ceramic parts require a much higher compacting force. If the ceramic parts are unable to have pressure transmit suitably for a uniform pressed density then isostatic pressing may be used. One of the most serious disadvantages of dry pressing lies in the difficulty in fabricating high quality large and complex-shaped ceramics or ceramics with a fine structure, which are required for various devices.

To resolve the problems associated with the conventional dry pressing, new wet forming techniques, such as gelcasting (Omatete et al., 1991), electrophoretic casting (Biesheuvel et al., 1999), hydrolysis assisted solidification (Novak et al., 2002), and direct coagulation casting (Graule et al., 1996), etc., are becoming increasingly attractive for advanced ceramic materials. Since in these techniques the ceramic powders are dispersed in a liquid medium and thoroughly mixed, wet forming techniques have the advantages of reducing some structure defects that are difficult to remove in dry pressed ceramic parts. Among the many wet forming techniques, aqueous gelcasting represents the latest improvements. In the technique, a high solids loading slurry obtained by dispersing the ceramic powders in the pre-mixed solution containing monomer and cross-linker is cast in a mold of the desired shape. When heated, the monomer and cross-linker polymerize to form a three-dimensional network structure, thus the slurry is solidified *in situ* and green bodies of the desired shape are obtained, which consists mainly of ceramic powders with a low polymer content. Gelcasting may be seen as a milestone in fabricating complex-shaped ceramic parts since it has initiated a new branch of research in ceramic processing due to its intriguing properties of near-net-shape forming, high green strength, and low binder concentration, etc. As a pressure free method, it can also be used in fabricating large ceramic components that have simple shapes. For example, to make a ceramic disc with a diameter of 20 cm by dry pressing, a load of hundreds of tons is required. In contrast, such a ceramic disc can be easily made by gelcasting via a simple ring-shaped mold at much less cost. Furthermore, gelcasting may be developed to fabricate complex-shaped (Cai et al., 2003) or fine-structured (Guo et al., 2003) ceramic parts that are rather difficult or even impossible to be formed by the conventional dry pressing method.

Since its invention in 1991 gelcasting has been widely used for the fabrication of structural ceramics, including Al_2O_3 (Young et al., 1991), SiC (Zhou et al., 2000), and Si_3N_4 (Huang et al., 2000), etc. Later, it was applied to a number of functional ceramics in a number of ways including $LaGaO_3$ (Zha et al., 2001), ZnO (Bell et al., 2004), and $Sr_{0.5}Ba_{0.5}Nb_2O_6$ (Chen et al., 2006), etc. The first journal report about the application of gelcasting to piezoelectric ceramics appeared at 2002 (Guo et al., 2003). Compared to structural ceramics, much less research about gelcasting of functional ceramics has been conducted so far.

Among the many ferroelectric ceramic materials, lead zirconate titanate (PZT) is the most widely used one owing to their superior piezoelectric, pyroelectric and dielectric properties. A fascinating feature of multicomponent ferroelectric ceramics is that their electrical properties can be modified by doping with acceptors and donors (Shaw et al., 2000). As a result, a series of PZT materials with tailored properties are commercially available. Unfortunately, this feature also leads to the problem of high sensitivity of the electrical properties of PZT to composition. In addition, PZT is commonly used with a composition close to the morphotropic phase boundary (MPB) at a Zr/Ti ratio of about 52/48, where properties such as piezoelectric coefficients, dielectric permittivity, and coupling factors are maximized and thus may be more sensitive to the composition (Noheda et al., 2006). On the other hand, since shape formation in gelcasting is achieved through *in situ* polymerization, organic additives are used in the premix solution. Also, addition of commercial surfactants that may have a complicated composition is indispensable, because successful fabrication of ceramics via gelcasting or other colloidal methods requires to prepare high solids loading ceramic slurry with still a low viscosity. Consequently, in order to apply gelcasting to the formation of ferroelectric ceramics such as PZT, it is necessary to remove the possible influence of the impurities introduced by the various additives on the electrical performance of the final products. This makes the problems more complicated than that of structural ceramics.

2. Colloidal chemistry and rheological properties of PZT suspensions

We first give a short introduction the gelcasting technique. The details are out of the focus of this chapter, and interested readers can refer to other papers. According to the liquid medium used, there are two types of gelcasting systems: aqueous and nonaqueous systems. In aqueous gelcasting deionized water is used as the medium, and acrylamide ($C_2H_3CONH_2$) may be seen as a prototype monomer. Generally, N,N-methylenebisacrylamide (($C_2H_3CONH)_2CH_2$, MBAM) N,N,N_0,N_0-Tetramethylethylenediamine (TEMED) and $(NH_4)_2S_2O_8$ are used as the cross-linker, the catalyst and the initiator, respectively. The gelcasting process of PZT is similar to that of previous studies. The flowchart of the gelcasting is shown in Figure 1. First, the PZT powders are added in the premix solution containing AM and MBAM and thoroughly mixed. After addition of initiator and catalyst, the slurry is de-aired in vacuum, then the slurry is cast into the mold with desired shape and heated around 60-80°C in a oven for several hours for polymerizatoin and drying. After demolding the green ceramic body is obtained.

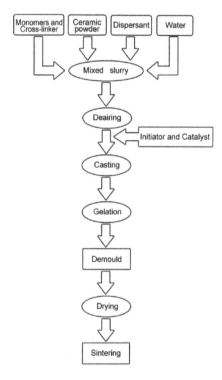

Figure 1. The flowchart of gelcasting

The colloidal and rheological properties of the PZT suspension are important issues that should be addressed first. Homogenous dispersion of the powder in the premix solution and stability of the suspension are determined mainly by attractive and repulsive forces between the particles in the system. The former generally arises from the van der Waals forces while the latter can arise from electrostatic repulsion or steric repulsion of surfactant materials absorbed on the particle surfaces (Israelachvili, 1992). Magnitude of the van der Waals forces is mainly determined by the nature of the particle surface and the solvent. While the repulsive forces can be modified over a wide range by dispersants.

Commercial Name	TAC	JN281	SP2	SGA
Producer	Beijing Chemical Reagent Co.	Beijing Pinbao Chemical Co.	Beijing Tongfang Chemical Co.	Beijing Tongfang Chemical Co.
Composition	Triammonium citrate	Poly (acrylic acid), NH_4^+ salt solution	Poly (acrylic acid-co-maleic acid), Na^+ salt solution	Poly (acrylic acid), Na^+ salt solution

Table 1. Composition and producer of the dispersants

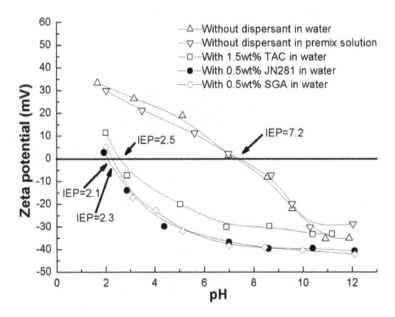

Figure 2. Effect of dispersants on the zeta potentials of the PZT suspensions at different pH values.

Polyelectrolyte dispersants are well known to be effective for various ceramic slurries due to both electrostatic and steric repulsions of the macromolecules. Here we show the effects of typical polyelectrolyte dispersants and a widely used organic surfactant triammonium cit-rate (TAC). The details are listed in Table 1. Electrostatic repulsion is dependent on the zeta potential (ζ) of the powders. The higher this value with the same polarity, the stronger the electrostatic repulsion between the particles. When close to the isoelectric point (IEP), the particles tend to flocculate. Zeta potentials of various PZT aqueous suspensions (0.06 vol % solids) at different pH values are shown in Figure 2. The zeta potential of pure PZT suspen-sion changes from 33 mV at pH = 1.7 to - 35.1 mV at pH = 11.9 with an IEP at about pH = 7.2, suggesting that neutral environment is disadvantageous for good dispersion. Addition of monomer only slightly decreases the relative value of the potential and has little effect on IEP, indicating that the uncharged AM molecules screen the charge of the PZT particles. With the addition of TAC, JN281 and SGA the IEP is moved to pH = 2.5, 2.1 and 2.3, respec-tively. In the range of neutral environment to pH = 12, the zeta potential is almost constant. The higher absolute potential values of suspensions with JN281 and SGA than that with TAC imply that polyelectrolyte dispersants are more effective as far as the electrostatic re-pulsion is concerned.

Figure 3 shows the viscosity as a function of the shear rate for different PZT slurries solids loading. Adjustment of pH values to either acid or basic conditions has little effect, while addition of dispersants can greatly decrease the viscosity. The high viscosities at the begin-ning indicate a 'Bingham' type behavior, which is followed by a shear thinning at low shear rates. Shear-thinning behavior can be attributed to a certain kind of rearrangement of the relative spatial disposition of the particles. For concentrated suspensions of hard solid parti-cles in Newtonian liquids, a flow-induced layered structure has been verified (Ackerson, 1990). Such a structure can provide a low resistance of the particle movement between dif-ferent layers under the shear flow. For the systems containing dispersants, when the shear rates increase to a critical value (γ_c) a shear-thickening behavior appears, indicating that the flow-induced structure is destroyed. It is clear that the polyelectrolyte dispersants are much more effective than TAC. In addition, PZT slurries in the premix solution have almost the same viscosity values as those in the pure water, suggesting that addition of AM (15 wt.%) has little effect on the viscosity. These and the zeta potential results indicate that the polye-lectrolyte dispersants work by both electrostatic and steric stabilization mechanism.

Generally, there is a complex nonlinear relationship between viscosity and solids volume fraction, which is closely related to many factors, including the continuous phase viscosity, particle-size distribution, and particle shape, etc. Influence of solids loading on the apparent viscosity is shown in Figure 4. At low solids loading the slurries show a low viscosity. The continuous addition of particles finally three-dimensional contact throughout the suspen-sion, making flow impossible. The particular solid phase volume at which this happens is called the maximum packing fraction Φ_m (Barnes & Hutton, 1989). Before adding dispersant the PZT suspension has a measurable viscosity at a solids loading slightly smaller than 32 vol.%. Adding a little more PZT powders leads to a 'solidified' slurry whose viscosity is im-possible to be measured by the normal rheometer. The Φ_m is thus determined to be 32 vol.%. Adding 0.8 and 1.5 wt.% of TAC increase Φ_m to about 47 and 53 vol.%, respectively. The lower Φ_m for a TAC concentration of 2.2 wt.% indicates that excess dispersant is harmful.

The relationship between the viscosity and the solids volume fraction for monodispersed suspension can be explained by the Krieger–Dougherty (K–D) model:

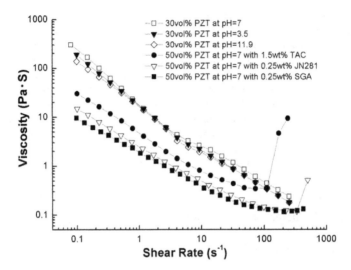

Figure 3. Influence of dispersants and pH value on the viscosity of the PZT suspensions

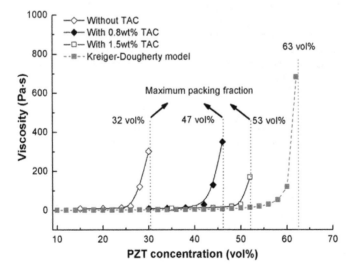

Figure 4. Influence of solids loading on the viscosities of the PZT suspensions

$$\eta = \eta_o (1 - \phi / \phi)^{-[\eta]\phi_m} \tag{1}$$

where η and η_o are the viscosity of the suspension and the solvent, respectively. The true volume fraction of the powder in the suspension is represented by f. The intrinsic viscosity $[\eta]$ is a function of particle geometry; a value of 2.5 is suitable for spherical particles. A maximum packing fraction of 0.63±0.002 is suitable for random close packing at low shear rate. The K–D model curve with $[\eta]$ = 2.5 and Φ_m=0.63 is plotted in the Figure. Although the experimental curves are somewhat deviated from the K–D model, they still show a similar shape: low viscosity at low solids loading and sharp increase at high solids loading. The discrepancy may partly be ascribed to the inhomogeneity of the particle size. In addition, particle flocculation will lead to a lower Φ_m because the flocs themselves are not close-packed (Starov et al., 2002) and they can trap part of the liquid phase, thus, leading to higher 'effective phase volume' and viscosity than those of the primary particles. The results clearly show the remarkable effect of dispersants in getting low viscosity ceramic slurry.

3. Electrical characterization and analysis of the doping effects

3.1. Electrical characterization for identifying the doping type

The piezoelectricity of PZT type materials originates from the displacement of Zr and Ti sublattices and the electrical properties of the materials can be dramastically affected by doping atoms. There are primarily two types of dopants for PZT, i.e. the donor type (soft type), and the acceptor type (hard type) (Jaffe et al., 1971). The former is mainly caused by substitution of higher valence ions for the A site Pb or B site Zr and Ti, and correspondingly higher piezoelectric coefficient (d_{33}), planar electromechanical coupling factor (K_p), relative permittivity (ε_r), loss tangent ($tg\delta$), P_r values and a lower mechanical quality factor (Q_m) value are obtained. The latter, which is caused by substitution of lower valence ions for the A or B site atoms, has contrary effects.

Sample	Average d_{33}	ε_r	$tg\delta$	K_p
Dry pressed	469 pC/N	1430	0.0311	0.621
G-TAC	468 pC/N	1369	0.0325	0.633
G-JN281	479 pC/N	1352	0.0401	0.675
G-SP2	403 pC/N	1285	0.0250	0.591
G-SGA	330 pC/N	1244	0.0202	0.541

Table 2. Comparison of densities and some electrical parameters of dry pressed and gelcast PZT samples.

Some electrical parameters of different soft-doped PZT samples are compared in Table 2, where G-TAC, G-JN281, G-SP2 and G-SGA represent the best available gelcast PZT samples with TAC, JN281, SP2 and SGA as the dispersants, respectively. The ε and $tg\delta$ are 1 kHz data under room temperature. The dry pressed sample is obtained under a pressure of ~ 80MPa. G-TAC shows similar electrical properties with those of the dry pressed one. This indicates that the organic species used in gelcasting, including the dispersant TAC, the monomer and the cross-linker, etc., have almost no doping effects. Compared to the dry pressed sample, G-SP2 and G-SGA show higher Q_m and decreased d_{33}, K_p, ε, $tg\delta$ and P_r values, while change of these parameters for G-JN281 is to the contrary. Thus, the data in reveal that SP2 and SGA induced evident 'hard' doped characteristics, while JN281 induced 'soft' doped characteristics. Except for the dispersant species, the samples were prepared under the same conditions. Hence, the different properties should be mainly attributed to the dispersants used. A 'fingerpint' of hard doping effect is the increased Q_m (Damjanovic, 1998). Since Na^+ is the main metal cation in SP2 and SGA, we neglect other difference and check the correlation between Na concentration (inspected by X-ray Fluorescence) and Q_m. As shown in Figure 5, Q_m roughly shows an increasing trend with Na mole fraction except for G-JN281. Then we intentionally introduced Na into the G-JN281 by adding NaOH to the corresponding slurry to increase its pH value to ~13. As expected, a higher Q_m is indeed obtained after adding Na to G-JN281. These reveal that the Na^+ ion introduced by the dispersants has a strong doping effect for gelcast PZT. From the valence and diameter (1.02Å) of Na^+, we deduce that it should substitute for the A site Pb^{2+} (1.18Å) as an acceptor dopant.

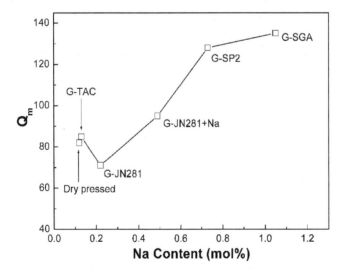

Figure 5. Illustration of the relationship between Na concentration (mole percentage in total metal elements) and Q_m of various PZT samples.

3.2. Complex impedance spectra

Figure 6(a), 6(b), 6(c) and 6(d) show the complex impedance spectra in the temperature range of 300~540 °C of the dry pressed sample, G-JN281, G-JN281+Na and G-SP2, respectively. The spectra of G-TAC is very similar with those of the dry pressed sample and thus not shown. The complex impedance spectra of all samples show only one Cole-Cole semicircle crossing the origin, which can be assigned with an equivalent circuit composed of a simple parallel RC element with a impedance that can be expressed as equation (2):

$$Z = \frac{R}{1 + (i\omega RC)^{1-n}},$$ (2)

where the parameter n characterizes the distribution width of the relaxation times around a mean value τ_0=RC (West et al., 1997). The semicircles of the samples show no obvious depression, i.e. the depression angle β (=$n\pi/2$) between the real axis and the line from the high frequency intercept to the centre of the circle is close to 0. Thus, n is close to 0, indicating a debye-like behavior with a single relaxation time (West, 1997; Cao, 1990). Then based on equation (2) the following relationship about the real (Z') and imaginary (Z") parts of the impedance can be derived:

$$\left(Z' - R/2\right)^2 + Z''^2 = \left(R/2\right)^2.$$ (3)

The resistance R derived from the diameter of the semicircle, and the capacitance C can be derived from the angular frequency ω at the peak of the circular arc based on the relationship ωRC=1. The C values of the samples are in the order of 10^{-10}-10^{-9} F, which can be associated with the intrinsic ferroelectric bulk (or grain) response.

Using the R values, the logarithm of conductivity of different samples as a function of reciprocal temperature is plotted in Figure 6(e). The conductivity σ of the ferroelectric bulk phase shows an Arrhenius type behavior that can be described by equation (4):

$$\sigma = \sigma_o \exp(\frac{-E_g}{kT}),$$ (4)

where σ_0, E_g, k and T are the pre-exponential factor, the activation energy, the Boltzmann's constant and the absolute temperature, respectively. The logarithm of equation (4) gives equation (5):

$$\ln \sigma_T = \ln \sigma_o - \frac{E_g}{k}(\frac{1}{T}) = -2.30259 \log \rho_T.$$ (5)

where ρ_T is the resistivity. Based on the sample geometry, the E_g values associated with the intrinsic bulk phase conduction of the samples were derived by using equation (5). The data are compared in Table 3.

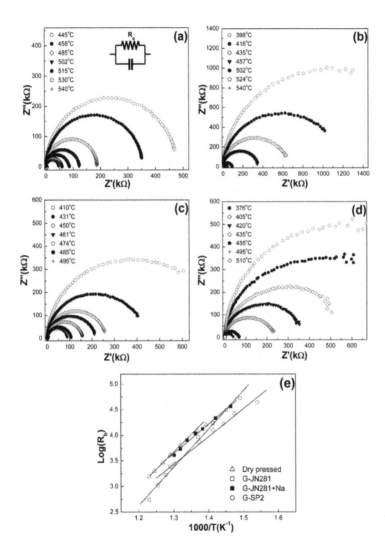

Figure 6. Complex impedance spectra of different PZT samples. (a) Dry pressed. (b). G-JN281. (c) G-JN281+Na. (d) G-SP2. (e) Arrhenius plot for the ferroelectric bulk resistivity of the samples derived from the impedance spectra.

Sample	Dry pressed	G-JN281	G-JN281+Na	G-SP2	G-SGA
E_g	1.24 eV	1.48 eV	1.14 eV	1.08 eV	1.07 eV

Table 3. Comparison of the conductivity activation energies of different PZT samples.

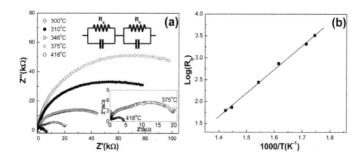

Figure 7. (a) Complex impedance spectra of G-SGA. The inset shows the magnified 375_oC and 418_oC curves. (b) Arrhenius plot of the ferroelectric bulk phase resistivity of the sample derived from the grain response of its impedance spectra.

The spectra of G-SGA shown in Figure 7(a), particularly the high temperature curves shown in the inset, consist of two semicircles, which can be assigned with an equivalent circuit composed of two parallel RC elements. This indicates the much different 'electrical micro-structure'[14] of G-SGA, which has the highest Na concentration. The capacitance obtained by fitting the first semicircle can also be attributed to the ferroelectric bulk response, while the second semicircle may be attributed to the grain boundary response due to the higher capacitance. Actually, one semicircle spectrum may also appear if different responses in a sample overlap as a result of their similar relaxation time. A closer check of Figure 6(d) indicates that the one semicircle impedance spectra of G-SP2 have a slightly higher depression angle than other spectra in Figure 6, consistent with its higher Na concentration (see Figure 5). These imply that the grain boundary response is gradually magnified with increasing Na concentration. The Arrhenius plot of the conductivity of the bulk phase response of G-SGA is shown in Figure 7 (b). The derived E_g value is also listed in Table 3.

The loss of PbO through its volatility causes oxygen vacancies in PZT, leading to a p-type conductivity (Barranco et al., 1999). The acceptor Na^+ incorporated by the dispersants can substitute for Pb^{2+}, which can be expressed as following:

$$Na_2O = 2Na_{pb}' + O_o + V_o \tag{6}$$

Thus, Na^+ ions replace Pb^{2+} ions and more oxygen vacancies are simultaneously created for charge compensation. Oxygen vacancies are the only lattice defects in the perovskite oxides that have a significant mobility, and the conductivity should be improved by the oxygen vacancy conduction mechanism via hopping of atoms in the oxygen octahedral network (Raymond & Smyth, 1996). This well explains the lower E_g data of G-SP2 and G-SGA than that of the dry pressed sample (Table 3), revealing again the hard doping effect of the Na-containing dispersants. The relatively higher activation energy of G-JN281 implies the presence of a certain donor-type impurity in G-JN281. Addition of NaOH into G-JN281 leads to a lower E_g value and further confirms the doping effect of Na.

3.3. Ferroelectric hysteresis loops

The ferroelectric (polarization-electric field) hysteresis loops of the PZT samples are shown in Figure 8. In Figure 8(a), G-SP2 and G-SGA show remnant polarization (P_r) of 32.7 $\mu C/cm^2$ and 16.7 $\mu C/cm^2$, respectively. Both values are smaller than the value of 39.8 $\mu C/cm^2$ of the dry pressed sample. The difference can also be interpreted by increased oxygen vacancies due to doping of Na. As afore mentioned, oxygen vacancies can move in the oxygen octahedral network. This may lead to a low stability of the $2Na_{pb}'-V_o$ defect dipoles. The defect dipoles tend to orient themselves along the polarization direction, resulting in stabilized ferroelectric domains. A stabilized domain wall structure in turn give rise to more difficult poling and depoling (switching) and a smaller P_r (Warren, et al., 1996; Lambeck & Jonker, 1978). The round open loop of G-SP2 implies a higher leakage current. The hysteresis loops of G-JN281 and G-JN281+Na in Figure 8 (b) also clearly demonstrates the effect of Na: decreased P_r and a loop with a rounder shape.

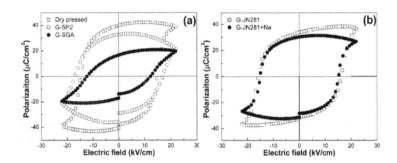

Figure 8. Ferroelectric hysteresis loops of the PZT samples.

4. Microstructural characterization

The XRD spectra of All samples (Figure 9) show typical perovskite structures (Soares et al., 2000; Hammer et al., 1998). The XRD patterns of G-TAC, G-SP2 and G-SGA are very similar to that of the dry pressed sample. While the peaks of the XRD pattern of G-JN281 shift very slightly to higher diffraction angles, indicating a contracted lattice cell. Considering the soft doped characteristics of G-JN281, the change in its XRD pattern may again be attributed to a certain unknown donor impurity ion introduced by the dispersant.

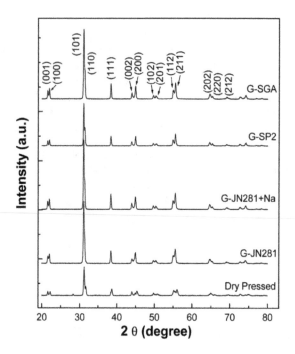

Figure 9. XRD patterns of different PZT samples.

As shown in Figure 10, although sintered under the same procedure, the samples exhibit rather different morphologies. SEM image of G-JN281+Na is not illustrated since it is very similar to that of G-JN281. Dry pressed sample shows an intergranular fracture surface and relative uniform grains with a diameter of 3~5 μm. G-JN281 shows a morphology very similar to that of the dry pressed one. G-SP2 also shows a basically intergranular fracture surface, but the grains are much larger with a diameter of 6~10 μm. Much different from other samples, G-SGA shows a transgranular fracture surface with the largest grains with a diameter of about 12 μm. Such a transgranular growth may result from chemical inhomogeneity and presence of intergranular phases or expanded grain boundary region. This is consistent with the two semicircle impedance spectra of G-SGA shown in Figure 7 (a).

In summary, electrical and structural characterization indicate that the Na$^+$ ion, which is the main cation in many widely used commercial dispersants (Xu et al., 1996; Tomasik et al., 2003), shows detrimental hard doping effects, leading to deteriorated electrical performance. Also, the impurity species introduced by dispersants or other additives may have a complicated influence on the electrical properties and microstructure of gelcast PZT samples. The results indicate that the specific doping effect, e.g. the change in electrical performance by the additives during processing, is a critical issue that should be paid special attention when applying gelcasting to the formation of electronic ceramics.

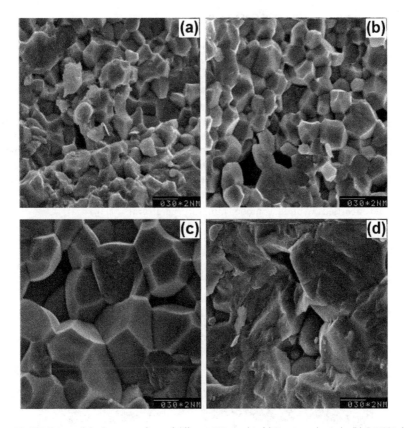

Figure 10. SEM images of the fracture surfaces of different PZT samples. (a) Dry pressed sample; (b) G-JN281; (c) G-SP2; (d) G-SGA.

5. Development of gelcasting for special-shaped PZT ceramics

Application of gelcasting to ferroelectric ceramics is not a mechanical copy of the technique to a different powder material. In addition to the demonstrated doping effect of the necessary additives, there are many other specific issues deserving further investigation. The many devices require the ferroelectric ceramic components to have various specific shapes (Scott, 2007). The shapes of the gelcast bodies are formed during gelation and drying process with the confinement of the molds. The two most critical issues about shape formation that have attracted more and more attention are probably the fracture (or crack growth) and deformation of the ceramics. No doubt that the former should be avoided, however, what is interesting is that the latter may even be used to form shapes that are difficult to formed by the molds. The two issues are briefly discussed in the following.

5.1. Factors affecting fracture and crack growth

Fracture or crack growth results from the competition of the strength of the gelcast body and the stress developed during gelation and drying. Many factors can affect fracture or crack growth, including the monomer concentration, monomer/cross-linker ratio, initiator concentration, initiator/catalyst ratio, gelation temperature and drying condition (humidity), etc (Ma et al., 2006). It is easy to be understood that too low a monomer concentration is insufficient to keep the three dimensional polymer network structure. A suitable monomer/cross-linker ratio is necessary to keep a strong gelcast body with still controllable stresses. Compared to other factors, initiator and catalyst seem to be more critical in controlling the gelcast body strength and stresses, since they can determine the 'joints' of the gel network. As shown in Figure 11. A slight decrease of the $(NH_4)_2S_2O_8$ concentration from the normal level causes a seriously broken green body. In addition to use of optimized premix solution, drying is a complicated process that should be carried out under contolled humidity (Barati et al., 2003).

Fracture or crack growth may also appear after sintering (Zheng et al., 2008), even if the shapes of the ceramics are well preserved after drying. During sintering the macromolecular gel network is destroyed and the stresses developed due to material densification may give rise to fracture or cracks in the ceramic body with decreased strength.

0.003 wt.% $(NH_4)_2S_2O_8$
(2/3 of the normal level)
20 wt.% AM solution

Figure 11. A fractured PZT-4 type green body caused by a lower initiator concentration

5.2. Deformation controllable gelcasting

Factors that affect crack growth discussed in the forgoing section also affect deformation. In most cases, deformation should be kept as smaller as possible. However, well controlled deformation can also be used to fabricate certain devices that use specially-shaped PZT as the active components. Here we show two examples: spherical PZT shell vibrator and PZT minitube.

As an effective non-invasive surgical tool, High intensity Focused sound (HIFU) has been used for the treatment of human tissues such as liver, kidney, breast, uterus and pancreas, and it is receiving growing interest (Wan, et al., 2008; Davies et al., 1998). As shown in Figure 12, some therapeutic HIFU transducers require to use spherical piezoelectric PZT shells as the active components (Saletes et al., 2011). Conventionally, such a spherical plate is fabricated by grinding both sides of a cylindrical plate. The mechanical method is time-consuming and wastes a lot of PZT material. Gelcasting of flat PZT plates generally uses a cylindrical ring mold and a flat bottom plate. During drying, the upper surface of the gelcast body is exposed to atmosphere while the other side is still berried until the whole body solidifies. Then we found an interesting phenomenon: the upper part of the gelcast body might dry and shrink first and a spherical plate was formed. By using suitable premix solution and under well controlled temperature and humidity, the plate edge bends up and a rather ideal spherically shaped PZT green plate can be formed. After sintering a spherical PZT plate was formed, which only needs to be grinded slightly to produce a homogenous spherical vibrator. Furthermore, because a lower humidity and a higher temperature can give rise to a larger deformation, the curvature radius can be controlled in a certain range under well controlled experimental conditions. So far the smallest focal length obtained for a plate with a diameter of 10 cm without further machining is around 15 cm. A PZT-8 plate with a diameter of 12 cm is shown in Figure 13.

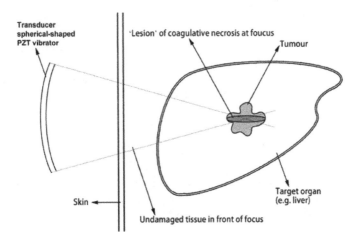

Figure 12. Schematic illustration of the therapeutic effect of a HIFU transducer

5.3. Fabrication of hollow spherical PZT shell

Thin-walled hollow sphere Omnidirectional Transducer has been used in hydrophones for many years (Li et al., 2010), which uses a hollow spherical PZT shell as the active component. Hollow spherical PZT shell is generally fabricated by Cold isostatic pressing, which requires complicated facilities. We show here that gelcasting can be developed to fabricate such Hollow spherical PZT shells by using a ball-shaped removable polymeric mold. First, the polymeric ball mold is fixed in a normal metal mold with a large spherical cavity, then the aqueous PZT premix slurry is poured into the mold. After gelation and drying, the whole mold containing the PZT green body and polymeric ball is removed by a careful thermal treatment. The most critical step in this method is thermal treatment, which solidifies the PZT slurry and melt the polymeric ball in due time. A PZT-4 hollow spherical shell with a focal length of ~ 200 mm, a diameter of ~ 120 mm and a wall thickness of 2 mm is also show in Figure 13.

Figure 13. Illustration of special-shaped gelcast PZT ceramics: spherical PZT plate and hollow spherical PZT shell.

In addition to the above mentioned advanced techniques, other gelcasting based ceramic fabrication or processing techniques have been developed in the authors' group, including spatter laser drilling techniques (Guo, 2004), and Rapid Prototyping of PZT bodies by combining gelcasting and Selective Laser Sintering (Guo, 2003), etc.

6. Conclusions

Comparison of various electrical properties and microstructures of various gelcast soft PZT samples with those of the dry pressed PZT one suggests that the Na^+ ion, which is the main cation in many widely used commercial dispersants, shows detrimental hard doping effects, leading to deteriorated electrical performance. The conclusion may be transferable to other doping ions like K^+, which is also contained in many commercial dispersants. Also due to the doping ions introduced by dispersants, the performance of the glecast PZT sample may also be improved as well. Considering that dispersants are indispensible in getting concentrated low viscosity ceramic slurries, the possible doping effect of metal ions or impurities introduced by the dispersants or other additives should be generally considered when applying gelcasting to forming multicomponent electronic ceramic materials whose electrical properties are sensitive to the composition.

In addition, we also demonstrate some advanced gelcasting techniques, including deformation controllable gelcasting of spherical PZT disc and gelcasting based hollow spherical PZT fabrication technique, etc. Application of gelcasting to PZT is not a mechanical copy of the technique to a different powder material. The results show that successful application of gelcasting to ferroelectric ceramics is not a mechanical copy of the technique to a different powder material.

Author details

Dong Guo[1] and Kai Cai[2]

1 Institute of Acoustics, Chinese Academy of Sciences, Beijing, China

2 Beijing Center for Chemical and Physical Analysis, Beijing Municipal Science and Technology Research Institute, Beijing, China

References

[1] Crsos, LE. (1996). Ferroelectric materials for electromechanical transducer applications. *Materials Chemistry and Physics*, 43, 108-115

[2] Setter, N.; Waser, R. (2000). Electroceramic materials. *Acta Materialia*, 48, 151-178

[3] Omatete, O.O.; Janney, M.A.; Strehlow, R.A. (1991). Gelcasting: a new ceramic forming process. *American Ceramic Society Bulletin*, 70, 1641-1649

[4] Biesheuvel, PM.; Verweij, H. (1999). Theory of Cast Formation in Electrophoretic Deposition. *Journal of the Amarican Ceramic Society*, 82, 1451-1455

[5] Novak, S.; Kosmac, T.; Krnel, K.; Drazic, G. (2002). Principles of the hydrolysis assisted solidification (HAS) process for forming ceramic bodies from aqueous suspension. *Journal of the European Ceramic Society*, 22, 289-295

[6] Graule, TJ.; Gauckler, LJ.; Baader, FH. (1996). Direct coagulation casting–a new green shaping technique.PT.1. processing principles. *Industrial Ceramics*, 16, 31-34

[7] Cai, K.; Guo, D.; Huang, Y.; Yang, JL. (2003). Solid freeform fabrication of alumina ceramic parts through a lost mould method. *Journal of the European Ceramic Society*, 23, 921-925

[8]   Guo, D.; Li, LT.; Gui, ZL.; Nan, CW. (2003). B-Solid state materials for advanced technology. *Materials Science and Engineering*, 99, 25-28

[9]   Young, A.C.; Omatete, O.O.; Janney, M. A.; Menchhofer, P.A.(1991). Gelcasting of Alumina. *Am.Ceram. Soc*, 74, 612-618

[10] Zhou, LJ.; Huang, Y.; Xie, ZP.; (2000). Gelcasting of concentrated aqueous silicon carbide suspension. *Journal of the European Ceramic Society*, 20, 85-90

[11] Huang, Y.; Ma, LG.; Tang, Q.; Yang, JL.; Xie, ZP.; Xu, XL. (2000). Surface oxidation to improve water-based gelcasting of silicon nitride. *Journal of Materials Science*, 35, 3519-3524

[12] Zha, SW.; Xia, CR.; Fang, XH.; Wang, HB.; Peng, DK.; Meng, GY. (2001). Processing and electrical properties of doped-LaGaO$_3$ by gelcasting. *Ceramics Intrernational*, 27, 649-654

[13] Bell, NS.; Voigt, JA.; Tuttle, BA. (2004). Colloidal processing of chemically prepared zinc oxide varistors. Part II: Near-net-shape forming and fired electrical properties. *Journal of Materials Research*, 19, 1341-1347

[14] Chen, W.; Kinemuchi, Y.; Watari, K.; Tamura, T.; Miwa, K. (2006). Preparation of Grain-Oriented Sr$_{0.5}$Ba$_{0.5}$Nb$_2$O$_6$ Ferroelectric Ceramics by Magnetic Alignment. *Journal o f the American Ceramic Society* , 89, 381-384

[15] Guo, D.; Cai, K.; Li, LT.; Gui, ZL. (2003). Application of gelcasting to the fabrication of piezoelectric ceramic parts. *Journal of the European Ceramic Society*, 23, 1131-1137

[16] Shaw, TM.; Trolier-McKinstry, S; McIntyre, PC.; (2000). The Properties of Ferroelectric Films at small dimensions. *Annual Review of Materials Science*, 30, 263-298

[17] Noheda, B.; Cox, DE. (2006). Bridging phases at the morphotropic boundaries of lead oxide solid solutions. *Phase Transitions*, 79, 5-20

[18] Israelachvili, J.; (1992). *Intermolecular & Surface Forces, 2nd ed.*

[19] Ackerson, B. J. (1990). Shear induced order and shear processing of model hard sphere suspensions. *J. Rheol*, 34, 553–590

[20] Barnes, H.A. (1989). Shear-thickening (dilatancy) in suspensions of nonaggregating solid particles dispersed in Newtonian liquids. *J. Rheol*, 33, 329–366

[21] Barnes, H.A.; Hutton,J.F.; Walers, K.; (1989). *An Introduction to Rheology, Elsevier, Oxford.*

[22] Starov, V.; Zhdanov, V.; Meireles M. et al. (2002) Viscosity of concentrated suspensions: influence of cluster formation. *Advances in Colloid and Interface Science* , 96, 279-293

[23] Jaffe, B.; Cook, W. (1971). *Piezoelectric Ceramics*. Academic Press,London/New York.

[24] Damjanovic, D. (1998). Ferroelectric, dielectric and piezoelectric properties of ferro-electric thin films and ceramics. Reports on Progress in Physics,61, 1267.

[25] West, A. R.; Sinclair, D. C. and Hirose, N. (1997). *J. Electroceram*, 65, 71

[26] Cao, W. Q.; and Gerhardt, R. (1990). *Solid State Ionics*, 42, 213

[27] Barranco, A. P.; Pinar, F. C.; Martinez, O. P. ; Guerra, J. D. ; and Carmenate,I. G. (1999). *J. Eur. Ceram. Soc*, 19, 2677

[28] Raymond, M. V.; Smyth, D. M. (1996). Defects and charge transport in perovskite fer-roelectrics. *J. Phys. Chem. Solids*, 57, 1507-1511

[29] Warren, W.L.; Pike, G.E.; Vanheusden,K.; Dimos, D.; Tuttle, B. A.; Robertson,J. (1996). Defect-dipole alignment and tetragonal strain in ferroelectrics. *J. Appl. Phys*, 79, 9250-9257

[30] Lambeck, P. V.; Jonker,G. H. (1978). *Ferroelectrics*, 22, 729

[31] Soares, MR.; Senos, AMR.; Mantas, PQ. (2000). Phase coexistence region and dielec-tric properties of PZT ceramics. *J. Eur. Ceram. Soc*, 20, 321-334

[32] Hammer, M.; Monty, C.; Endriss, A. (1998). Correlation between surface texture and chemical composition in undoped, hard, and soft piezoelectric PZT ceramics. *Journal of the American Ceramic Society*, 81, 721-724

[33] Xu, ZH.; Ducker, W.; Israelachvili, J. (1996). *Langmuir*, 12, 2263-2270

[34] Tomasik, P.; Schilling, CH.; Jankowiak, R.; Kim, JC. (2003). The role of organic disper-sants in aqueous alumina suspensions. *Journal of the European Ceramic Society*, 23, 913-919

[35] Scott, J. F. (2007). *Science*, 315, 954-959

[36] Ma, LG.; Huang, Y.; Yang, JL.; Le, HR.; Sun, Y. (2006). Control of the inner stresses in ceramic green bodies formed by gelcasting. *Ceramics International*, 32, 93-98

[37] Barati, A.; Kokabi, M.; Famili, N. (2003). Modeling of liquid desiccant drying method for gelcast ceramic parts. *Ceramics International*, 29, 199-207

[38] Zheng, ZP.; Zhou, DX.; Gong, SP. (2008). Studies of drying and sintering characteris-tics of gelcast $BaTiO_3$-based ceramic parts. *Ceramics International*, 34, 551-555

[39] Wan, Yayun.; Ebbini, Emad S. (2008). Transactions on Ultrasonics Ferroelectrics and frequency Control. *IEEE* 55, 1705-1718

[40] Davies, BL.; Chauhan, S.; Lowe, M. (1998). International Conference on Medical Im-age Computing and Computer-Assisted Intervention. *Lecture Notesin Computer Sci-ence*, 1496, 386-396

[41] Saletes, Izella.; Gilles, Bruno.; Bera, Jean-Christophe. (2011). Promoting inertial cavi-tation by nonlinear frequency mixing in a bifrequency focused ultrasound beam. *Ul-trasonics*,51, 94-101

[42] Li, XF.; Peng, XL.; Lee, KY. (2010). The static response of functionally graded radially polarized piezoelectric spherical shells as sensors and actuators. *Smart Materials & Structures*, 19, 3

[43] Guo, D.; Li, LT.; Cai, K.; Gui, ZL.; Nan, CW. (2004). Rapid Prototyping of Piezoelectric Ceramics via Selective Laser Sintering and Gelcasting. *Journal of the American Ceramic Society*, 87, 17-22

[44] Guo, D.; Cai, K.; Huang, Y.; Li, LT. (2003). A novel anti-spatter and anti-crack laser drilling technique: application to ceramics. *Applied Physics A:materials Science & Processing* , 76, 1121-1124

Doping-Induced Ferroelectric Phase Transition and Ultraviolet-Illumination Effect in a Quantum Paraelectric Material Studied by Coherent Phonon Spectroscopy

Toshiro Kohmoto

Additional information is available at the end of the chapter

1. Introduction

There has been significant interest in a quantum paraelectric material strontium titanate ($SrTiO_3$), and its lattice dynamics and unusual dielectric character have been extensively studied. In low temperatures, its dielectric constant increases up to about 30 000. The dielectric constant increases extraordinarily with decreasing temperature while the paraelectric phase is stabilized by quantum fluctuations without any ferroelectric phase transition even below the classical Curie temperature T_c=37 K [1]. In $SrTiO_3$, a ferroelectric transition is easily induced by a weak perturbation such as an uniaxial stress [2], an isotopic substitution of oxygen 18 for oxygen 16 [3], and an impurity doping [4,5].

$SrTiO_3$ has a perovskite structure as shown in Fig. 1(a) and is known to undergo a structural phase transition at T_c=105 K [6]. The cubic (O_h) structure above T_c, where all phonon modes are Raman forbidden, changes into the tetragonal (D_{4h}) structure below T_c, where Raman-allowed modes of symmetries A_{1g} and E_g appear [7]. The phase transition is due to the collapse of the Γ_{25} mode at the R point of the high-temperature cubic Brillouin zone. Below T_c, the R point becomes the Γ point of the D_{4h} phase. The phase transition is characterized by the softening of phonons at the R point and concomitant doubling of the unit cell.

The distortion consists of an out-of-phase rotation of adjacent oxygen octahedra in the (100) planes [6]. The order parameter for the phase transition is inferred to be the angle of rotation of the oxygen octahedra. Only a small rotation of the oxygen octahedra is involved for the transition. The rotation angle for the oxygen octahedra varies from ~2° of arc near 0 K down to zero at T_c=105 K; the transition is second order. At liquid nitrogen temperature, the rotation angle is about 1. 4° and the linear displacement of the oxygen ions about their high-temperature equilibrium positions is less than 0.003 nm. This oxygen octahedron motion can be described as a rotation only as a first approximation; the oxygen ions actually remain on the faces of each cube and therefore increase in separation from the titanium. Because the (100) planes are equivalent in the cubic phase, the distortion produces domains below Tc in which the [100], [010], or [001] axis becomes the unique tetragonal c axis.

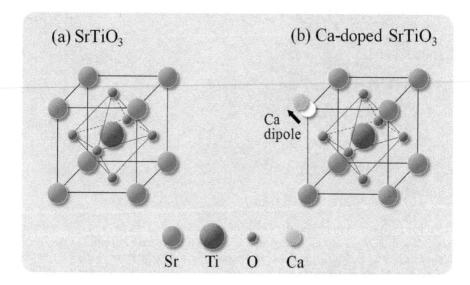

Figure 1. (a) Perovskite structure in SrTiO₃. (b) Doped Ca ions are substituted for the Sr ions in Ca-doped SrTiO₃.

According to the measurement of dielectric constants, $Sr_{1-x}Ca_xTiO_3$ undergoes a ferroelectric transition above the critical Ca concentration x_c=0. 0018, where doped Ca ions are substituted for the Sr ions [8–11] as shown in Fig. 1(b). The cubic structure above the structural phase-transition temperature T_{c1} changes in to the tetragonal structure below T_{c1} and into the rhombohedral structure below the ferroelectric transition temperature T_{c2}. The structural phase transition at T_{c1} is also antiferrodistortive as in pure SrTiO₃ at 105 K [8]. As in the case of impurity systems, Li-doped KTaO₃ and Nb-doped KTaO₃ [4], off-centered impurity ions are supposed. Their polarized dipole moments show a ferroelectric instability below the ferroelectric transition temperature [4]. In the case of Ca-dopedSrTiO₃, a spontaneous polarization occurs along [110] directions in the c plane, where the tetragonal (D_{4h}) symmetry is lowered to C_{2v} [8]. With decreasing temperature a ferroelectric ordering process dominates, that is, due to the thermal growth of the polarization clouds surrounding the off-center Ca^{2+} dipoles [8]. The system behaves like a super paraelectric as the ferroelectric nano-ordered regions contain disordered cluster like regions. The investigation by x-ray and neutron diffractions and first-principles calculations [12] suggests that polar instabilities originating from the off-center displacements of

Ca^{2+} ions are not likely to directly polarize the host matrix by an electrostatic mechanism. Instead, the possible role of random fields in inducing the presence of disordered polar clusters was suggested, which is similar to polar nanoregions in relaxor materials.

Recently, a gigantic change in the dielectric constant by an ultraviolet (UV) illumination was discovered [13,14], and a deeper interest has been taken in $SrTiO_3$ again. The origin of the giant dielectric constants, however, has not yet been clarified. In Ca-doped $SrTiO_3$, it was reported that a UV illumination causes a ferroelectric peak shift of the dielectric constant toward the lower temperature side [11]. In several ferroelectric materials such as $BaTiO_3$ [15], SbSI [16], and oxygen-isotope-substituted $SrTiO_3$ [17], the T_c reduction under a UV illumination has been observed.

The optical information on the dielectric response is usually obtained from the experiments of Raman scattering or infrared spectroscopy. The usefulness of the investigation of low-frequency dielectric response by observing coherent phonons have also been demonstrated by the time-resolved study of the dynamics of phonons [18] and phonon polaritons [19]. At low frequencies this technique is very sensitive and provides a very good signal-to-noise ratio as compared to the conventional frequency-domain techniques while at higher frequencies a better performance will be achieved by using the conventional techniques. Therefore the coherent phonon spectroscopy and the conventional frequency-domain techniques can be considered to be complementary methods for the investigation of the dielectric response.

The observed signal of the Raman scattering [20] in $SrTiO_3$ is very weak because the distortion from cubic structure in the low-temperature phase is very small. The intensity of the first-order Raman signal is of the same order of magnitude with many second-order Raman signals, and then a background-free signal of the first-order Raman scattering cannot be observed.

Under a UV illumination, $SrTiO_3$ and Ca-doped $SrTiO_3$ show a broadband luminescence in the visible region originated from a relaxed excited state [21]. The coherent phonon spectroscopy is not sensitive to the luminescence and a powerful technique to investigate UV-illumination effects in paraelectric materials as compared to the Raman-scattering measurement because in Raman-scattering experiments, it is not easy to separate Raman-scattering signals from the luminescence.

In the present study, ultrafast polarization spectroscopy is used to observe the coherent optical phonons in pure and Ca-doped $SrTiO_3$, which are generated by femtosecond optical pulses through the process of impulsive stimulated Raman scattering [22,23]. Time-dependent linear birefringence induced by the generated coherent phonons is detected as a change of the polarization of probe pulses. High detection sensitivity of $\sim 10^{-5}$ in polarization change has been achieved in our detection system. Damped oscillations of coherent phonons in $SrTiO_3$ were observed below the structural phase-transition temperature (T_c=105 K), and temperature dependences of the phonon frequency and the relaxation rate are measured [24]. The mechanism of the phonon relaxation is discussed by using a population decay model, in which an optical phonon decays into two acoustic phonons due to an harmonic phonon-phonon coupling.

The doping-induced ferroelectric phase transition in Ca-doped $SrTiO_3$ is investigated by observing the birefringence and coherent phonons [25]. In the birefringence measurement, the structural and the ferroelectric phase-transition temperatures are examined. In the observation of coherent phonons, the soft phonon modes related to the structural (T_{c1}=180 K) and the ferroelectric (T_{c2}=28 K) phase transitions are studied, and their frequencies are obtained from the observed coherent phonon signals. The behavior of the softening toward each phase-transition temperature and the UV-illumination effect on the ferroelectric transition are discussed. In addition to the ferroelectric phase transition at T_{c2}=28 K, another structural deformation at 25 K is found. A shift of the ferroelectric phase-transition temperature under the UV illumination and a decrease in the phonon frequencies after the UV illumination are found. We show the approach in the time domain is very useful for the study of the soft phonon modes and their UV-illumination effect in dielectric materials.

2. Experiment

The experiments are performed on single crystals of pure $SrTiO_3$ and Ca-doped $SrTiO_3$ with the Ca concentration of x=0.011. $SrTiO_3$ was obtained commercially (MTI Corporation) and Ca-doped $SrTiO_3$ was grown by the floating zone method [11]. The thickness of the sample is 1. 0 and 0.5 mm for the pure and Ca-doped crystals, respectively. In Ca-doped$SrTiO_3$, the structural phase-transition ($O_h{\rightarrow}D_{4h}$) temperature, T_{c1}=180 K, is obtained from the result of the birefringence measurement in section 3. The ferroelectric phase-transition temperature, T_{c2}=28 K, was determined by the measurement of dielectric constants [11]. The value of x was determined by the empirical relation of Bednorz and Müller [5].

Schematic diagram of the experiment of polarization spectroscopy is shown in Fig. 2. The change in optical anisotoropy (birefringence) is detected by a polarimeter as the change in the polarization of the probe light (ellipticity). In the birefringence measurement, the birefringence generated by the lattice deformation is detected by a continuous-wave (cw) probe beam with no pump beam. In the coherent phonon spectroscopy, the transient birefringence due to the coherent phonons generated by a pump pulse is detected by a probe pulse.

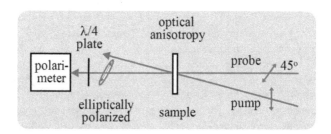

Figure 2. Schematic diagram of the experiment of polarization spectroscopy.

2.1. Birefringence measurement

In the birefringence measurement, the linearly polarized probe beam is provided by a Nd:YAG laser (532 nm, cw) and is perpendicular to the (001) surface of the sample.

The construction of the polarimeter is shown in Fig. 3. The polarimeter [26,27] detects the rotation of polarization plane of a light beam. A linearly-polarized beam is split by a polarized beam splitter (PBS) and incident on the two photodiodes (PD) whose photocurrents are subtracted at a resistor (R). When the polarized beam splitter is mounted at an angle of 45° to the plane of polarization of the light beam, the two photocurrents cancel. If the plane of polarization rotates, the two currents do not cancel and the voltage appears at the resistor.

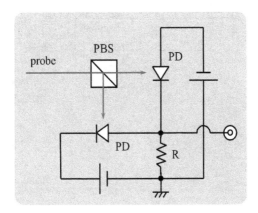

Figure 3. Construction of the polarimeter.

In the present experiment, the birefringence generated by the lattice deformation is detected as the change in polarization of the probe beam using a quarterwave plate and the polarimeter. The birefringence generated in the sample changes the linear polarization before transmission to an elliptical polarization after transmission. The linearly-polarized probe beam is considered to be a superposition of two circularly-polarized components which have the opposite polarizations and the same intensities. The generated birefringence destroys the intensity balance between the two components. The two circularly-polarized beams are transformed by the quaterwave plate to two linearly-polarized beams whose polarizations are crossed each other, and the unbalance of circular polarization is transformed to the unbalance of linear polarization or the rotation of polarization plane. This rotation is detected by the polarimeter as the signal of the lattice deformation.

2.2. Observation of coherent phonons

Coherent phonons are observed by ultrafast polarization spectroscopy with the pump-probe technique. The experimental setup for coherent phonon spectroscopy is shown in Fig. 4. Coherent phonons are generated by femtosecond optical pulses through the process of impulsive

stimulated Raman scattering [22,23], and are detected by monitoring the time-dependent anisotropy of refractive index induced by the pump pulse. The pump pulse is provided by a Ti:
sapphire regenerative amplifier whose wavelength, pulse energy, and pulse width at the sample are 790 nm, 2 μJ, and 0.2 ps, respectively. The probe pulse is provided by an optical para-
metric amplifier whose wavelength, pulse energy, and pulse width are 690 nm, 0. 1 μJ, and 0.2
ps, respectively. The repetition rate of the pulses is 1kHz. The linearly polarized pump and
probe beams are nearly collinear and perpendicular to the (001) surface of the sample, and are
focused on the sample in a temperature-controlled refrigerator. The waist size of the beams at
the sample is about 0.5 mm.

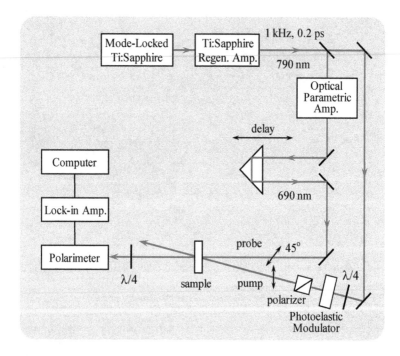

Figure 4. Experimental setup for coherent phonon spectroscopy.

The induced anisotropy of refractive index is detected by the polarimeter with a quarter-wave
plate as the polarization change in the probe pulse. The plane of polarization of the probe pulse
is tilted by 45° from that of the pump pulse. The two different wavelengths for the pump and
probe pulses and pump-cut filters are used to eliminate the leak of the pump light from the input of the polarimeter. The time evolution of the signal is observed by changing the optical delay between the pump and probe pulses. The pump pulse is switched on and off shot by shot by
using a photoelastic modulator, a quarter-wave plate, and a polarizer, and the output from the
polarimeter is lock-in detected to improve the signal-to-noise ratio.

The source of UV illumination is provided by the second harmonics (380 nm, 3. 3 eV) of the output from another mode-locked Ti: sapphire laser, whose energy is larger than the optical band gap of $SrTiO_3$ (3.2 eV). Since the repetition rate of the UV pulses is 80 MHz, this UV illumination can be considered to be continuous in the present experiment, where the UV-illumination effect appearing more than one minute after is studied.

3. Birefringence measurement

Figure 5 shows the temperature dependences of the change in birefringence in $SrTiO_3$ and Ca-doped $SrTiO_3$ between 4.5 and 250 K, where the polarization plane of the probe light is along the [110]and [100]axes. In $SrTiO_3$, a change in birefringence appears below T_c=105 K, which is the temperature of the structural phase transition, and is increased as the temperature is decreased. In Ca-doped $SrTiO_3$, large changes in birefringence come out at T_{c1}=180 K, which is the temperature of the structural phase transition, and at T_{c2}=28 K, which is the temperature of the ferroelectric phase transition. The change in birefringence is increased as the temperature is decreased from T_{c1} to T_{c2}, as well as the case in $SrTiO_3$, and shows a increase in the gradient for both axes around T_{c2}. Below T_{c2}, another kind of lattice distortion is added, and the peaks of Δn due to the competition between the two kinds of lattice distortion appear. In both $SrTiO_3$ and Ca-doped$SrTiO_3$, a cusp on the birefringence curve appears around T_c=105 K and T_{c1}=180 K, respectively, because of the fluctuation associated with the second-order structural phase transition [28].

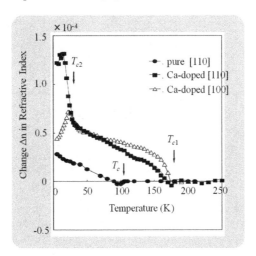

Figure 5. Temperature dependences of the change in birefringence between 4. 5 and 250 K in $SrTiO_3$ and Ca-doped $SrTiO_3$. The polarization plane of the probe light for $SrTiO_3$ is along the [110] axis (solid circles). That for Ca-doped $SrTiO_3$ is along the [1 10] axis (solid squares) and along the [100] axis (open triangles).

4. Coherent phonon spectroscopy in SrTiO$_3$

4.1. Angular dependence of the coherent phonon signal

Figure 6 shows the transient birefringence in SrTiO$_3$ observed at 6 K for the 0° pumping, where the polarization direction of the pump pulse is parallel to the [100] axis of the crystal. Vertical axis is the ellipticity η in electric-field amplitude of the transmitted probe pulse. At zero delay, a large signal due to the optical Kerr effect, whose width is determined by the pulse width, appears. After that, damped oscillations of coherent phonons are observed as shown in Fig. 6(b), where the vertical axis is enlarged by 50 from that of Fig. 6(a). The change of the polarization for the oscillation amplitude of the coherent phonon signal is 4×10^{-4} of the electric-field amplitudeof the probe pulse, which corresponds to the change $\Delta n = 7\times10^{-8}$ of the refractive index. In our detection system, polarization change of $\sim10^{-5}$ in the electric field amplitude can be detected.

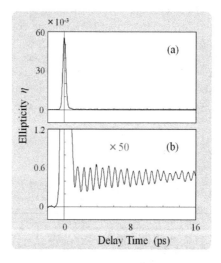

Figure 6. Transient birefringence in SrTiO$_3$ observed at 6 K for the 0° pumping. The vertical axis is the ellipticity η in electric-field amplitude of the transmitted probe pulse. (a) A large signal due to the optical Kerr effect appears at zero delay. (b) Damped oscillations of coherent phonons follow after the Kerr signal. The vertical axis of (b) is enlarged by 50 from that of (a).

Angular dependence of the coherent phonon signal in SrTiO$_3$ observed at 6 K is shown in Fig. 7(a), where the angle between the [100] axis of the crystal and the polarization direction of the pump pulse is 0°, 15°, and 45°. The angle between the polarization directions of the pump and probe pulses is fixed to 45°. The 0.7 ps period signal for the 0° pumping disappears for the 45°pumping, where the 2.3 ps period signal appears. The Fourier transform of the coherent phonon signals in Fig. 7(a) is shown in Fig. 7(b). Oscillation frequency of the signal for the 0° pumping is 1.35 THz, and that for the 45° pumping is 0.4 THz. For other pumping angles both frequency components coexist in the coherent phonon signal. From the oscillation frequencies the 1.35 THz component is considered to correspond to the A_{1g} mode, and the 0.4 THz component to the E_g mode [7].

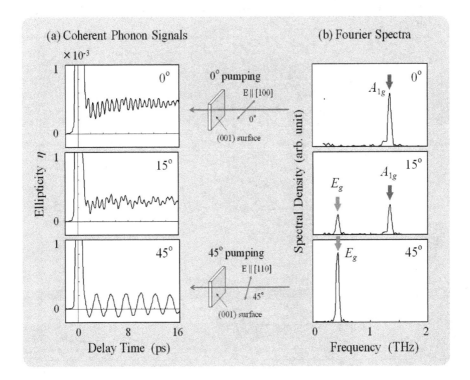

Figure 7. (a) Coherent phonon signals in SrTiO$_3$ observed at 6 K for the 0°, 15°, and 45° pumping, where the angle between the [100] axis of the crystal and the polarization direction of the pump pulse is changed. (b) Fourier transform of the coherent phonon signals in (a).

In addition to the oscillation signal there exists a dc component. The dc component has a maximum amplitude for the 0° pumping and minimum amplitude for the 45° pumping. However, the creation mechanism is not clear at present. In the following we pay attention to the oscillation component.

4.2. Temperature dependence of the coherent phonon signal

The temperature dependence of the coherent phonon signal in SrTiO$_3$ observed for the 0° pumping, which corresponds to the A_{1g} mode, is shown in Fig. 8(a). The oscillation period and the relaxation time of coherent phonons at 10 K are 0.7 and 12 ps. As the temperature is increased, the oscillation period becomes longer and the relaxation time becomes shorter. At

T_c=105 K, the phase transition point, the oscillation disappears. Above T_c no signal of coherent phonons is observed. The temperature dependence of the coherent phonon signal in SrTiO₃observed for the 45° pumping, which corresponds to the E_g mode, is shown in Fig. 8(b). The oscillation period and the relaxation time of coherent phonons at 10 K are 2.3 and 45 ps. Similar behavior to that of the A_{1g} mode was observed as the temperature was increased.

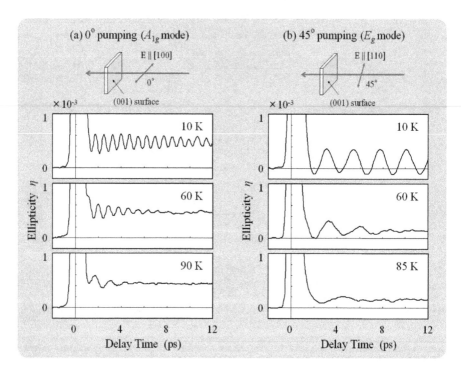

Figure 8. Temperature dependence of the coherent phonon signals in SrTiO₃ observed (a) for the 0° pumping which corresponds to the A_{1g} mode and (b) for the 45° pumping which corresponds to the E_g mode.

4.3. Phonon frequencies

The observed coherent phonon signal $S(t)$ is expressed well by the damped oscillation

$$S(t) = Ae^{-\gamma t}\sin \omega t , \tag{1}$$

where ω is the oscillation frequency and γ is the relaxation rate. This sine-type function is expected for phonons induced by impulsive stimulated Raman scattering [22,23]. The temperature dependence of the oscillation frequency in SrTiO₃ obtained from the observed coherent phonon signal below T_c is shown in Fig. 9. The diamonds are the oscillation frequency for the A_{1g} mode, and the squares are that for the E_g mode. As the temperature is increased from 6 K, the oscillation frequencies decrease and approach zero at the phase transition temperature T_c for both modes. This result is consistent with the temperature dependence of phonon frequency observed by Raman scattering [7]. The solid curves describe a temperature dependence of the form $\omega \propto (T_c - T)^n$. The experimental results for the temperature region between 50 K and T_c are explained well by $n=0.4$ for both modes, while those below 40 K deviate from that form.

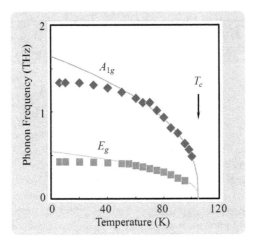

Figure 9. Temperature dependence of the oscillation frequency in SrTiO$_3$ obtained from the coherent phonon signal below T_c. The diamonds are the oscillation frequency for the A_{1g} mode, and the squares are that for the E_g mode. The solid curves describe a temperature dependence of the form $\omega \propto (T_c - T)^n$, where T_c=105 K and n=0.4 for both modes.

The intensity of the first-order Raman-scattering signal in SrTiO$_3$ is very weak and is of the same order of magnitude as many second-order Raman-scattering signals because the distortion from the cubic structure in the low-temperature phase is very small. Then observation of a background-free signal of first-order Raman scattering is not easy, and information on the relaxation, or the spectral width, is not given in a study of Raman scattering [20]. By the present method of coherent phonon spectroscopy in the time domain, on the other hand, background-free damped oscillations can be observed directly, and the oscillation frequency and the relaxation rate can be obtained accurately.

4.4. Relaxation rates

The temperature dependence of the relaxation rate in SrTiO$_3$ obtained from the observed coherent phonon signal below T_c is shown in Fig. 10. The diamonds in Fig. 10(a) are the relaxation rate for the A_{1g} mode and the squares in Fig. 10(b) are that for the E_g mode. The relaxation rates increase as the temperature is increased.

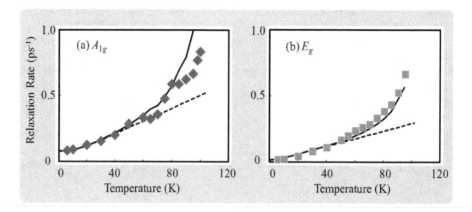

Figure 10. Temperature dependence of the relaxation rate in SrTiO$_3$ obtained from the coherent phonon signal below T_c. (a) The diamonds are the relaxation rate for the A_{1g} mode, and (b) the squares are that for the E_g mode. The solid curves show the theoretical curves including the frequency change obtained from Eq. (2).

In general, relaxation of coherent phonons is determined by population decay (inelastic scattering) and pure dephasing (elastic scattering). In metals, pure dephasing due to electron-phonon scattering, which depends on the hot electron density, contributes to the phonon relaxation [29]. In dielectric crystals, the relaxation process of the coherent phononis considered to be dominated by the population decay due to the anharmonic phonon-phonon coupling [30-32], rather than pure dephasing. According to the anharmonic decay model [30], the relaxation of optical phonons in the center of the Brillouin zone is considered to occur through two types of decay process, the down-conversion and up-conversion processes. In a down-conversion process, the initial ω_0 phonon with wave vector $k \cong 0$ decays into two lower-energy phonons ω_i and $\omega_{j'}$ with opposite wave vectors k and $-k$, which belong to the i branch and the j branch of the phonon. Energy and wave-vector conservation is given by $\omega_0 = \omega_{ik} + \omega_{j-k}$. In an up-conversion process, the initial excitation is scattered by a thermal phonon (ω_{jk}) into a phonon of higher energy (ω_{jk}), where $\omega_0 + \omega_{ik} = \omega_{jk}$. The down-conversion process can be realized either for $i = j$ (overtone channel), or for $i \neq j$ (combination channel), depending on the phonon band structure of the material, while the up-conversion process contains only the combination channel and has no overtone channel. The combination channel is less likely, because three frequencies of phonons and three phonon branches have to be concerned and stringent limitations are imposed by the energy and wave-vector conservation. The overtone channel, on the other hand, is more likely because two (an optical and an acoustic) phonon branches are concerned, and the energy and wave-vector conservation are necessarily satisfied by two acoustic phonons with the same frequency and opposite wave vectors, if the frequency maximum of the acoustic branch is higher than half the frequency of the initial optical phonon.

Here we consider the down-conversion process in which an optical phonon decays into two acoustic phonons with half the frequency of the optical phonon and with opposite wave vec-

tors. Schematic diagram of this down-conversion process is shown in Fig. 11. The temperature dependence of the relaxation rate γ of the coherent phonon is given by [30,31]

$$\gamma = \gamma_0\left(1 + \frac{2}{\exp\left[\hbar(\omega_0/2)/k_B T\right] - 1}\right)$$

(2)

where ω_0 is the frequency of the optical phonon, and k_B is the Boltzmann constant.

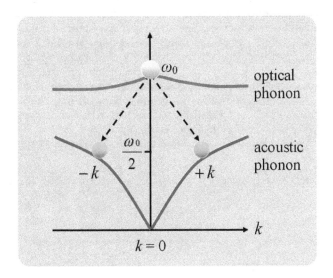

Figure 11. Schematic diagram of the down-conversion processe (overtone channel) in the an harmonic decay model
of optical phonons. The initial ω_0 optical phonon with wave vector $k \cong 0$ decays into two $\omega_0/2$ acoustic phonons, with
opposite wave vectors k and $-k$.

In ordinary materials the temperature dependence of the phonon frequency is small, and a
theoretical curve with a fixed value of phonon frequency fits the experimental data well. In
SrTiO$_3$, however, the phonon frequencies are changed greatly as the temperature is increased; thus a frequency change has to be considered. The solid curves in Fig. 10 show the
theoretical curves including the frequency change obtained from Eq. (2) with $\gamma_0 = 8.0 \times 10^{10}$ s^{-1}
for the A$_{1g}$ mode and 1.5×10^{10} s^{-1} for the E$_g$ mode, where the observed phonon frequencies in
Fig. 9 are used for each temperature. The broken curves show the theoretical curves in the
case of no frequency change. As is seen in Fig. 10, the solid curves explain well the experimental data.

Deviation of the experimental data near T_c from the solid curve may be caused by the effect
of the phase transition. However, the relaxation rate just around the phase transition point
cannot be obtained in the present experiment, and the relation between the temperature-dependent relaxation rate and the structural phase transition is not clear.

5. Coherent phonon spectroscopy in Ca-doped SrTiO₃

5.1. Angular dependence of the coherent phonon signal

The angular dependence of the coherent phonon signal in Ca-doped $SrTiO_3$ observed at 50 K is shown in Fig. 12(a), where the angle between the [100] axis of the crystal and the polarization plane of the pump pulse is $0°$, $25°$, and $45°$. The angle between the polarization planes of the pump and probe pulses is fixed to $45°$. Vertical axis is the ellipticity in electric field amplitude of the transmitted probe. At zero delay, a large signal due to the optical Kerr effect, whose width is determined by the laser-pulse width, appears. After that, damped oscillations of coherent phonons are observed. For the $0°$ pumping a 0.5 ps period signal appears while it disappears for the $45°$ pumping and a 2 ps period signal appears. The Fourier transform of the coherent phonon signals in Fig. 12(a) is shown in Fig. 12(b). The oscillation frequency of the signal for the $0°$ pumping is 1.9 THz and that for the $45°$ pumping is 0.5 THz. For other pumping angles both frequency components coexist in the coherent phonon signal. These phonon modes are the soft modes related to the structural phase transition at $T_a = 180$ K. From the oscillation frequencies the 1.9 THz component corresponds to the A_{1g} mode, and the 0.5 THz components to the E_g mode, which are assigned from the modes of pure $SrTiO_3$ [7] and from that of $Sr_{1-x}Ca_xTiO_3$ (x=0.007) [8].

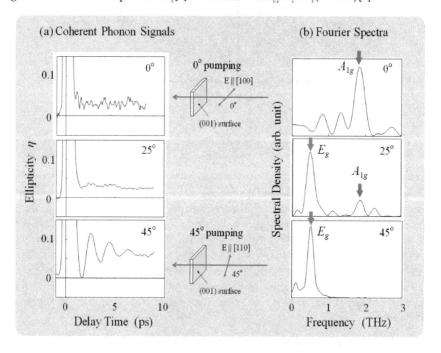

Figure 12. (a) Coherent phonon signals in Ca-doped SrTiO₃ observed at 50 K for the 0°, 25°, and 45° pumping, where the angle between the [100] axis of the crystal and the polarization direction of the pump pulse is changed. (b) Fourier transform of the coherent phonon signals in (a).

5.2. Temperature dependence of the coherent phonon signal

The temperature dependence of the coherent phonon signal in Ca-doped $SrTiO_3$ observed for the 45° pumping, which corresponds to the E_g mode, is shown in Fig. 13. At 6 K some oscillation components, which have different frequencies, are superposed. As the temperature is increased, the number of the oscillation components is decreased, the oscillation period becomes longer and the relaxation time becomes shorter. At T_{c1}=180 K, which is the structural phase-transition temperature obtained from the birefringence measurement, no signal of coherent phonons is observed.

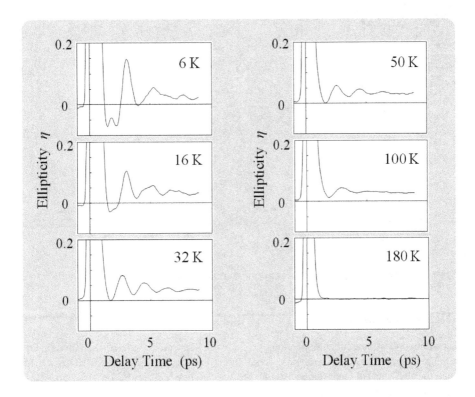

Figure 13. Temperature dependence of the coherent phonon signal in Ca-doped SrTiO3 observed for the 45° pumping.

The temperature dependence of the coherent phonon signal in Ca-doped $SrTiO_3$ observed for the 25° pumping are shown in Figs. 14 and 15. Figure 14(a) shows the coherent phonon signals below the ferroelectric phase-transition temperature T_{a} and Fig. 15(a) shows those above T_{a}. Figure 14(b) shows the Fourier transform of the coherent phonon signals in Fig. 14(a), where the peaks with arrows 1, 2, and 3 correspond to the ferroelectric phonon modes [8]. Figure 15(b) shows the Fourier transform of the coherent phonon signals in Fig. 15(a), where modes 1, 2, and 3 disappear but the A_{1g} and E_g modes related to the structural phase-transition remain.

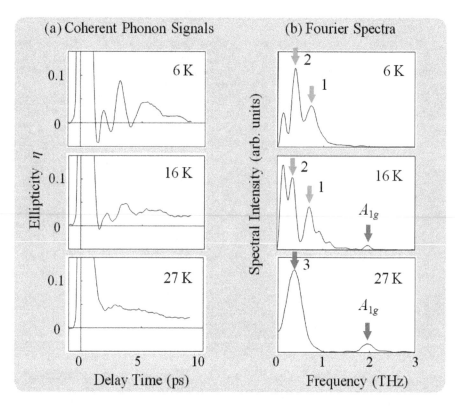

Figure 14. (a) Temperature dependence of the coherent phonon signal in Ca-doped SrTiO$_3$ observed for the 25° pumping below the ferroelectric phase-transition temperature T_{c2}. (b) Fourier transform of the coherent phonon signals in (a).

5.3. Phonon frequencies

Each component of the coherent phonon signal is expressed well by damped oscillations in Eq. (1). The temperature dependence of the oscillation frequencies in Ca-doped SrTiO$_3$ obtained from the observed coherent phonon signals below T_a is shown in Fig. 16. The solid circles are the oscillation frequencies for the A_{1g} and E_g modes. The triangles are that for modes 1, 2, and 3 which are related to the ferroelectric phase transition. As for the A_{1g} and E_g modes, as the temperature is increased from 6 K, the oscillation frequencies decrease and approach to zero at the structural phase-transition temperature T_a for both modes. This result is consistent with the temperature dependence of phonon frequency observed by Raman scattering [7] and coherent phonons in section 4.3 for pure SrTiO$_3$ except for the phase-transition temperature. The broken curves describe a temperature dependence of the form $\omega \propto (T_c - T)^n$. The experimental results for the temperature region between T_a and T_a are explained well by n=0.5 for both modes. Below the ferroelectric phase-transition temperature $T_{a'}$ another mode appears at 0.9 THz. It is

considered that the doubly degenerate E_g mode is split into two components under the tetrago-
nal-to-rhombohedral lattice distortion.

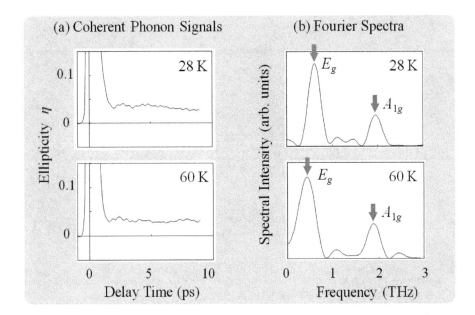

Figure 15. (a) Temperature dependence of the coherent phonon signal in Ca-doped SrTiO$_3$ observed for the 25°
pumping above the ferroelectric phase-transition temperature T_a. (b) Fourier transform of the coherent phonon signals in (a).

The temperature dependence of the phonon frequencies, which are related to the ferroelectric phase transition, obtained from the observed coherent phonon signals below T_a is shown
in Fig. 17, where only the frequencies of the reproducible peaks in the Fourier spectrum are
plotted. While the lowest peaks at 6 and 16 K in Fig. 14, for example, may be the third mode
observed in the Raman experiment [8], the frequencies of the peaks with poor reproducibility are not plotted in Figs. 16 and 17. Under dark illumination, two phonon modes 1 and 2
are softened toward about 25 K and degenerate into mode 3, which seems to be softened toward the ferroelectric phase-transition temperature T_a=28 K. The lift of degeneracy for mode
3 below T_a suggests that another structural deformation occurs at 25 K. In the Raman-scattering experiment [8], three modes deriving from the TO$_1$ mode has been observed. The
three modes do not all split off at the same temperature; the highest-energy component
splits off at the ferroelectric transition temperature while the other two split off at the temperature about 3 K lower. These two temperatures may correspond to the two temperatures,
28 and 25 K, observed in our experiment, and the temperature differences are nearly equal
to each other. The existence of the second structural deformation at 25 K is consistent with
the Raman-scattering data.

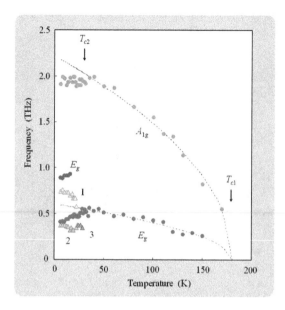

Figure 16. Temperature dependence of the phonon frequencies in Ca-doped SrTiO3 obtained from the coherent phonon signals below T_{c1}. The solid circles are the oscillation frequencies for the A_{1g} and E_g modes. The broken curves describe a temperature dependence of the form $\omega \propto (T_c - T)^n$, where T_c=180 K and n=0.5 for both modes. The triangles are that for modes 1, 2, and 3 related to the ferroelectric phase transition.

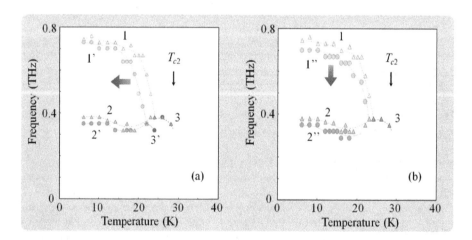

Figure 17. Temperature dependence of the phonon frequencies (a) under dark (1, 2, 3) and under UV (1',2',3') illumination below T_{c1}, which correspond the phonon modes related to the ferroelectric phase transition, and (b)under dark (1, 2, 3) and after UV(1'',2'',3'') illumination, where the intensity of the UV illumination is 7 mW/mm².

5.4. UV-illumination effect

In order to examine the UV-illumination effect, two types of measurements, under the UV illumination and after the UV illumination, are carried out. The temperature dependence of the phonon frequencies under the UV illumination is shown in Fig. 17(a), where the intensity of the UV illumination is 7 mW/mm^2. As is seen in Fig. 17(a), the temperature, toward which the two modes 1 and 2 are softened and degenerate into mode 3, shifts to the lower temperature side. The temperature shift due to the UV-illumination effect is ~3 K.

Doped Ca ions behave as permanent dipoles and ferroelectric clusters are formed around Ca dipoles with high polarizability of the host crystal. The ferroelectric transition is caused by the ordering of randomly distributed Ca dipoles. The UV-illumination-induced T_a reduction is related to the screening of the internal macroscopic field by UV-excited carriers. Under the UV-illumination non equilibrium carriers appear, which are captured by traps and screen the polarization field. The ordering is prevented by the photo excited carriers. Thus, the screening effect due to the UV-excited carriers weakens the Coulomb interaction between dipoles, and the transition temperature is decreased.

The theoretical T_a reduction under the UV illumination ΔT_a is given by [11]

$$\frac{\Delta T_{c2}}{T_{c2}(0)} = 1 - \left(1 + \lambda a + \frac{\gamma}{4\pi}\lambda^2 a^2\right)\exp\left(-\lambda a\right), \tag{3}$$

where $T_a(0)$ is the transition temperature before the UV illumination, λ is the inverse of the screening length, a is the mean separation between the dipoles, and γ is the number of the nearest-neighbor dipoles. The expression of λ is presented by $\lambda = \sqrt{ne^2/\varepsilon\varepsilon_0 k_B T}$, where n is the carrier concentration which depends on the UV-illumination intensity and ε is the relative dielectric constant. In the present case, the domain is large enough and the dipole-dipole interaction is expressed as a simple Coulomb interaction. The UV-illumination-intensity dependence of the T_a reduction for Sr$_{1-x}$Ca$_x$TiO$_3$ (x=0. 011) was analyzed by using Eq. (3). Assuming that the carrier concentration is proportional to the UV-illumination intensity, the fitting result reproduced well the experimental result obtained in the measurement of dielectric constants [11]. Here we estimate the inverse λ of the screening length at the UV-illumination intensity of 7 mW/mm^2. As a result of the measurement of dielectric constant [11], the carrier concentration is n=2. 5× 10^{17} cm^{-3} at 3 mW/mm^2. From this value we can estimate the carrier concentration to be n=5. 8× 10^{17} cm^{-3} at 7 mW/mm^2. The value of the inverse of the screening length is obtained as

$$\lambda = \sqrt{\frac{ne^2}{\varepsilon\varepsilon_0 k_B T}} = \sqrt{\frac{8\pi n a_B E_{1s}}{\varepsilon k_B T}} = 1.1 \times 10^7 \text{ cm}^{-1}, \tag{4}$$

where a_B and E_{1s} are Bohr radius and the energy of the hydrogen atom in the 1s ground state, respectively, and we use the dielectric constant ε=4 in the visible region. Substituting the value of λ into Eq. (3), the shift of the transition temperature can be estimated to be ΔT_a~8 K. This estimation is not inconsistent with the observed values of the temperature shift in the experiment of coherent phonons, in which the value of the observed transition temperature shift is ~3 K.

The temperature dependence of the phonon frequencies after the UV illumination is shown in Fig. 17(b), where theUV illumination of 7 mW/mm² is on before the coherent phonon measurement but is off during the measurement. In this case, on the other hand, the shift of the softening temperature for modes 1 and 2 is not clear. The phonon frequencies for modes 1 and 2 are decreased while the coherent phonon signal for mode 3 is not observed. The relaxation time of the UV-illumination induced carriers is on the order of milliseconds below 30 K for SrTiO₃ [21]. As is seen in Fig. 17(b), however, it is suggested that the UV-illumination effect, frequency decrease for modes 1 and 2 and disappearance of mode-3 signal, remains for at least several minutes, even if the UV illumination is switched off, although the T_d shifting effect disappears immediately.

6. Summary

We observed coherent phonons in pure and Ca-doped SrTiO₃ using ultrafast polarization spectroscopy to study the ultrafast dynamics of soft phonon modes and their UV-illumination effect. Coherent phonons are generated by linearly polarized pump pulses. The time-dependent linear birefringence induced by the generated coherent phonons is detected asa change of the polarization of the probe pulses. A high detection sensitivity of ~10⁵ in polarization change, which corresponds to the change $\Delta n = 2 \times 10^9$ of the refractive index for a 1 mm sample, has been achieved in our detection system.

In SrTiO₃, damped oscillations of coherent phonons for the A_{1g} and E_g modes, which contribute to the structural phase transition at 105 K, were observed. The temperature dependences of the frequency and the relaxation rate of the observed coherent phonons were measured. Softening of the phonon frequencies was observed. The phonon relaxation is explained well by a decay model of a frequency-changing phonon, in which the optical phonon decays into two acoustic phonons due to the anharmonic phonon-phonon coupling.

We observed the temperature dependences of the birefringence and the coherent phonon signal to investigate the doping-induced ferroelectric phase transition in Ca-dopedSrTiO₃ with the Ca concentration of x=0.011. In the birefringence measurement, it was confirmed that the structural phase-transition temperature is T_1=180 K. Coherent phonons were observed by using ultrafast polarization spectroscopy. The damped oscillations of coherent phonons for the A_{1g} and E_g modes, which contribute to the structural phase transition at T_1=180 K, and for the modes 1, 2, and3, which contribute to the ferroelectric phase transition at T_d=28 K, were observed. The phonon frequencies were obtained from the observed signals of coherent phonons, and their softening toward each phase-transition temperature was observed. Another structural deformation at 25 K was found in addition to the ferroelectric phase transition at T_d. It was found that the UV illumination causes the shift of the ferroelectric phase-transition temperature toward the lower temperature side, and the temperature shift is ~3 K. The decrease in the phonon frequencies after the UV illumination suggests the UV-illumination effect remains even if the UV illumination is switched off.

It was shown that the coherent phonon spectroscopy in the time domain is a very useful approach to study the soft phonon modes and their UV-illumination effect in dielectric materials.

Acknowledgement

We would like to thank Dr. Y. Koyama for experimental help and Dr. Y. Yamada and Prof. K. Tanaka for providing us the samples of Ca-doped $SrTiO_3$.

Author details

Toshiro Kohmoto

Graduate School of Science, Kobe University, Japan

References

[1]   K. A. Müller and H. Burkard, Phys. Rev. B 19, 3593 (1979).

[2]   H. Uwe and T. Sakudo, Phys. Rev. B 13, 271 (1976).

[3]   M. Itoh, R. Wang, Y. Inaguma, T. Yamaguchi, Y.-J. Shan, and T.Nakamura, Phys. Rev. Lett. 82, 3540 (1999).

[4]   B. E. Vugmeister and M. P. Glinchuk, Rev. Mod. Phys. 62, 993 (1990).

[5]   J. G. Bednorz and K. A. Müller, Phys. Rev. Lett. 52, 2289 (1984).

[6]   J. F. Scott, Rev. Mod. Phys. 46, 83 (1974), and references therein.

[7]   P. A. Fleury, J. F. Scott, and J. M. Worlock, Phys. Rev. Lett. 21,16 (1968).

[8]   U. Bianchi, W. Kleemann, and J. G. Bednorz, J. Phys.: Condens. Matter 6, 1229 (1994).

[9]   U. Bianchi, J. Dec, W. Kleemann, and J. G. Bednorz, Phys. Rev. B 51, 8737 (1995).

[10] Bürgel, W. Kleemann, and U. Bianchi, Phys. Rev. B 53, 5222 (1995).

[11] Y. Yamada and K. Tanaka, J. Phys. Soc. Jpn. 77, 5 (2008).

[12] G. Geneste and J.-M. Kiat, Phys. Rev. B 77, 174101 (2008).

[13] M. Takesada, T. Yagi, M. Itoh, and S. Koshihara, J. Phys. Soc. Jpn. 72, 37 (2003).

[14] T. Hasegawa, S. Mouri, Y. Yamada, and K. Tanaka, J. Phys. Soc. Jpn. 72, 41 (2003).

[15] G. Godefroy, P. Jullien, and L. Cai, Ferroelectrics 13, 309 (1976).

[16] S. Ueda, I. Tatsuzaki, and Y. Shindo, Phys. Rev. Lett. 18, 453 (1967).

[17] Y. Yamada and K. Tanaka, J. Lumin. 112, 259 (2005).

[18] T. P. Dougherty, G. P. Wiederrecht, K. A. Nelson, M. H. Garrett,H. P. Jensen, and C. Warde, Science 258, 770 (1992).

[19] H. J. Bakker, S. Hunsche, and H. Kurz, Rev. Mod. Phys. 70, 523 (1998).

[20] W. G. Nilsen and J. G. Skinner, J. Chem. Phys. 48, 2240 (1968).

[21] T. Hasegawa, M. Shirai, and K. Tanaka, J. Lumin. 87-89, 1217 (2000).

[22] Y.-X. Yan, E. B. Gamble, Jr., and K. A. Nelson, J. Chem. Phys.83, 5391 (1985).

[23] G. A. Garrett, T. F. Albrecht, J. F. Whitaker, and R. Merlin, Phys. Rev. Lett. 77, 3661 (1996).

[24] T. Kohmoto, K. Tada, T. Moriyasu, and Y. Fukuda,Phys. Rev. B 74, 064303 (2006).

[25] Y. Koyama, T. Moriyasu, H. Okamura, Y. Yamada, K. Tanaka, and T. Kohmoto, Phys. Rev. B 81, 024104 (2010).

[26] T. Kohmoto, Y. Fukuda, M. Kunitomo, and K. Isoda, Phys. Rev. B 62, 579 (2000).

[27] R. V. Jones, Proc. R. Soc. London, Ser. A 349, 423 (1976).

[28] E. Courtens, Phys. Rev. Lett. 29, 1380 (1972).

[29] K. Watanabe, N. Takagi, and Y. Matsumoto, Phys. Rev. Lett. 92,057401 (2004).

[30] F. Vallée, Phys. Rev. B 49, 2460 (1994).

[31] M. Hase, K. Mizoguchi, H. Harima, S. I. Nakashima, and K. Sakai, Phys. Rev. B 58, 5448 (1998).

[32] M. Hase, K. Ishioka, J. Demsar, K. Ushida, and M. Kitajima, Phys. Rev. B 71, 184301 (2005).

Permissions

The contributors of this book come from diverse backgrounds, making this book a truly international effort. This book will bring forth new frontiers with its revolutionizing research information and detailed analysis of the nascent developments around the world.

We would like to thank Dr. Aimé Peláiz Barranco, for lending her expertise to make the book truly unique. She has played a crucial role in the development of this book. Without her invaluable contribution this book wouldn't have been possible. She has made vital efforts to compile up to date information on the varied aspects of this subject to make this book a valuable addition to the collection of many professionals and students.

This book was conceptualized with the vision of imparting up-to-date information and advanced data in this field. To ensure the same, a matchless editorial board was set up. Every individual on the board went through rigorous rounds of assessment to prove their worth. After which they invested a large part of their time researching and compiling the most relevant data for our readers. Conferences and sessions were held from time to time between the editorial board and the contributing authors to present the data in the most comprehensible form. The editorial team has worked tirelessly to provide valuable and valid information to help people across the globe.

Every chapter published in this book has been scrutinized by our experts. Their significance has been extensively debated. The topics covered herein carry significant findings which will fuel the growth of the discipline. They may even be implemented as practical applications or may be referred to as a beginning point for another development. Chapters in this book were first published by InTech; hereby published with permission under the Creative Commons Attribution License or equivalent.

The editorial board has been involved in producing this book since its inception. They have spent rigorous hours researching and exploring the diverse topics which have resulted in the successful publishing of this book. They have passed on their knowledge of decades through this book. To expedite this challenging task, the publisher supported the team at every step. A small team of assistant editors was also appointed to further simplify the editing procedure and attain best results for the readers.

Our editorial team has been hand-picked from every corner of the world. Their multi-ethnicity adds dynamic inputs to the discussions which result in innovative outcomes. These outcomes are then further discussed with the researchers and contributors who give their valuable feedback and opinion regarding the same. The feedback is then collaborated with the researches and they are edited in a comprehensive manner to aid the understanding of the subject.

Apart from the editorial board, the designing team has also invested a significant amount of their time in understanding the subject and creating the most relevant covers. They scrutinized every image to scout for the most suitable representation of the subject and create an appropriate cover for the book.

The publishing team has been involved in this book since its early stages. They were actively engaged in every process, be it collecting the data, connecting with the contributors or procuring relevant information. The team has been an ardent support to the editorial, designing and production team. Their endless efforts to recruit the best for this project, has resulted in the accomplishment of this book. They are a veteran in the field of academics and their pool of knowledge is as vast as their experience in printing. Their expertise and guidance has proved useful at every step. Their uncompromising quality standards have made this book an exceptional effort. Their encouragement from time to time has been an inspiration for everyone.

The publisher and the editorial board hope that this book will prove to be a valuable piece of knowledge for researchers, students, practitioners and scholars across the globe.

List of Contributors

Kaoru Miura
Canon Inc., Tokyo, Japan

Hiroshi Funakubo
Tokyo Institute of Technology, Yokohama, Japan

Ashok Kumar
National Physical Laboratory, Council of Scientific and Industrial Research (CSIR), New Delhi, India
Department of Physics and Institute for Functional Nano materials, University of Puerto Rico, San Juan, Puerto Rico

Margarita Correa, Nora Ortega, Salini Kumari and R. S. Katiyar
Department of Physics and Institute for Functional Nano materials, University of Puerto Rico, San Juan, Puerto Rico

Shu-Tao Ai
School of Science, Linyi University, Linyi, People's Republic of China

Desheng Fu
Division of Global Research Leaders, Shizuoka University, Johoku, Naka-ku, Hamamatsu, Japan

Hiroki Taniguchi and Mitsuru Itoh
Materials and Structures Laboratory, Tokyo Institute of Technology, Nagatsuta, Yokohama, Japan

Shigeo Mori
Department of Materials Science, Osaka Prefecture University, Sakai, Osaka, Japan

Zhigao Hu, Yawei Li, Kai Jiang and Ziqiang Zhu
Key Laboratory of Polar Materials and Devices, Ministry of Education, Department of Electronic Engineering, East China Normal University, Shanghai, People's Republic of China

Junhao Chu
Key Laboratory of Polar Materials and Devices, Ministry of Education, Department of Electronic Engineering, East China Normal University, Shanghai, People's Republic of China
National Laboratory for Infrared Physics, Shanghai Institute of Technical Physics, Chinese Academy of Sciences, Shanghai, People's Republic of China

A. Peláiz-Barranco, F. Calderón-Piñar, O. García-Zaldívar and Y. González-Abreu
Physics Faculty–Institute of Science and Technology of Materials, Havana University, San Lázaro y L, Vedado, La Habana, Cuba

Martun Hovhannisyan
"ENI" Institute of Electronic Materials LTD, Armenia

Hideo Kimura, Qiwen Yao and Lei Guo
National Institute for Materials Science, Sengen 1-2-1, Tsukuba, Japan

Hongyang Zhao
National Institute for Materials Science, Sengen 1-2-1, Tsukuba, Japan
Shanghai Institute of Ceramics, Chinese Academy of Sciences, Shanghai, China

Xiaolin Wang and Zhenxiang Cheng
Institute for Superconducting and Electronics Materials, University of Wollongong, Innovation Campus, Fairy Meadow, Australia

A.K. Bain and Prem Chand
Department of Physics Indian Institute of Technology Kanpur, Kanpur-208016, INDIA

Akira Onodera and Masaki Takesada
Department of Physics, Faculty of Science, Hokkaido University, Sapporo, Japan

Dong Guo
Institute of Acoustics, Chinese Academy of Sciences, Beijing, China

Kai Cai
Beijing Center for Chemical and Physical Analysis, Beijing Municipal Science and Technology, Research Institute, Beijing, China

Toshiro Kohmoto
Graduate School of Science, Kobe University, Japan

Printed in the USA
CPSIA information can be obtained
at www.ICGtesting.com
JSHW011451221024
72173JS00005B/1032